MONOGRAPHS IN STATISTICAL PHYSICS AND THERMODYNAMICS

Volume 1
Non-Equilibrium Statistical Mechanics
by I. PRIGOGINE

Volume 2
Thermodynamics
With quantum statistical illustrations
by P. T. LANDSBERG

Volume 3
Ionic Solution Theory
Based on cluster expansion methods
by HAROLD L. FRIEDMAN

MONOGRAPHS IN STATISTICAL PHYSICS AND THERMODYNAMICS

Editor: I. PRIGOGINE
Professor of Physical Chemistry and of Theoretical Physics
Université Libre, Brussels, Belgium

VOLUME 3

IONIC SOLUTION THEORY

BASED ON CLUSTER EXPANSION METHODS

HAROLD L. FRIEDMAN
IBM Research Center
Yorktown Heights, New York

INTERSCIENCE PUBLISHERS 1962

a division of John Wiley & Sons, New York • London

Library of Congress Catalog Card Number 62-18931

PRINTED IN THE UNITED STATES OF AMERICA

Preface

The cluster theories of statistical mechanics, and in particular the cluster theory of ionic solutions, have now reached so high a stage of development, and promise such numerous applications, that there is a need for a single, reasonably complete, unified exposition for students of the subject. This book is intended primarily to fill this need for readers whose knowledge of statistical mechanics extends to the configuration integral. The methods of statistical thermodynamics are used as far as possible and hence a certain familiarity with chemical thermodynamics will also be helpful. For the rest, the book has been written to be of service to readers who are not following a course of lectures in the subject; to this end an effort has been made to be particularly complete with bibliographic references and cross references in the text. Advanced mathematical methods have been used freely wherever they seem to facilitate the proceedings. The underlying mathematical theorems are mostly proved only formally here, with references made to more rigorous treatments.

It has been found necessary to devote the first half of the book to the cluster theory of non-ionic systems, so the title is a little misleading as far as the content of the book is concerned. As usual the Table of Contents is a more reliable guide.

Many of the results in this book derive from the published work of Professor J. E. Mayer and various parts of the theory are usually referred to elsewhere as *the* Mayer theory or *the* X-Mayer theory. If this practice had been followed here Mayer's name would be prominent in the title of the book and in most of the section headings. It was finally decided that it would be more helpful to the reader to use designations that are more adjectival than genitive and to indicate the origins of the various parts of the theory in a less casual way in the text.

I am very grateful to Professor Thor Bak who read and criticized part of the manuscript, and to several of my colleagues at the I.B.M. Research Center, in particular J. Greyson, W. R. Heller, D. Jepson,

D. Mattis, and E. W. Montroll, for providing criticism, discussion, and suggestions.

I also must express my deepest appreciation to Professor I. Prigogine for the inspiration, guidance, and encouragement of my initial efforts in this field and even for suggesting that I undertake to write this book.

HAROLD L. FRIEDMAN

Contents

1. Preliminaries · · · · · · · · · · · · · · · · 1

1. Introduction · · · · · · · · · · · · · · · 1
2. Rudimentary Derivation of the Cluster Theories · · 4
3. Notation and Some Mathematical Topics · · · · · 17
4. Potentials of Average Force · · · · · · · · · · 34
5. Cluster Expansions · · · · · · · · · · · · 41

2. The Cluster Theory of Non-Ionic Systems · · · · 67

6. Thermodynamics. Volume as an Independent Variable · · · · · · · · · · · · · · · · · 67
7. The Theory of the Grand Partition Function · · · 72
8. The Density Expansion of the Excess Free Energy · 88
9. Spatial Correlation Functions · · · · · · · · 93
10. Some Applications of the Cluster Theory · · · · 100

3. Cluster Theory of Ionic Solutions · · · · · · · · 115

11. Introduction · · · · · · · · · · · · · · · 115
12. Some Mathematical Aspects · · · · · · · · · 125
13. The Excess Free Energy of Ionic Systems · · · · 147
14. Spatial Correlation Functions for Ionic Solutions · · 167
15. Questions of Convergence · · · · · · · · · · 173

4. Application to Ionic Systems · · · · · · · · · · 191

16. Thermodynamics. Pressure as an Independent Variable · · · · · · · · · · · · · · · · · 191
17. Solutions of a Single Electrolyte · · · · · · · 208
18. Mixed Electrolyte Solutions · · · · · · · · · 225

Appendix. Table of $J(L, K)$ · · · · · · · · · · · 249

Index of Notation · · · · · · · · · · · · · · · 255

Subject Index · · · · · · · · · · · · · · · · 263

CHAPTER 1

Preliminaries

1. Introduction

The study of electrolyte solutions has occupied a central place in chemistry ever since the time of van't Hoff and Arrhenius. The primary reason for this is certainly the large number of interesting and useful chemical processes that involve ions in solution. Another reason develops when one tries to understand these processes in terms of the structure of the solution and the energetics and mechanisms of chemical reactions. Then it is soon found that the long-range forces among the ions have a powerful effect on the observed properties of the solution, so much so that it becomes far more difficult than in non-ionic systems to arrive at a molecular interpretation of the observations. Many studies of electrolyte solutions, this one among others, are motivated by the wish to understand the effects of the long-range forces and their interplay with the effects of other molecular interactions.

Almost all of the statistical mechanical studies of electrolyte solutions treat the idealization in which all but the long-range forces are reduced to a minimum. This is the *primitive model*, comprising ions represented by hard spheres with charges at their centers and a solvent represented by an ideal dielectric fluid. However, even the observations on real systems chosen to conform as closely as possible to the primitive model are so complicated that in early developments three distinct but complementary theories were proposed for the interpretation of various aspects of the observations. Each of these theories continues to have an important influence.

Bronsted's theory of specific ion interaction[1] (1920) is a phenomenological theory, rather than statistical, that is based on the approximation that chemical interaction between ions is limited to the interaction of ions of opposite sign. Its basis is that two ions of the same sign are less likely to approach each other closely enough for short-range interactions to manifest themselves. This has been helpful in the interpretation of the observations of mixed electrolyte solutions at high dilution, as exemplified in Bronsted's own work.

The Debye-Hückel theory of ionic interaction (1923) was the first statistical theory of electrolyte solutions. In its application to extremely dilute solutions it has been remarkably successful and has served as the basis for theories of various transport properties of aqueous solutions, from Onsager's theory of electrolytic conductance (1927) to Helfand and Kirkwood's theory of thermal diffusion in ionic solutions (1960), but unfortunately it is limited in validity to exceedingly low concentrations, almost always below the experimental concentration range. Furthermore, it does not seem to be possible to amend the Debye-Hückel method to extend it to more general models nor to higher concentrations.

Bjerrum's theory of ion association (1926) allows one to estimate the number of ion pairs in an electrolyte solution. A pair is two ions of opposite sign whose separation is less than some arbitrary distance. Such a pair is supposed not to contribute to the electrical conductivity of the solution while all unpaired ions contribute as free ions modified by the ion atmosphere according to the Onsager theory. Although there has been some difficulty with the specification of how close two ions have to be to constitute a pair, Bjerrum's theory and its later development by Fuoss have served for the interpretation of a tremendous body of conductivity data, especially for electrolyte solutions at low concentrations in organic solvents. It has also been applied to some manifestations of ion pairing in other than conductivity measurements, and as a guide to when one should expect ion pairs to be formed in significant amounts in various systems.

Although these theories have led to valuable insight into the nature of ionic solutions there are many interesting systems for which they offer only the most uncertain guide to understanding. Among the equilibrium systems for which this is true are simple electrolyte solutions in the preparative concentration range, mixed electrolyte solutions, and systems exhibiting "salting out," the weak interaction of an electrolyte with a non-electrolyte other than the solvent. Furthermore, since all three theories deal with ion interactions one should hope to deduce all of these results from a single satisfactory theory. This has yet to be done and indeed most later theories have dealt only with the Debye-Hückel aspect: solutions of a single strong electrolyte.

The limiting form of the Debye-Hückel theory at low concentrations can be derived in several ways which may be expected to yield correct results at moderate concentrations as well. These are the theories of Kramers,[6] Kirkwood,[7] Kirkwood and Poirier,[8] Bogoliubov,[9] and Mayer.[10] In each case the mathematical difficulty encountered in proceeding beyond the Debye-Hückel limiting law stage is very formidable and only for Mayer's cluster theory of ionic solutions have the calculations been brought to sufficient completion to provide a basis for a comparison of the calculations from various models with the observations on solutions in the range of moderate concentrations.

The difficulties which remain for Mayer's cluster theory calculations are largely in obtaining an appropriate mathematical description of the model, a description in terms of the forces among two or three ions at various configurations in the solvent. The potentials corresponding to these forces are the *direct potentials* on which the cluster theory operates to yield thermodynamic functions and other average properties of solutions of finite concentration at equilibrium. Therefore, the theory derived here, although in an important sense complete in itself, will in general require elaborate supplementary calculations of the direct potentials when it is applied to a given model for a solution. Such calculations lie outside the scope of this book.

Several other topics which one might expect to be treated here are only mentioned in passing, for example, the application of cluster methods to strictly quantum mechanical systems. The cluster theories of surface properties and phase equilibria have only been referred to. An effort has yet to be made to apply the cluster theory results in calculating the transport properties of ionic solutions. Finally, there is almost no discussion of non-cluster ionic solution theories. All of these topics have been omitted here because of limitations of the author's competence or limitations of space, rather than lack of interest.

Although the derivation of the cluster ionic solution theory seems to be rigorous there is one feature that may possibly limit its usefulness to solutions of low concentration. That is, all of the results are expressed as infinite series, which, although mathematically convergent, may not converge fast enough for reliable numerical computation with the first few terms unless the ionic concentration is low. The direct mathematical study of this problem is difficult; an easier approach may be to compare experimental quantities with values calculated from a model using only the first few terms of the series. Because of the dearth of calculated direct potentials, except for the primitive model, only tentative comparisons can be made at this time. These do seem to be satisfactory.

The plan of this book is to give the various statistical mechanical derivations in as compact a form as possible. Each is preceded by detailed discussions of the mathematical procedures and followed by some applications of the statistical mechanical results. This program begins in Section 3 and continues to Section 10 for non-ionic systems, and then begins in Section 11 for ionic systems. Graph-topological methods are used throughout, after an elementary introduction in Section 2 and a more detailed treatment in Section 5. A special mathematical topic of a different kind is discussed in Section 15—a study of the convergence of the series for ionic systems. The convergence question itself is rather academic, however it is found that in many numerical computations with the theory questions are raised that may be answered by the methods of that section.

Notes and References

1. J. N. Bronsted, *J. Am. Chem. Soc.* **44**, 877 (1922).
2. P. Debye and E. Hückel, *Physik. Z.* **24**, 185 (1923).
3. L. Onsager, *Physik. Z.* **28**, 277 (1927).
4. E. Helfand and J. G. Kirkwood, *J. Chem. Phys.* **32**, 857 (1960).
5. N. Bjerrum, *Kgl. Danske Videnskab. Selskab, mat.-fys. Medd.* **7**, No. 9 (1926).

6a. H. A. Kramers, *Proc. Roy. Acad. (Amsterdam)* XXX, 145 (1927).
6b. T. H. Berlin and E. W. Montroll, *J. Chem. Phys.* **20**, 75 (1952).

7. J. G. Kirkwood, *Chem. Revs.* **19**, 275 (1936).
8. J. G. Kirkwood and J. C. Poirier, *J. Phys. Chem.* **58**, 591 (1954).
9. M. N. Bogoliubov. Described in *Statistical Physics*, by L. D. Landau and E. M. Lifshitz (English Edition: Pergamon Press Ltd., London, 1958), Section 74.
10. J. E. Mayer, *J. Chem. Phys.* **18**, 1426 (1950).

2. Rudimentary Derivation of the Cluster Theories

The most rigorous and general derivations of the cluster theories make abundant use of mathematical techniques that are likely to be unfamiliar to the reader. He is therefore in danger of losing all feeling for the logical structure of the theory as he works through the individual steps. In order to minimize this danger we present here a skeletonized version of the theory.

In this version most of the mathematical problems do appear, but in a less general and more familiar form than in the general theory. This is accomplished partly by considering only the simplest models for a system of interacting particles and partly by considering only the first terms of the various expansions. It is therefore to be emphasized that it is not safe to base any conclusions about the range of validity of cluster theories in general on the limitations of this rudimentary derivation.

The theory in this section does not employ either the grand partition function or the correlation functions, all of which are used in the more general theory. However these topics are more physical than mathematical and it seems unnecessary to show at this stage how they fit into the theory.

Non-Ionic Systems

The following derivation is similar in some respects to a short derivation of the cluster theory given by Brout.[1]

We consider a system of N identical monatomic molecules in a vessel of volume $\mathcal{V}$ at temperature T. It is further assumed that when the molecules are instantaneously at a configuration represented by the coordinates $\{N\}$ (i.e., $\{N\}$ is a set of $3N$ center-of-mass coordinates) then the energy $U(\{N\})$ of the system may be accurately calculated by summing over the pairwise interaction potentials,

$$U(\{N\}) = \sum_{\text{pairs}} u(r_{ij}) \tag{2.1}$$

where r_{ij} is the separation of the pair ij in the given configuration.

As is well known the thermodynamic properties of such a system may be expressed in terms of the configuration integral,

$$Z(N,\mathcal{V},T) \equiv \int \exp\left[-U(\{N\})/kT\right] d\{N\} \tag{2.2}$$

The characteristic maneuver of the cluster theories is to introduce a *cluster function* defined as

$$\gamma_{ij} \equiv \exp\left[-u(r_{ij})/kT\right] - 1 \tag{2.3}$$

into this equation for Z.‡ This leads to the expansion

‡ In the derivation of Brout, however, another principle is used for expansion of the configuration integral and the cluster function is formed by a summation procedure at a later stage.[1]

$$\exp\,[-U(\{N\})/kT] = 1 + \sum_{\text{pairs}} \gamma_{ij} + \sum_{\text{triples}} [\gamma_{ij}\gamma_{jk} + \gamma_{ij}\gamma_{jk}\gamma_{ik}] + \sum_{\text{quadruples}} [\gamma_{ij}\gamma_{kl} + \cdots] + \cdots \tag{2.4}$$

which converts (2.2) into a sum of integrals.

The first of these terms of Z is simply

$$\int_{\mathcal{V}} d\{N\} = \mathcal{V}^N \tag{2.5}$$

All of the pairwise terms in (2.4) may be integrated to give, collectively, $N(N - 1)/2$ terms of the form

$$\mathcal{V}^{N-2} \int_{\mathcal{V}} \gamma_{ij}\, d\{ij\} \tag{2.6}$$

because the integration of each particle *not* represented as a subscript in the integrand may be done independently of the others.

This indicates the basic procedure. For the rest it is convenient to proceed with the term-by-term integration of (2.4) in a more systematic fashion. First we describe the terms of (2.4) as corresponding to *graphs* in a way that is particularly suited to this problem.

The first term, 1, is represented as a graph consisting of only N vertices (points). This set of vertices forms the *skeleton* of each of the graphs representing the higher terms as well (Fig. 2.1), but the graphs of the higher terms contain *bonds* which connect vertices of the skeleton. Each of the terms in the sum over pairs is represented by a graph with a single bond, representing γ_{ij}, connecting a pair of vertices that correspond to molecules i and j. This procedure is readily extended to allow the representation of the higher terms as shown in Fig. 2.1.

The important characteristic of these graphs is not their geometry (shape and size, position and orientation) but only their *topology*. The topology of a graph is the scheme of connections among its vertices and bonds. If for a moment we think of a graph as having geometric characteristics which reflect a given $\{N\}$ we see that the integration (2.2) corresponds to the changing of the geometry of the graph over the whole range that is consistent with the molecules being in the same volume. In other words two graphs with the same topology but

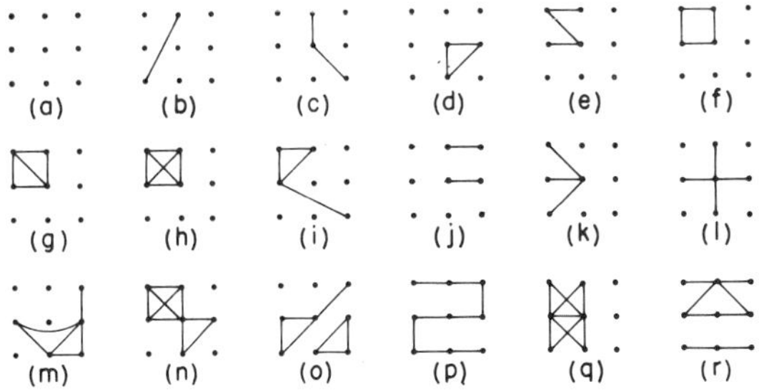

Fig. 2.1. Graphical representation of some of the first terms of the cluster expansion of exp $[-U(\{N\})/kT]$ for $N = 9$.

different geometry merely correspond to the same term of the cluster expansion (2.4) for two different values of $\{N\}$, hence for two different stages in the integration.

In addition to the topology as defined above there is one other characteristic of the graphs that is important. This has to do with possible distinguishability of the vertices. The graphs of Fig. 2.1 are all distinguishable from each other even if the vertices are not numbered, colored, or labeled in any way. These graphs are distinguishable even if the skeleton consists of unlabeled vertices. (Of course after a graph is drawn on such a skeleton by inserting bonds some of the vertices may then be distinguishable topologically from others because they intersect different numbers of bonds, or for more subtle reasons. We cannot then say that the vertices of the graph are indistinguishable but only unlabeled.)

To illustrate the role played by graphs with labeled vertices we consider equation (2.4) for $N = 9$. In this case there will be 252 (i.e., $9 \times 8 \times 7/2$) terms of the same topology as Fig. 2.1(c). On the other hand if we number the vertices of the skeleton of Fig. 2.1 from 1 to 9 then we may draw 252 graphs that are distinguishable from each other but which would all be identical to Fig. 2.1(c) if we erased the labels. By making our numbering of the vertices of the skeleton correspond to the numbering of the molecules (this number is implied by writing γ_{ij} different from γ_{jk}) we make a one-to-one correspondence between the terms of the cluster expansion and the graphs on skeletons of labeled vertices. However integration is equivalent to erasing the labels on the vertices because all terms that correspond to graphs that are mutually distinguishable only because the vertices are labeled give the same result on integration, as we saw for the γ_{ij} terms of (2.4). (This statement must be modified for multicomponent systems.) This is the origin of the combinatorial problems which appear in the cluster theories.

Now we may describe the *cluster expansion*, equation (2.4), of $\exp\left[-U(\{N\})/kT\right]$ in graph-topological terms. There is one term in the cluster expansion corresponding to each graph of γ bonds on a skeleton of N vertices provided that we count all graphs that are distinguishable from each other by topological criteria if the vertices of the skeleton are numbered from 1 to N. Then we note again that any group of graphs that are distinguishable from each other if and only if their vertices are labeled corresponds to a group of terms of the cluster expansion that gives the same integral.

Based on these considerations the cluster expansion of the configuration integral may be written as

$$Z(N,\mathcal{V},T) = \mathcal{V}^N \sum_m \binom{N}{m} \mathcal{V}^{-m} \sum_\tau A_{m,\tau} I_{m,\tau}, \qquad m = 0, 2, 3, 4, 5, \cdots, N-1, N \tag{2.7}$$

The binomial coefficient $\binom{N}{m}$ is the number of ways of dividing a collection of N molecules into two groups, one of m and the other of $N - m$ molecules, with no concern for the order within each group. The second sum in the equation is over the graphs of distinguishable topology τ on a skeleton of m unlabeled vertices. Only those graphs may be included here in which each vertex intersects at least one bond. The unconnected vertices of Fig. 2.1 give, by integration, the factor $\mathcal{V}^{N-m}$ in equation 2.7. The combinatorial factor $A_{m,\tau}$ is the number of distinguishable ways of numbering the vertices of the graph m,τ from 1 to m. Alternatively it is the number of distinguishable graphs on a skeleton of m numbered vertices that become identical to m,τ if the labels on the vertices are erased. Finally $I_{m,\tau}$ is the integral corresponding to the graph m,τ.

In many cases $I_{m,\tau}$ may be expressed as a product of integrals corresponding to simpler graphs. The simplest examples are

$$I_{4,a} = \int_{\mathcal{V}} \gamma_{ij}\gamma_{kl}\, d\{ijkl\} = \left[\int_{\mathcal{V}} \gamma_{ij}\, d\{ij\}\right]^2 \tag{2.8}$$

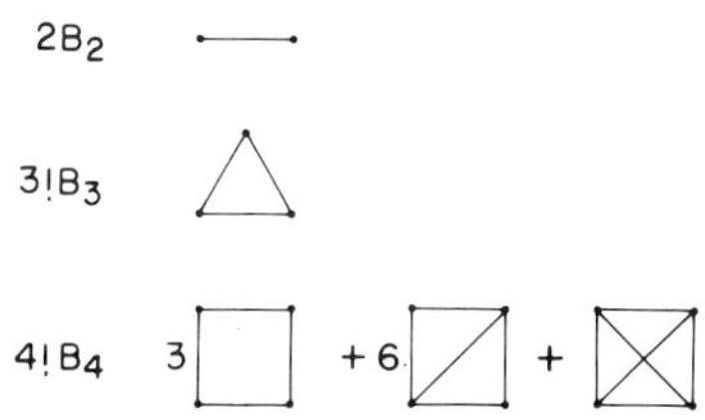

Fig. 2.2. Graphical representation of the first few irreducible cluster integrals.

which corresponds to Fig. 2.1(j) and

$$I_{3,a} = \int_{\mathcal{V}} \gamma_{ij}\gamma_{jk}\, d\{ijk\} \tag{2.9}$$

which corresponds to Fig. 2.1(c). The integration of I_{3a} may be performed by first holding j fixed and then integrating over i and k. When this is done the integrations over i and k are independent and so we have

$$\begin{aligned} I_{3a} &= \int_{\mathcal{V}} d\{j\} \left[\int_{\mathcal{V}} \gamma_{ij}\, d\{i\}\right]\left[\int_{\mathcal{V}} \gamma_{jk}\, d\{k\}\right] \\ &= \int_{\mathcal{V}} d\{j\} \left[\int_{\mathcal{V}} \gamma_{ij}\, d\{i\}\right]^2 \end{aligned} \tag{2.10}$$

Because γ_{ij} depends only on the separation of i and j, the integration over $\{j\}$ is trivial and only produces a factor $\mathcal{V}$. Hence we have

$$I_{3,a} = \left[\int_{\mathcal{V}} \gamma_{ij}\, d\{ij\}\right]^2 \Big/ \mathcal{V}. \tag{2.11}$$

It is convenient to define special symbols for cluster integrals which appear here and which cannot be factored in either of the ways illustrated. They are called *irreducible* cluster integrals. The first few are listed here and represented graphically in Fig. 2.2.

$$2!\beta_2 \equiv \int_{\mathcal{V}} \gamma_{ij}\, d\{ij\} \tag{2.12}$$

$$3!\beta_3 \equiv \int_{\mathcal{V}} \gamma_{ij}\gamma_{jk}\gamma_{ik}\, d\{ijk\} \tag{2.13}$$

$$4!\beta_{4a} \equiv \int_{\mathcal{V}} \gamma_{ij}\gamma_{jk}\gamma_{kl}\gamma_{il}\, d\{ijkl\} \tag{2.14}$$

$$4!\beta_{4b} \equiv \int_{\mathcal{V}} \gamma_{ij}\gamma_{jk}\gamma_{kl}\gamma_{il}\gamma_{ik}\, d\{ijkl\} \tag{2.15}$$

$$4!\beta_{4c} \equiv \int_{\mathcal{V}} \gamma_{ij}\gamma_{jk}\gamma_{kl}\gamma_{il}\gamma_{ik}\gamma_{jl}\, d\{ijkl\} \tag{2.16}$$

$$\beta_4 \equiv 3\beta_{4a} + 6\beta_{4b} + \beta_{4c} \tag{2.17}$$

As an example of the use of this terminology we have

$$I_{4a} = 4\beta_2^2 \qquad \text{and} \qquad I_{3a} = 4\beta_2^2/\mathcal{V}$$

for equations (2.8) and (2.11), respectively.

In Table 2.1 we show an analysis of the first terms of the cluster expansion of Z based on the foregoing considerations. The first four columns of the table pertain to matters that we have already discussed. Before considering the last column we must examine the behaviour of the product

$$\mathcal{V}^{-m}\binom{N}{m} = N[N-1]\cdots[N-m+1]/\mathcal{V}^m m! \tag{2.18}$$

in the limit as $\mathcal{V} \to \infty$ at fixed $c \equiv N/\mathcal{V}$. This limiting process corresponds to considering larger and larger systems of the same composition, a familiar procedure to assess bulk thermodynamic properties. For any fixed m we have in this limit

$$\mathcal{V}^{-m}\binom{N}{m} \to c^m/m! \tag{2.19}$$

that is, this function becomes independent of the volume at large $\mathcal{V}$. On the other hand $I_{m,\tau}$ evidently may be a function of the volume at large $\mathcal{V}$. An exact analysis is unnecessary here. We need only note that a term of $Z\mathcal{V}^{-N}$ such as

$$\mathcal{V}^{-3}\binom{N}{3} 3[4\beta_2^2/\mathcal{V}]$$

certainly becomes negligible compared to

$$\mathcal{V}^{-4}\binom{N}{4} 3[4\beta_2^2]$$

TABLE 2.1. The cluster expansion of $Z(N,\mathcal{V},T)$.

m	τ,m	$A_{\tau,m}$	$I_{\tau,m}$	Non-negligible term of Z
0	none	1	1	1
2		1	$2\beta_2$	$c^2\beta_2$
3		3	$4\beta_2^2/\mathcal{V}$	
3		1	$6\beta_3$	$c^3\beta_3$
4		3	$4\beta_2^2$	$c^4\beta_2^2/2$
4		12	$8\beta_2^3/\mathcal{V}^2$	
4		4	$8\beta_2^3/\mathcal{V}^2$	
4		12	$12\beta_2\beta_3/\mathcal{V}$	
4		3	$24\beta_{4a}$	
4		6	$24\beta_{4b}$	$c^4\beta_4$ (for the three rows)
4		1	$24\beta_{4c}$	
5		30	$8\beta_2^3/\mathcal{V}$	
5		60	$16\beta_2^4/\mathcal{V}^3$	
5		60	$16\beta_2^4/\mathcal{V}^3$	
5		10	$12\beta_2\beta_3$	$c^5\beta_2\beta_3$
⋮				
6		15	$8\beta_2^3$	$c^6\beta_2^3/6$
⋮				

as $\mathcal{V} \to \infty$. In the same way every other term in which $I_{m,\tau}$ may be expressed as a product of β_n's with $\mathcal{V}$ in the denominator may be shown to be negligible compared to some other term in which the integral may be expressed as a product of the same β_n's, but without the powers of $\mathcal{V}$ in the denominator. So we arrive at the non-negligible terms listed in the last row of the table. It is apparent that these are the first terms of the expansion of

$$\mathcal{V}^{-N}Z = \exp\left(\sum_n{}'' c^n\beta_n\right) = 1 + \sum_n{}'' c^n\beta_n + \left[\sum_n{}'' c^n\beta_n\right]^2/2 + \cdots \tag{2.20}$$

where $\sum''$ indicates that the terms for $n = 0$ and 1 are omitted from the summation.

Now we use this result with the general relation

$$\mathcal{V}^{-N}Z = \exp\left(-\mathfrak{F}^{\text{ex}}/kT\right) \tag{2.21}$$

where $\mathfrak{F}^{\text{ex}}$ is the free energy of the real gas minus that of the equivalent ideal gas at the same c and T. Comparing (2.21) with (2.20) we see that

$$-\mathfrak{F}^{\text{ex}}/kT = \sum{}'' c^n\beta_n \tag{2.22}$$

In deducing this we have let $\mathcal{V} \to \infty$ and therefore $\mathfrak{F}^{\text{ex}}$ is infinite as well. This inelegance can be avoided at the expense of an increase in the complexity of the derivation, but it can be handled almost as well by dividing through by $\mathcal{V}$ at this stage. We define

$$\mathfrak{S} \equiv -\mathfrak{F}^{\text{ex}}/kT\mathcal{V} \tag{2.23}$$

$$B_n \equiv \beta_n/\mathcal{V} \tag{2.24}$$

These functions remain finite as $\mathcal{V} \to \infty$ at constant c (at least in the absence of Coulomb forces between the molecules, as discussed below) and so we have as the basic cluster expansion for thermodynamic properties,

$$\mathfrak{S} = \sum{}'' c^n B_n \tag{2.25}$$

where, strictly speaking, B_n is the limiting value of $\beta_n/\mathcal{V}$ as $\mathcal{V} \to \infty$.

As an example of the use of this expansion we note that thermodynamic manipulation of (2.23) leads to the relation of $\mathfrak{S}$ to the pressure P (cf. Section 6),

$$P/kT = c + \partial(\mathfrak{S}/c)/\partial(1/c) \tag{2.26}$$

This equation applied to (2.25) leads to the virial equation

$$P/kT = c - \sum{}'' c^n[n - 1]B_n \tag{2.27}$$

Therefore $[1 - n]B_n$ is the statistical mechanical expression for the nth virial coefficient. We recall that β_n , and hence B_n too, is defined in terms of the forces acting between pairs of molecules. By other thermodynamic manipulations equation (2.25) leads to expressions for all of the thermodynamic properties of systems of finite concentration in terms of the interactions of two molecules in the absence of all other molecules.

In Section 8 we present the derivation of the same equation for much more general systems. The main difference in the two methods of derivation is that the combinatorial problems are much more severe in the more powerful derivation.

Ionic Systems

Equation (2.25) is a useful form for non-ionic systems but we find that if we attempt to calculate the B_n for an ionic gas they all diverge. Let us assume, for example, that we have a one-component monatomic ionic gas with the pairwise potential

$$u(r) = \varepsilon^2/r \tag{2.28}$$

where ε is the electronic charge. Then

$$\begin{aligned} 2B_2 &= \lim_{\mathcal{V}\to\infty} \int_{\mathcal{V}} \gamma_{ij}\, d\{ij\}/\mathcal{V} \\ &= \lim_{R\to\infty} \int_0^R \gamma(r) 4\pi r^2\, dr \\ &= \lim_{R\to\infty} \int_0^R [e^{-\varepsilon^2/rkT} - 1] 4\pi r^2\, dr \end{aligned} \tag{2.29}$$

where R is the radius of the containing vessel, assumed to be spherical. We may investigate this integral by expanding the exponential. Let

$$\lambda \equiv 4\pi\varepsilon^2/kT \tag{2.30}$$

Then

$$2B_2 = \sum_{n\geq 1} [-\lambda]^n \lim_{R\to\infty} \int_0^R [4\pi r]^{1-n} r\, dr/n! \tag{2.31}$$

and we see that the terms for $n < 4$ lead to divergence of the sum as $R \to \infty$. The integrals on terms for $n \geq 3$ also diverge at the lower limit even for finite R. However the integral in (2.29) does not diverge at this limit, so the effect of *this* divergence of the integral must cancel in the summation in (2.31). This observation is a clue to the evaluation of $\mathfrak{S}$ for the ionic system, for it suggests that by an appropriate summation procedure the divergences at $R \to \infty$ may be shown to cancel too, because the configuration integral itself is convergent. The following procedure for achieving this is due to Mayer.[3] The short derivation presented here is similar to that given by Montroll and Ward.[4]

Before beginning the derivation we have one other observation to make. The Debye-Hückel limiting law expression for the thermodynamic function $\mathfrak{S}$ is $\kappa^3/12\pi$ where κ is defined by the equation

$$\kappa^2 \equiv \lambda c \tag{2.32}$$

Therefore the leading term in an expansion of $\mathfrak{S}$ in powers of c must have c to the 3/2 power. This is certainly inconsistent with (2.25) *unless* the B_n are infinite.

Now to introduce Mayer's summation procedure we expand every γ_{ij} in the B_n of (2.25) just as we expanded the γ_{ij} in B_2 above:

$$\gamma_{ij} = \sum_{n\geq 1} [-\lambda/4\pi r]^n/n! \tag{2.33}$$

In terms of the graphical representations of the B_n this has the effect of dividing every graph in B_n into an infinite number of graphs with $1/r$ bonds instead of γ bonds. This is shown as the expansion from left to right in Fig. 2.3.

Having made the expansion we must now decide in which order to collect the terms. There is an infinite number of possibilities, one of which is the reverse of the expansion we have just used to get this array from the configuration integral. If convergence conditions are satisfied every procedure for collecting the terms must give the same result, but all will not be equally convenient. The series and the integrals in (2.25) do converge if $\mathcal{V}$ is finite, corresponding to a

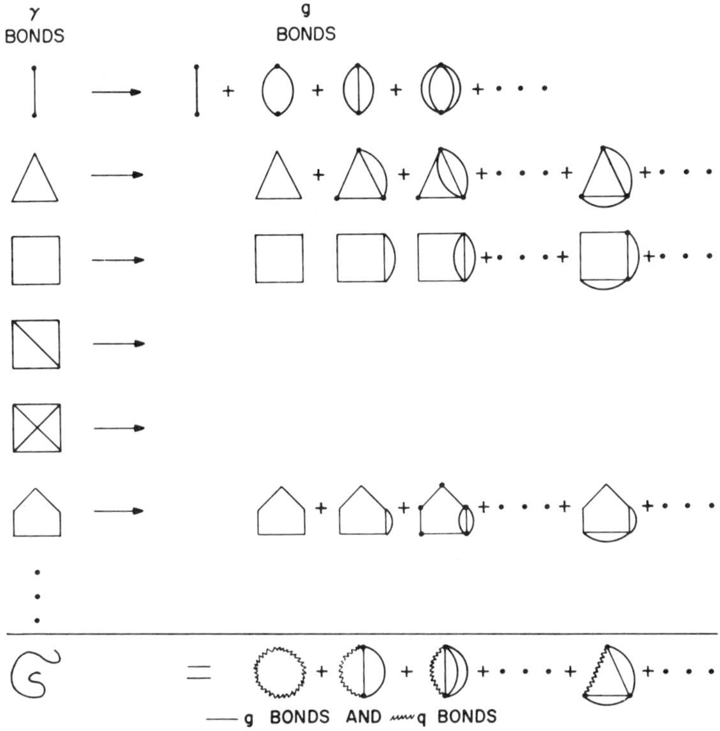

Fig. 2.3. Illustration of Mayer's summation procedure.

physical system. The conventional mathematical procedure would be to pass to the limit of infinite volume only after the divergences as $\mathcal{V} \rightarrow \infty$ are made to cancel by regrouping terms. These aspects are omitted from this short derivation.

We note that if we compare two graphs of Fig. 2.3 with the same number of vertices but with different numbers of bonds then the one with fewer bonds corresponds to the more divergent integral. This is illustrated by the discussion of equation (2.31). As noted above it is the divergent parts of the B_n which contribute to the limiting law. Therefore we sum the most divergent integrals first, using as a criterion the number of bonds minus the number of vertices.

We shall neglect the integral on the first graph, the one corresponding to $-\lambda/4\pi r$, because it has a coefficient of zero in electrically neutral systems. Its appearance here reflects the energy that the system has because of its net charge, $N\varepsilon$, and this should indeed increase without bound as the volume of the system increases at constant c.

The next most divergent integrals come from the graphs in Fig. 2.3 which are simple cycles of $1/r$ bonds. The sum over this series is $\mathfrak{S}_c$.

$$\mathfrak{S}_c = \sum_n{}'' \frac{[-c\lambda]^n}{n!} \frac{[n-1]!}{2} \int g_{12}g_{23}\cdots g_{n-1,n}g_{n,1}\, d\mathbf{r}_{12}\, d\mathbf{r}_{13}\cdots d\mathbf{r}_{1,n} \tag{2.34}$$

where

$$g_{ij} \equiv 1/4\pi r_{ij} \tag{2.35}$$

$\mathbf{r}_{ij}$ is the vector from particle 1 to j, and the integration is over the infinite volume (over the infinite range of $3(n-1)$ Cartesian coordinates).

Now let us define

$$p_n(r_{1,n+1}) \equiv \int g_{12}g_{23}\cdots g_{n,n+1}\, d\mathbf{r}_{12}\, d\mathbf{r}_{13}\cdots d\mathbf{r}_{1,n} \tag{2.36}$$

and

$$q(r) \equiv \sum{}' [-\kappa^2]^{n-1} p_n(r) \tag{2.37}$$

where κ is the function of concentration defined in (2.32), and where $\sum'$ is the sum for all $n \geq 1$. Then it is readily verified that

$$\mathfrak{S}_c = \int_0^\kappa \kappa\, d\kappa \lim_{r\to 0} [g(r) - q(r)] \tag{2.38}$$

Next we take the Fourier transform of both sides of (2.37). (This procedure is discussed in detail in Section 12. Here we only show in a formal way that it leads to a closed expression for $q(r)$.) The transform is

$$\begin{aligned} \tilde{q}(t) &\equiv \int q(r) \exp(i\mathbf{r}\cdot\mathbf{t})\, d\mathbf{r} \\ &= \sum{}' [-\kappa^2]^{n-1} [\tilde{g}(t)]^n \end{aligned} \tag{2.39}$$

where we have used the convolution theorem to deduce that the transform of p_n is the product of the transforms of the factors $g(r)$ in its integrand. We have the transform (Section 12)

$$\tilde{g}(t) = \int g(r) \exp(i\mathbf{t}\cdot\mathbf{r})\, d\mathbf{r} = t^{-2} \tag{2.40}$$

and so

$$\tilde{q}(t) = \kappa^{-2}\sum{}' [-\kappa^2/t^2]^n = [\kappa^2 + t^2]^{-1} \qquad (2.41)$$

Therefore

$$q(r) = [2\pi]^{-3} \int \tilde{q}(t) \exp(-i\mathbf{r}\cdot\mathbf{t})\, d\mathbf{t} = e^{-\kappa r}/4\pi r \qquad (2.42)$$

This result may be substituted in (2.38) and the limit taken to get

$$\mathfrak{S}_c = \kappa^3/12\pi$$

the Debye-Hückel limiting law for $\mathfrak{S}$. The higher terms in $\mathfrak{S}$ are obtained by summing in order the other vertical series in Fig. 2.3. The result is that every such summation results in a graph with the same topology as the simplest graph of the series but with a $q(r)$ bond in place of the chain of $g(r)$ bonds that increases in length in the series. The function $q(r)$ is essentially the Debye potential, and so we see that Mayer's summation procedure results in replacing the Coulomb potential by the Debye potential in the various cluster integrals. When this is done the integrals no longer diverge, but also they are no longer independent of the composition of the system.

The details of the derivations of the higher terms in the series for $\mathfrak{S}$ are not given here because the combinatorial aspects are complicated even for our hypothetical one-component ionic system.

Notes and References

1. R. Brout, *Phys. Rev.* **115,** 824 (1959).
2. R. Fowler and E. A. Guggenheim, *Statistical Thermodynamics*, (Cambridge University Press, New York, 1952).
3. J. E. Mayer, *J. Chem. Phys.* **18,** 1426 (1950).
4. E. W. Montroll and J. C. Ward, *Phys. Fluids* **1,** 55 (1958).

3. Notation and Some Mathematical Topics

In this section we describe a special system of notation for some of the variables in statistical mechanics. The notation is admittedly complicated and it takes some effort to read and write with ease, but its use offers the following rewards: (1) The equations for one-component systems may be written more concisely than if the same information is presented in standard mathematical notation. (2) The equations for multicomponent systems have exactly the same form as those for one-component systems. (3) In the derivations there are some mathematical operations which are difficult to describe, let alone perform, in standard notation and which become as easy as more familiar procedures if done in the specialized notation.

The essential ideas of the notation were introduced in statistical mechanics by Mayer and Montroll[1] and subsequently generalized by McMillan and Mayer[2] to apply to multicomponent systems. Some extensions were suggested by Meeron[3] and still others are introduced below. It is interesting to note that some aspects of the notation are also used in mathematical works on problems of many variables.[4, 5]

In the following discussion the foundation of the specialized notation in the theory of sets is emphasized. This is done with the intention of clarifying the mathematical operations with the notation and to provide a systematic way to describe the various concepts involved. Here we employ only a few elementary concepts of the theory of sets, which is one of the most general and powerful developments of modern mathematics.[6] We also note that more sophisticated treatments of most of the combinatorial problems discussed here and in Section 5 may be found in Riordan's book.[5]

The Set Notation

A set is a collection of *elements*. The elements can be anything at all: similar or dissimilar to each other, real or ideal, simple or complex (i.e., each element may itself be a set). Such a set, with no restrictions, we shall term a general set.

We define here several operations with general sets. Let **A** and **B** be two such sets. Then

$$\mathbf{A} \cup \mathbf{B}$$

called the *union* of **A** and **B** is the set of all elements that are in **A** or **B** or both. The relation that **A** is a *subset* of **B**, every element of **A** being also an element of **B**, is expressed by

$$\mathbf{A} \subseteq \mathbf{B}$$

On the other hand, if **A** is a *proper subset* of **B**,

$$\mathbf{A} \subset \mathbf{B}$$

then every element of **A** is an element of **B**, but there is at least one element of **B** that is not also an element of **A**.

If $\mathbf{A} \subset \mathbf{B}$, then we define the difference,

$$\mathbf{B} \setminus \mathbf{A}$$

as the set of all elements that are in **B** but not in **A**. In order to have the difference defined for all pairs of sets we define the *empty set* as the set containing no elements at all. Then if $\mathbf{A} \not\subset \mathbf{B}$, **A** is not a proper subset of **B**, we have $\mathbf{B} \setminus \mathbf{A}$ equal to the empty set. Finally, if we want to specify that a is an element of **A** we write

$$a \in \mathbf{A}$$

Now we shall discuss some special sets. Consider, as an example of a system that one may encounter in statistical mechanics, a vessel containing a fluid composed of n_1 molecules of species 1, n_2 molecules of species 2, $\cdots$, n_σ molecules of species σ, where σ is the total number of species which may be present. We may regard the n molecules,

$$n \equiv n_1 + n_2 + \cdots + n_\sigma$$

as elements of a set. Now there are three other sets closely related to this set of molecules in the vessel:

First we define the *composition set*, whose elements are the numbers of molecules of each species. We write this definition as

$$\mathbf{n} \equiv n_1, n_2, \cdots, n_\sigma \tag{3.1}$$

where the commas separate the elements of the set. It must be emphasized that the elements of this set are non-negative integers, not molecules, and that the number of elements is σ. If in a particular state of the system there are no molecules of species s in the vessel, then the corresponding element of **n**, namely n_s, is zero in that state. In the particular state of the system in which there are no molecules in the vessel, **n** is *not* the empty set, but is a set of σ elements, all zero:

$$\mathbf{n} = 0, 0, \cdots, 0 \equiv \mathbf{0}$$

The *concentration* set, **c**, is closely related to **n**. We define the concentration (particle number density) of species s in our system as

$$c_s \equiv n_s/\mathcal{V} \tag{3.2}$$

where $\mathcal{V}$ is the volume of the vessel.[7] The concentration set,

$$\mathbf{c} \equiv c_1, c_2, \cdots, c_\sigma \tag{3.3}$$

contains σ elements, each a non-negative real number. If some of the species are absent from the vessel in a particular state the corresponding c_s are zero.

The *coordinate set*, $\{\mathbf{n}\}$, is defined as the set each of whose elements specifies the coordinates of one molecule in the vessel. We write

$$\{\mathbf{n}\} = \{1_1\}, \{2_1\}, \cdots, \{n_1\}, \{1_2\}, \{2_2\}, \\ \cdots, \{n_2\}, \cdots, \{1_\sigma\}, \{2_\sigma\}, \cdots, \{n_\sigma\} \tag{3.4}$$

where $\{m_s\}$ is a complete set of coordinates of the mth molecules of species s. Note that according to this definition a particular coordinate of the mth molecule, say the x coordinate of its center of mass, is not an element of $\{\mathbf{n}\}$ because each element of $\{\mathbf{n}\}$ is a *complete* set of coordinates of one molecule. (We shall, however, at a later stage, divide the coordinate set according to the equation, $\{\mathbf{n}\} = \{\mathbf{n}\}_s \cup \{\mathbf{n}\}_i$ where each element of $\{\mathbf{n}\}_s$ is the set of spatial (center-of-mass) coordinates of one molecule and each element of $\{\mathbf{n}\}_i$ is the set of internal coordinates of one molecule.) We note also that the concept of a coordinate set implies that all of the molecules are distinguishable, even those of the same species, because they occupy different coordinates. Configurations in which two molecules have the same coordinates are negligible because of the repulsive forces between molecules at small separations.

Now suppose we have two coordinate sets, $\{\mathbf{n}\}$ and $\{\mathbf{m}\}$ which are *disjoint*, that is, with no elements in common. Then if

$$\{\mathbf{t}\} = \{\mathbf{n}\} \cup \{\mathbf{m}\}$$

there is a relation among the corresponding composition sets which we shall write as

$$\begin{aligned} \mathbf{t} &= \mathbf{n} + \mathbf{m} \\ &\equiv n_1 + m_1, n_2 + m_2, \cdots, n_\sigma + m_\sigma \end{aligned} \tag{3.5}$$

Another possibility is that we have $\{\mathbf{m}\} \subseteq \{\mathbf{n}\}$. This corresponds to a relation between the corresponding composition sets which we shall write

$$\mathbf{m} \leq \mathbf{n}$$

which is an abbreviation for

$$m_1 \leq n_1, m_2 \leq n_2, \cdots, m_\sigma \leq n_\sigma$$

If $\{\mathbf{m}\} \subseteq \{\mathbf{n}\}$ and

$$\{\mathbf{d}\} \equiv \{\mathbf{n}\} \setminus \{\mathbf{m}\}$$

then there is a relation among the composition sets,

$$\begin{aligned}\mathbf{d} &= \mathbf{n} - \mathbf{m} \\ &\equiv n_1 - m_1, n_2 - m_2, \cdots, n_\sigma - m_\sigma\end{aligned} \tag{3.6}$$

If $\{\mathbf{m}\} = \{\mathbf{n}\}$ then $\{\mathbf{d}\} = \{\mathbf{n}\} \setminus \{\mathbf{m}\}$ is the empty set but $\mathbf{d} = \mathbf{n} - \mathbf{m}$ is not empty, but has σ elements, all zero.

These definitions are made in such a way as to make the operations with composition sets parallel those with coordinate sets. However this is not true of the following definitions, which are essentially abbreviations made in such a way that the equations for multicomponent systems will have the same form as those for one-component systems.

$$n \equiv n_1 + n_2 + \cdots + n_\sigma \tag{3.7}$$

$$c \equiv c_1 + c_2 + \cdots + c_\sigma \tag{3.8}$$

$$\mathbf{n}! \equiv n_1! n_2! \cdots n_\sigma! \qquad \text{(product of factorials)} \tag{3.9}$$

$$\mathbf{c}^{\mathbf{n}} \equiv c_1^{n_1} c_2^{n_2} \cdots c_\sigma^{n_\sigma} \qquad \text{(product of exponentials)} \tag{3.10}$$

We also define the chemical potential set,

$$\boldsymbol{\mu} \equiv \mu_1, \mu_2, \cdots, \mu_\sigma \tag{3.11}$$

The elements of $\boldsymbol{\mu}$ are the chemical potentials of the σ species. In this case each element is a real number, positive, zero, or negative. We also have

$$\boldsymbol{\mu} \cdot \mathbf{n} \equiv \mathbf{n} \cdot \boldsymbol{\mu} \equiv n_1\mu_1 + n_2\mu_2 + \cdots + n_\sigma\mu_\sigma \tag{3.12}$$

It is apparent that some, but not all, of the operations defined for the sets $\mathbf{n}$, $\mathbf{c}$, and $\boldsymbol{\mu}$ are like those with vectors in σ dimensions. It must be emphasized that n, c, $\mathbf{n}!$, $\mathbf{c}^{\mathbf{n}}$, and $\boldsymbol{\mu} \cdot \mathbf{n}$ are not sets, but real numbers.[8] Note also that according to our definitions we have

$$\mathbf{c}^{\mathbf{n}+\mathbf{m}} = \mathbf{c}^{\mathbf{n}}\mathbf{c}^{\mathbf{m}} \tag{3.13}$$

$$\mathbf{m}!\mathbf{n}! \neq [\mathbf{n} + \mathbf{m}]! = [n_1 + m_1]![n_2 + m_2]! \cdots [n_\sigma + m_\sigma]! \tag{3.14}$$

and

$$\{\mathbf{n} + \mathbf{m}\} = \{\mathbf{n}\} \cup \{\mathbf{m}\} \tag{3.15}$$

$$\{\mathbf{n} - \mathbf{m}\} = \{\mathbf{n}\} \setminus \{\mathbf{m}\} \tag{3.16}$$

TABLE 3.1. Classification of some sets.

Name of set	Symbol	Element
general set	**A, B**	unrestricted
coordinate set	{**n**}, {**m**}	coordinates of one molecule
composition set	**n, m**	non-negative integer
partition	**p**	non-negative integer
tree	**t**	non-negative integer
concentration set	**c**	non-negative real number
fugacity set	**z**	non-negative real number
activity set	**a**	non-negative real number
chemical potential set	$\boldsymbol{\mu}$	real number

In Table 3.1 we list the sets we have introduced and some others that will be introduced later. The symbols listed do not represent all of the symbols that will be used for each kind of set. The classification is according to the nature of the elements.

Miscellaneous Conventions

Brackets. Ten-point lightface braces are used only to represent coordinate sets, as $\{\mathbf{n}\}$. Ten-point lightface parentheses are used in mathematical expressions only to indicate functional dependence, as in $y = f(x)$. All brackets, [ ], and the larger styles of braces and parentheses are used in the usual way to represent mathematical grouping.

Range of summation. The same conventions are used for continued products.

$\sum f(\mathbf{n})$ = sum of $f(\mathbf{n})$ for all $n \geq 0$.

$\sum' f(\mathbf{n})$ = sum of $f(\mathbf{n})$ for all $n \geq 1$.

$\sum'' f(\mathbf{n})$ = sum of $f(\mathbf{n})$ for all $n \geq 2$.

$\sum_{i \in \mathbf{n}} f(i)$ = sum of $f(i)$ for every i that is an element of $\mathbf{n}$.

$\sum_{\mathbf{m} \leq \mathbf{n}} f(\mathbf{m})$ = sum of $f(\mathbf{m})$ for all $\mathbf{m} \leq \mathbf{n}$.

$\sum_{\{\mathbf{m}\} \subseteq \{\mathbf{n}\}} f(\{\mathbf{m}\})$ = sum of $f(\{\mathbf{m}\})$ for every $\{\mathbf{m}\}$ that is a subset of $\{\mathbf{n}\}$.

$\sum_{\mathbf{p}]\{\mathbf{n}\}} f(\mathbf{p})$ = sum of $f(\mathbf{p})$ for all $\mathbf{p}$ that are partitions of $\{\mathbf{n}\}$. See p. 28 ff.

$\sum_{\mathbf{p}]\mathbf{n}} f(\mathbf{p})$ = sum of $f(\mathbf{p})$ for all $\mathbf{p}$ that are partitions of $\mathbf{n}$. See p. 28 ff.

$\sum_{\mathbf{t}]\{\mathbf{n}\}} f(\mathbf{t})$ = sum of $f(\mathbf{t})$ for all $\mathbf{t}$ that are trees of $\{\mathbf{n}\}$. See p. 48 ff.

$\sum_{\mathbf{t}]\mathbf{n}} f(\mathbf{t})$ = sum of $f(\mathbf{t})$ for all $\mathbf{t}$ that are trees of $\mathbf{n}$. See p. 48 ff.

Integration

$$\int_{\mathcal{V}} f(\{\mathbf{n}\})\, d\{\mathbf{n}\}$$

is the integral of $f\{\mathbf{n}\}$ over the configuration space formed by the continuum of values of $\{\mathbf{n}\}$. The range includes all $\{\mathbf{n}\}$ that correspond to the molecules of the set $\mathbf{n}$ being within the volume $\mathcal{V}$. If $\mathcal{V}$ is omitted, the range is understood to be infinite.

To illustrate the notation we give here two equations in both the standard notation and in the set notation.

In the standard notation the configuration integral for a multicomponent system is

$$Z(n_a, n_b, \cdots, \mathcal{V}, T) = \int_{\mathcal{V}} \exp\,[-U(R_{a1}, R_{a2}, \cdots, R_{an_a}, R_{b1}, R_{b2}, \cdots)/kT]\, dR_{a1}\, dR_{a2} \cdots dR_{an_a}\, dR_{b1}\, dR_{b2} \cdots$$

where

$$R_{im} \equiv x_{im}, y_{im}, z_{im}, \text{ internal coordinates}$$

and

$$dR_{im} = dx_{im} dy_{im} dz_{im}\, d\,[\text{internal coordinates}]$$

In the present notation we have

$$Z(\mathbf{n}, \mathcal{V}, T) = \int_{\mathcal{V}} \exp\,[-U(\{\mathbf{n}\})/kT]\, d\{\mathbf{n}\}$$

The grand partition function is, in standard notation,

$$\Xi(z_a, z_b, \cdots z_\sigma, \mathcal{V}, T)$$

$$= \sum_{n_a \geq 0} \sum_{n_b \geq 0} \cdots \sum_{n_\sigma \geq 0} [z_a^{n_a}/n_a!][z_b^{n_b}/n_b!] \cdots [z_\sigma^{n_\sigma}/n_\sigma!] \times Z(n_a, n_b, \cdots, n_\sigma, \mathcal{V}, T)$$

and in the present notation,

$$\Xi(\mathbf{z}, \mathcal{V}, T) = \sum [\mathbf{z}^{\mathbf{n}}/\mathbf{n}!] Z(\mathbf{n}, \mathcal{V}, T)$$

In the remainder of Section 3 several mathematical theorems that are useful in the derivation of the cluster theories are given. It is hoped that the study of these theorems will give the reader sufficient exercise in the use of the specialized notation to enable him to read the following sections with ease.

Two Generalizations of the Binomial Theorem

A generalized binomial theorem is

$$[\mathbf{x} + \mathbf{y}]^{\mathbf{a}} = \sum_{\mathbf{b} \leq \mathbf{a}} \binom{\mathbf{a}}{\mathbf{b}} \mathbf{x}^{\mathbf{b}} \mathbf{y}^{\mathbf{a}-\mathbf{b}} \tag{3.17}$$

We assume that the elements of **a** are non-negative integers. The definition of the generalized binomial coefficient is

$$\binom{\mathbf{a}}{\mathbf{b}} \equiv \prod_{s=1}^{\sigma} \binom{a_s}{b_s} = \prod_{s=1}^{\sigma} a_s!/b_s![a_s - b_s]! \tag{3.18}$$

To derive (3.17), we expand the left side as follows. The first step is based on the definition, (3.10).

$$\begin{aligned} [\mathbf{x} + \mathbf{y}]^{\mathbf{a}} &= \prod_s [x_s + y_s]^{a_s} = \prod_s \sum_{b_s=0}^{a_s} \binom{a_s}{b_s} x_s^{b_s} y_s^{a_s - b_s} \\ &= [y_1^{a_1} + a_1 y_1^{a_1-1} x_1 + \cdots][y_2^{a_2} + a_2 y_2^{a_2-1} x_2 + \cdots] \cdots \end{aligned} \tag{3.19}$$

Every product of σ factors, one being a term of the first sum of $a_1 + 1$ terms, the next being a term of the second sum of $a_2 + 1$ terms, $\cdots$, and the last being a term of the σth sum of $a_\sigma + 1$ terms, corresponds to exactly one set,

$$\mathbf{a} - \mathbf{b} = a_1 - b_1, a_2 - b_2, \cdots, a_\sigma - b_\sigma$$

and hence, since **a** is given, to one set, **b**. Furthermore, every **b** that satisfies the condition

$$\mathbf{b} \leq \mathbf{a}$$

corresponds to one of the terms obtained by multiplying out the last member of (3.19), and no term corresponds to a **b** that is not a subset of **a**. Finally, every term of this expansion is of the form,

$$\binom{\mathbf{a}}{\mathbf{b}} \mathbf{x}^{\mathbf{b}} \mathbf{y}^{\mathbf{a}-\mathbf{b}}$$

and so (3.17) is obtained.

Another generalization of the binomial theorem that we shall require is the multinomial theorem,

$$[x_1 + x_2 + \cdots + x_\sigma]^n = n! \sum \mathbf{x}^{\mathbf{n}}/\mathbf{n}! \tag{3.20}$$

where the sum is over all composition sets **n** such that $\sum n_s = n$. Proofs are given by Mayer and Mayer[9] and Riordan.[5]

Theorems on Sequences of Series

Let **m** and **n** be composition sets and let **x** be a set of continuous variables such that $\mathbf{x}^{\mathbf{n}}$ is defined.

Let $A(\mathbf{m},\mathbf{x})$ and $B(\mathbf{m},\mathbf{x})$ be functions that are defined for some definite range of **x** and assume further that for $n > N$, where N is some positive integer, $0 = A(\mathbf{m} + \mathbf{n}, \mathbf{x}) = B(\mathbf{m} + \mathbf{n}, \mathbf{x})$.

Then if the A functions are related to the B functions by the series,

$$A(\mathbf{m}, \mathbf{x}) = \sum_{n=0}^{N} [\mathbf{x}^{\mathbf{n}}/\mathbf{n}!]B(\mathbf{m} + \mathbf{n}, \mathbf{x}) \tag{3.21}$$

the inverse of this series is

$$B(\mathbf{m}, \mathbf{x}) = \sum_{n=0}^{N} [[-\mathbf{x}]^{\mathbf{n}}/\mathbf{n}!]A(\mathbf{m} + \mathbf{n}, \mathbf{x}) \tag{3.21$'$}$$

where

$$[-\mathbf{x}]^{\mathbf{n}} \equiv [-x_1]^{n_1}[-x_2]^{n_2} \cdots [-x_\sigma]^{n_\sigma} = [-1]^n \mathbf{x}^{\mathbf{n}} \tag{3.22}$$

This inversion is proved by direct substitution in the following way.

$$B(\mathbf{m}, \mathbf{x}) = \sum [[-\mathbf{x}]^{\mathbf{n}}/\mathbf{n}!] \sum [\mathbf{x}^{\mathbf{r}}/\mathbf{r}!]B(\mathbf{m} + \mathbf{n} + \mathbf{r}, \mathbf{x})$$

We define $\mathbf{u} \equiv \mathbf{n} + \mathbf{r}$ and change indices accordingly to obtain,

$$B(\mathbf{m}, \mathbf{x}) = \sum_{u=0}^{2N} S_{\mathbf{u}} B(\mathbf{m} + \mathbf{u}, \mathbf{x}) \tag{3.22$'$}$$

For $u > N$, the B coefficients on the right vanish. For $u \leq N$ we have

$$\begin{aligned} S_{\mathbf{u}} &= \sum_{\mathbf{n}+\mathbf{r}=\mathbf{u}} \frac{[-\mathbf{x}]^{\mathbf{n}}[\mathbf{x}]^{\mathbf{r}}}{\mathbf{n}!\mathbf{r}!} \\ &= \frac{1}{\mathbf{u}!} \sum_{\mathbf{n}\leq\mathbf{u}} \binom{\mathbf{u}}{\mathbf{n}} [-\mathbf{x}]^{\mathbf{n}} \mathbf{x}^{\mathbf{u}-\mathbf{n}} = [\mathbf{x} - \mathbf{x}]^{\mathbf{u}}/\mathbf{u}! \end{aligned} \tag{3.23}$$

where, in the last equality, we use (3.17). Equation (3.23) shows that $S_{\mathbf{u}}$ vanishes except in the case that $u = 0$, when it reduces to unity. Therefore the only non-vanishing term on the right of (3.22)$'$ is $B(\mathbf{m},\mathbf{x})$ and the inversion, (3.21)$'$, is proved.

A special case of such sequences of series is defined by the additional equation,

$$B(\mathbf{m},\mathbf{x}) \equiv A(\mathbf{m},\mathbf{0}). \tag{3.24}$$

Then if

$$A(\mathbf{m},\mathbf{x}) = \sum_{n=0}^{N} [\mathbf{x}^{\mathbf{n}}/\mathbf{n}!] A(\mathbf{m} + \mathbf{n},\mathbf{0}) \tag{3.25}$$

the inverse is

$$A(\mathbf{m},\mathbf{0}) = \sum_{n=0}^{N} [[-\mathbf{x}]^{\mathbf{n}}/\mathbf{n}!] A(\mathbf{m} + \mathbf{n}, \mathbf{x}) \tag{3.25$'$}$$

In this special case there is also a *generalization* of the original series, (3.25),

$$A(\mathbf{m}, \mathbf{x} + \mathbf{y}) = \sum_{n=0}^{N} [\mathbf{x}^{\mathbf{n}}/\mathbf{n}!] A(\mathbf{m} + \mathbf{n}, \mathbf{y}) \tag{3.25$''$}$$

This generalization is derived in the following way. We write (3.25) in the form,

$$A(\mathbf{m}, \mathbf{x} + \mathbf{y}) = \sum [[\mathbf{x} + \mathbf{y}]^{\mathbf{n}}/\mathbf{n}!] A(\mathbf{m} + \mathbf{n}, \mathbf{0}). \tag{3.26}$$

Now (3.25)$'$ is substituted in this to obtain

$$\begin{aligned} &A(\mathbf{m},\mathbf{x} + \mathbf{y}) \\ &= \sum_{n} [[\mathbf{x} + \mathbf{y}]^{\mathbf{n}}/\mathbf{n}!] \sum_{r} [[-\mathbf{x}]^{\mathbf{r}}/\mathbf{r}!] A(\mathbf{m} + \mathbf{n} + \mathbf{r},\mathbf{x}) \end{aligned} \tag{3.27}$$

We define $\mathbf{u} = \mathbf{n} + \mathbf{r}$ and change indices in (3.27) to obtain

$$A(\mathbf{m},\mathbf{x} + \mathbf{y}) = \sum_{u=0}^{2N} S_{\mathbf{u}} A(\mathbf{m} + \mathbf{u},\mathbf{x}) \qquad (3.28)$$

The A coefficients on the right vanish for all $u > N$. For $u \leq N$ we have

$$S_{\mathbf{u}} = \sum_{\mathbf{n}+\mathbf{r}=\mathbf{u}} [[\mathbf{x} + \mathbf{y}]^{\mathbf{n}}/\mathbf{n}!][[-\mathbf{x}]^{\mathbf{r}}/\mathbf{r}!] = (1/\mathbf{u}!)[\mathbf{x} + \mathbf{y} - \mathbf{x}]^{\mathbf{u}} = \mathbf{y}^{\mathbf{u}}/\mathbf{u}!$$

With this substitution, (3.28) reduces to the same form as (3.25)″.

These equations are evidently valid for any integer N in the range from zero to infinity. For infinite N, equation (3.25)″ is similar to Taylor's series if $A(\mathbf{m},\mathbf{x}) \equiv \partial^{\mathbf{m}} A/\partial \mathbf{x}^{\mathbf{m}}$. However the theorems on sequences of series do not depend on the continuity of the functions and their derivatives in the range of variables in which the expansions are made. The relations among functions that are described here are similar to some that are known in the theory of interpolation.[10]

Some other forms of the inversion theorem for sequences of series are listed here. The first equation defines the sequence of series, the second equation is the inverse. If the B functions are defined in a way analogous to (3.24) then in each case except (3.33) a generalization, analogous to (3.25)″, can also be obtained.

$$A(m,x) = \sum_{n=0}^{N} [x^n/n!] B(m + n, x)$$
$$B(m,x) = \sum_{n=0}^{N} [[-x]^n/n!] A(m + n,x) \qquad (3.29)$$

$$A(\{\mathbf{m}\},\mathbf{x}) = \sum_{n=0}^{N} [\mathbf{x}^{\mathbf{n}}/\mathbf{n}!] \int_{\mathcal{V}} B(\{\mathbf{m} + \mathbf{n}\},\mathbf{x})\, d\{\mathbf{n}\}$$
$$B(\{\mathbf{m}\},\mathbf{x}) = \sum_{n=0}^{N} [[-\mathbf{x}]^{\mathbf{n}}/\mathbf{n}!] \int_{\mathcal{V}} A(\{\mathbf{m} + \mathbf{n}\},\mathbf{x})\, d\{\mathbf{n}\} \qquad (3.30)$$

$$A(\mathbf{N},\mathbf{m},\mathbf{x}) = \sum_{\mathbf{n}\leq\mathbf{N}} [\mathbf{x}^{\mathbf{n}}/\mathbf{n}!] B(\mathbf{N},\mathbf{n} + \mathbf{m},\mathbf{x})$$
$$B(\mathbf{N},\mathbf{m},\mathbf{x}) = \sum_{\mathbf{n}\leq\mathbf{N}} [[-\mathbf{x}]^{\mathbf{n}}/\mathbf{n}!] A(\mathbf{N},\mathbf{n} + \mathbf{m},\mathbf{x}) \qquad (3.31)$$

$$A(\mathbf{N},\mathbf{x}) = \sum_{\mathbf{n}\leq\mathbf{N}} [\mathbf{x}^{\mathbf{n}}/\mathbf{n}!] B(\mathbf{N} - \mathbf{n},\mathbf{x})$$
$$B(\mathbf{N},\mathbf{x}) = \sum_{\mathbf{n}\leq\mathbf{N}} [[-\mathbf{x}]^{\mathbf{n}}/\mathbf{n}!] A(\mathbf{N} - \mathbf{n},\mathbf{x}) \qquad (3.32)$$

$$A(\{\mathbf{N}\}) = \sum_{\{\mathbf{n}\}\subseteq\{\mathbf{N}\}} B(\{\mathbf{N} - \mathbf{n}\})$$

$$B(\{\mathbf{N}\}) = \sum_{\{\mathbf{n}\}\subseteq\{\mathbf{N}\}} [-1]^n A(\{\mathbf{N} - \mathbf{n}\}) \tag{3.33}$$

Equation (3.29) follows from (3.21) on letting $\sigma \rightarrow 1$, but may also be proved directly in the same way used for (3.21). This and some related equations were used by Mayer[11] in the theory of one-component systems. Equations (3.30) and (3.31) may readily be derived[2] in the same way as (3.21). Equation (3.32) may be derived from (3.31) by first considering as a special case of the latter,

$$A(\mathbf{N},\mathbf{m},\mathbf{x}) \rightarrow A(\mathbf{N} - \mathbf{m},\mathbf{x})$$

and

$$B(\mathbf{N},\mathbf{m},\mathbf{x}) \rightarrow B(\mathbf{N} - \mathbf{m},\mathbf{x}) \qquad \text{for} \quad \mathbf{m} \leq \mathbf{N}$$

and then letting $m \rightarrow 0$.

To derive (3.33) we begin with (3.32) and consider the special case in which each element of **N** is either 0 or 1. This corresponds to the case in which **N** is a set of numbers of molecules but all of the molecules are distinguishable from each other, or the case in which the elements of **N** are numbers

$$n_{\{1_1\}}, n_{\{2_1\}}, \cdots, n_{\{N_1\}}, n_{\{1_2\}}, n_{\{2_2\}}, \cdots, n_{\{N_2\}}, \cdots$$

of elements of a coordinate set {**N**} (compare with equation (3.4)). In this case there is a one-to-one correspondence between the elements of **N** and of {**N**}, and we may write {**N**} in place of **N** in the equations. In this case, also, $\mathbf{n}! = 1$. We also let every $x_s \rightarrow 1$ and the result is (3.33). This inversion, which is related to the problem of the definition of the components of the potentials of average force, has been derived in a different way by McMillan and Mayer.[2]

Partitions

A *covering*, **B**, of a set **A** is a set such that every element of **A** is an element of **B**. Obviously if **B** is a covering of **A**, **A** is a subset of **B**. This may be expressed by the notation $\mathbf{A} \subseteq \mathbf{B}$ or $\mathbf{B} \supseteq \mathbf{A}$. We shall have particular use for two kinds of coverings, *partitions* which are described here and *trees* which are described in Section 5.

If **B** is a partition of **A** then **B** consists of several subsets, $\mathbf{B}_1$, $\mathbf{B}_2$, $\mathbf{B}_3$, $\cdots$ such that

i) $\mathbf{B}_1 \cup \mathbf{B}_2 \cup \mathbf{B}_3 \cup \cdots = \mathbf{A}$

and

ii) No element of one subset, $\mathbf{B}_i$, is an element of another, $\mathbf{B}_j$. The first condition specifies that **B** is a covering of **A**, and also that no element of **B** is not an element of **A**. The second condition specifies that the subsets of **B** are disjoint. The partition concept is readily represented geometrically if the elements of a set are represented as dots on a sheet (Fig. 3.1). This representation is geometric, but a moment's reflection makes it clear that the *meaning* is topological: the coordinates of the dots on the sheet have no significance, nor does the two dimensional nature of our representation.

The partition concept may be directly applied to coordinate sets. For example if

$$\{\mathbf{a}\} \cup \{\mathbf{b}\} \cup \{\mathbf{c}\} = \{\mathbf{n}\}$$

and if $\{\mathbf{a}\}$, $\{\mathbf{b}\}$, and $\{\mathbf{c}\}$ are disjoint, then the three subsets comprise a partition of $\{\mathbf{n}\}$. For purposes of analysis we now define a *partition set* having one element for each subset of $\{\mathbf{n}\}$. Each such element may be either 0 or 1. The partition in the preceding example corresponds to the partition set,

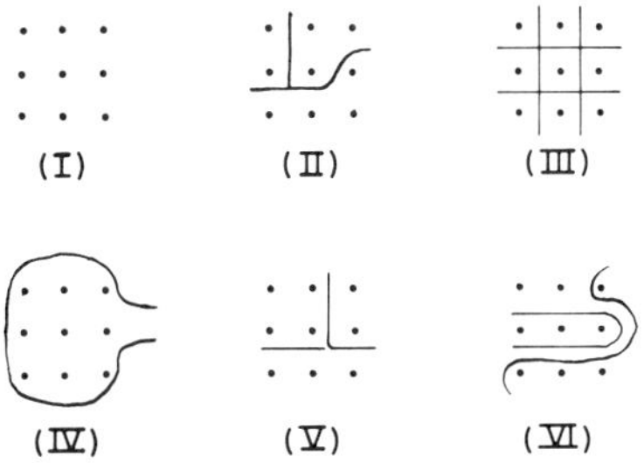

Fig. 3.1. (I) represents a set of 9 elements. (II) represents a partition of the set into three subsets. (III) is a partition of the set into 9 subsets, of one element each. This partition closely resembles the original set but so does the "trivial" partition represented in (IV), a partition of the set into one subset of 9 elements. Whether the partitions (V) and (VI) are distinguishable from each other and from (III) depends on whether the elements of the original set are all distinguishable from each other.

$$\mathbf{p} = 1_{\{a\}}, 1_{\{b\}}, 1_{\{c\}}, 0_{\{d\}}, 0_{\{e\}}, \cdots$$

and the correspondence is one to one: each partition set also defines a particular partition. We shall write a partition set for a coordinate set in the form

$$\mathbf{p}]\{\mathbf{n}\}$$

in order to suggest a relation to the standard notation for a covering, $\mathbf{B} \supseteq \mathbf{A}$.

Now we proceed to define a partition of a composition set, $\mathbf{n}$, and the corresponding partition set, $\mathbf{p}]\mathbf{n}$. If $\{\mathbf{a}\}$, $\{\mathbf{b}\}$, $\{\mathbf{c}\}$ is a partition of $\{\mathbf{n}\}$ then $\mathbf{a},\mathbf{b},\mathbf{c}$ is a partition of $\mathbf{n}$. In general if

$$\sum \mathbf{n}_i = \mathbf{n}$$

then the subsets $\mathbf{n}_1, \mathbf{n}_2, \cdots$ in the summand comprise a partition of $\mathbf{n}$. The partition set $\mathbf{p}]\mathbf{n}$ has one element for each distinguishable subset of $\mathbf{n}$, the element being the number of times this subset appears in the summand:

$$\mathbf{p} = p_{\mathbf{n}_1}, p_{\mathbf{n}_2}, \cdots$$

The elements of $\mathbf{p}]\mathbf{n}$ are thus non-negative integers. As a result, if $\mathbf{p}]\mathbf{n}$ then[12]

$$\sum_{\mathbf{n}_i \leq \mathbf{n}} \mathbf{n}_i p_{\mathbf{n}_i} = \mathbf{n} \qquad (3.34)$$

The importance of the partition sets in analysis lies in the fact that the operations which have been defined for composition sets are also defined for partition sets. In the sequel we shall simply call $\mathbf{p}]\{\mathbf{n}\}$ and $\mathbf{p}]\mathbf{n}$ partitions when no ambiguity can result from this usage.

As elementary examples of some $\mathbf{p}]\mathbf{n}$, let $\mathbf{n} = 5$, a set of five identical elements. Then $\sigma = 1$. The subsets of 5 are 1, 2, 3, 4, and 5 so any $\mathbf{p}]5$ is a set of the form

$$\mathbf{p} = p_1, p_2, p_3, p_4, p_5$$

and we may make Table 3.2.

Partitions in the sense illustrated by this example are important in many statistical problems and have been very thoroughly investigated.[5, 13]

In the following section we discuss a problem in which partitions of more general composition sets ($\sigma > 1$) appear in a natural way.

TABLE 3.2. Example of partition notation.

Partition	$\mathbf{p}]5$
1, 1, 1, 1, 1	5, 0, 0, 0, 0
1, 1, 3	2, 0, 1, 0, 0
1, 2, 2	1, 2, 0, 0, 0
1, 4	1, 0, 0, 1, 0
5	0, 0, 0, 0, 1

It will be convenient to employ the following definitions:

$$\sum_{\mathbf{p}]\{\mathbf{n}\}} a_{\mathbf{p}} = \text{sum of } a_{\mathbf{p}} \text{ for all partitions of } \{\mathbf{n}\}$$

$$\sum_{\mathbf{p}]\mathbf{n}} a_{\mathbf{p}} = \text{sum of } a_{\mathbf{p}} \text{ for all partitions of } \mathbf{n}$$

Cumulants and Moments

We consider here the relations among the coefficients in the equation

$$\varphi(x) = \sum_n x^n m_n/n! = \exp\left[\sum_n{}' x^n k_n/n!\right]$$

This problem is well known in the theory of the statistical treatment of observations[14] in which $\varphi(x)$, called the characteristic function, is the Fourier transform of a distribution, m_n is the nth *moment* of the distribution, and k_n is the nth *cumulant* or Thiele seminvariant. (Kendall[14] explains why the name, cumulant, is preferred.) In the statistical theory the moments can be used to characterize a distribution, but for some applications the cumulants are more satisfactory for the same purpose. An equation of the same form, but apparently with entirely different meaning, is encountered in the cluster theories of statistical mechanics, and the problem of relating the cumulants to the moments was there solved independently of the much earlier investigations of Thiele. It was first pointed out by Ono[15] that the mathematical problem is the same in both cases. Hill[16] discusses the other methods which have been used for the solution of the problem for one-component systems.

We shall consider the problem here only in multivariate or multicomponent form, the terms being appropriate to the statistical theory of observations and to the cluster theories, respectively. We write the equation in the form

$$\sum \mathbf{x}^{\mathbf{n}} M_{\mathbf{n}} = \exp\left[\sum{}' \mathbf{x}^{\mathbf{n}} K_{\mathbf{n}}\right] \tag{3.35}$$

and shall call $M_{\mathbf{n}}$ the moment and $K_{\mathbf{n}}$ the cumulant for set $\mathbf{n}$, although these definitions differ by $\mathbf{n}!$ from those used previously. The relation of moments to cumulants for the multivariate case is also treated by Kendall[14]; a different derivation is given by Meeron,[3] who was the first to use the multivariate cumulant-moment relation in statistical mechanics. The proof we give follows Kendall's.

We expand the right side of (3.35) as follows.

$$\begin{aligned}\sum \mathbf{x}^{\mathbf{n}} M_{\mathbf{n}} = \prod{}' \exp\left(\mathbf{x}^{\mathbf{m}} K_{\mathbf{m}}\right) = \prod_{\mathbf{m}} [1 + \mathbf{x}^{\mathbf{m}} K_{\mathbf{m}} \\ + [1/2!][\mathbf{x}^{\mathbf{m}} K_{\mathbf{m}}]^2 + \cdots + [1/p_{\mathbf{m}}!][\mathbf{x}^{\mathbf{m}} K_{\mathbf{m}}]^{p_{\mathbf{m}}} + \cdots]\end{aligned} \tag{3.36}$$

We now equate coefficients of $\mathbf{x}^{\mathbf{n}}$ in this equation. A given product, characterized by $\mathbf{m}_1, \mathbf{m}_2, \cdots$ is found only once among the terms obtained by multiplying out the right side of (3.36). Therefore the coefficient of $\mathbf{x}^{\mathbf{n}}$ on the right is obtained simply by summing over all sets, $\mathbf{m}_1, \mathbf{m}_2, \cdots$ that are partitions of $\mathbf{n}$. So we obtain

$$M_{\mathbf{n}} = \sum_{\mathbf{p}]\mathbf{n}} \prod_{\mathbf{m} \leq \mathbf{n}} K_{\mathbf{m}}^{p_{\mathbf{m}}}/p_{\mathbf{m}}! \tag{3.37}$$

By an obvious generalization of the set-exponent notation (3.10) we may write (3.37) more concisely as

$$M_{\mathbf{n}} = \sum_{\mathbf{p}]\mathbf{n}} \mathbf{K}^{\mathbf{p}}/\mathbf{p}! \tag{3.38}$$

We see that in a sense the moment-cumulant relation *generates* partitions.

We shall not need the inverse of (3.38) but its derivation is included as an additional exercise with the set notation. We take the logarithm of (3.35) to obtain, for $M_0 = 1$,

$$\begin{aligned}\sum{}' \mathbf{x}^{\mathbf{n}} K_{\mathbf{n}} = \ln\left(\sum \mathbf{x}^{\mathbf{m}} M_{\mathbf{m}}\right) &= \sum_a{}' [-1]^{a+1} [\sum_m{}' \mathbf{x}^{\mathbf{m}} M_{\mathbf{m}}]^a / a \\ &= \sum{}' [[-1]^{a+1}/a] a! \sum_{p=a} \prod_{\mathbf{m}} [\mathbf{x}^{\mathbf{m}} M_{\mathbf{m}}]^{p_{\mathbf{m}}}/p_{\mathbf{m}}!\end{aligned} \tag{3.39}$$

The condition $p = a$ means $\sum p_{\mathbf{m}} = a$ for each term. Compare with equation (3.20). The condition that any one of the products in the last member of (3.39) contain **x** to exactly the **n** power is

$$\sum \mathbf{m}p_{\mathrm{m}} = \mathbf{n}$$

or, **p**]**n**. We add together all such products and so obtain

$$\begin{aligned} K_{\mathrm{n}} &= \sum_{\mathbf{p}]\mathbf{n}} [-1]^{p+1}[p-1]! \prod_{\mathrm{m}} [M_{\mathrm{m}}^{p_{\mathrm{m}}}/p_{\mathrm{m}}!] \\ &= \sum_{\mathbf{p}]\mathbf{n}} [-1]^{p+1}[p-1]!\mathbf{M}^{\mathbf{p}}/\mathbf{p}! \end{aligned} \qquad (3.40)$$

For multicomponent systems an equation equivalent to (3.38) was first published by Fuchs.[17] The inverse, (3.40) was first given by McMillan and Mayer.[2] Kendall's[13] equations for multivariate statistics are in somewhat less general form.

Notes and References

1. J. E. Mayer and E. Montroll, *J. Chem. Phys.* **9,** 2 (1941).
2. W. G. McMillan and J. E. Mayer, *J. Chem. Phys.* **13,** 276 (1945).
3. E. Meeron, *J. Chem. Phys.* **27,** 1238 (1957).
4. L. Schwartz, *Theorie des Distributions* (Hermann & Cie, Paris, 1957).
5. J. Riordan, *Introduction to Combinatorial Analysis* (John Wiley & Sons, Inc., New York, 1958).
6. An excellent introduction in workbook form is given by D. J. Aiken and C. A. Beseman, *Modern Mathematics* (McGraw-Hill Book Company, Inc., New York, 1959). A more general account is given by E. Kamke, *Theory of Sets* (Dover Publications, New York, 1950). For more sophistication see P. S. Aleksandrov, *Combinatorial Topology* (Graylock Press, Rochester, N. Y., 1956), Vol. I.
7. We shall always use script capital letters to represent extensive thermodynamic variables.
8. Although it would be confusing to do so here, it is possible to regard ordinary real numbers as sets. See Kamke, *Theory of Sets* (Dover Publications, New York, 1950).
9. J. E. Mayer and M. G. Mayer, *Statistical Mechanics*, (John Wiley & Sons, Inc., New York, 1940).
10. C. Jordan, *Calculus of Finite Differences* (Chelsea Publishing Company, New York, 1950), second ed., pp. 8–10.
11. J. E. Mayer, *J. Chem. Phys.* **10,** 629 (1942).
12. We define $\mathbf{m}p_{\mathrm{m}} = \mathbf{m} + \mathbf{m} + \dots + \mathbf{m}$, with p_{m} terms on the right. If m_1, m_2, ... are the elements of **m** then $\mathbf{m}p_{\mathrm{m}} = p_{\mathrm{m}}m_1, p_{\mathrm{m}}m_2, \dots$. This is like the multiplication of a vector by a scalar.
13. H. Gupta, *Tables of Partitions*, (Cambridge University Press, New York, 1958).
14. M. G. Kendall, *The Advanced Theory of Statistics* (Charles Griffin & Co., Ltd., London, 1948), Vol. I, Sections 3.11–3.13 and 3.28–3.29.

15. S. Ono, *J. Chem. Phys.* **19,** 504 (1941).
16. T. L. Hill, *Statistical Mechanics* (McGraw-Hill Book Company, Inc., New York, 1956) Chapter 5.
17. K. Fuchs, *Proc. Roy. Soc.* (*London*)**A179,** 408 (1942).

4. Potentials of Average Force

There remain a few specialized mathematical topics to explore before beginning the derivation of the cluster theories, but these topics, much more than those covered in Section 3, seem excessively abstract if completely removed from their physical origins. Therefore we shall briefly describe a physical situation that is appropriate to these topics and then return to the mathematics in Section 5.

We begin with potentials of average force, which are closely related to distribution functions and correlation functions. The underlying theory and the wide applicability of all these concepts have been discussed by Hill.[1]

Let us consider a closed system of interacting molecules of composition set **N**. For simplicity we assume that the system is in equilibrium and that external fields are absent. For a classical system the potential energy is then a function of $3N$ spatial [center-of-mass] coordinates of the molecules and of a number of internal coordinates that depends on the structure of the molecules.

Let $\{\mathbf{N}\}$ represent the *spatial* coordinates, $\{\mathbf{N}\}_i$ the internal coordinates,[2] and $U(\{\mathbf{N}\},\{\mathbf{N}\}_i)$ the potential energy for a particular configuration. Then the probability of this configuration is proportional to

$$\exp\left[-U(\{\mathbf{N}\},\{\mathbf{N}\}_i)/kT\right]$$

and the probability of a definite configuration specified only by the spatial coordinates, $\{\mathbf{N}\}$, is proportional to

$$\int \exp\left[-U)\{\mathbf{N}\},\{\mathbf{N}\}_i)/kT\right] d\{\mathbf{N}\}_i \equiv \exp\left(-U(\{\mathbf{N}\})/kT\right) \int d\{\mathbf{N}\}_i \tag{4.1}$$

We regard (4.1) as the definition of $U(\{\mathbf{N}\})$ for a classical system. We shall often abbreviate it as $U_{\mathbf{N}}$ or write it out more fully as $U_{\mathbf{N}}(\{\mathbf{N}\})$, according to what seems appropriate. For a quantum mechanical system the corresponding definition is the Slater sum,[2, 3, 4]

$$\exp[-U(\{\mathbf{N}\})/kT] = N!\Lambda^{3N} \sum_i |\Psi_i(\{\mathbf{N}\})|^2 e^{-E_i/kT} \quad (4.2)$$

where the sum is over quantum states of the system defined by $\mathbf{N}$ and $\mathcal{V}$, the volume. The function $\Psi_i(\{\mathbf{N}\})$ is the eigenfunction of the ith quantum state at spatial coordinates $\{\mathbf{N}\}$. E_i is the energy of this state while Λ is a normalization factor. It is clear in both cases that $U_{\mathbf{N}} = U_{\mathbf{N}}(\{\mathbf{N}\})$, unlike $U(\{\mathbf{N}\},\{\mathbf{N}\}_i)$ or E_i, is a function of the temperature and so is not simply a potential energy. The temperature dependence of $U_{\mathbf{N}}$ is negligible for systems of monatomic molecules or non-polar diatomic molecules over a wide temperature range but for more complex molecules it may be very important.

The physical meaning of $U_{\mathbf{N}}$ can be established in the following way. We differentiate both sides of (4.1) with respect to the spatial coordinates of one molecule of the set, say molecule ω. This operation will be represented as ∇_ω. We have

$$\nabla_\omega \exp[-U(\{\mathbf{N}\})/kT] = -[1/kT]\nabla_\omega(U(\{\mathbf{N}\}))\exp[-U(\{\mathbf{N}\})/kT]$$

$$= -[1/kT]\int \nabla_\omega U(\{\mathbf{N}\},\{\mathbf{N}\}_i)\exp[-U(\{\mathbf{N}\},\{\mathbf{N}\}_i)/kT] \cdot d\{\mathbf{N}\}_i \Big/ \int d\{\mathbf{N}\}_i \quad (4.3)$$

The force acting on a particle ω in a completely specified configuration is the vector

$$\mathbf{F}_\omega = -\nabla_\omega U(\{\mathbf{N}\},\{\mathbf{N}\}_i)$$

and hence,

$$\bar{\mathbf{F}}_\omega \equiv -\nabla_\omega U(\{\mathbf{N}\}) = \frac{\int \mathbf{F}_\omega \exp[-U(\{\mathbf{N}\},\{\mathbf{N}\}_i)/kT]\,d\{\mathbf{N}\}_i}{\int \exp[-U(\{\mathbf{N}\},\{\mathbf{N}\}_i)/kT]\,d\{\mathbf{N}\}_i} \quad (4.4)$$

is the *average* force on the particle ω in the configuration, $\{\mathbf{N}\}$, where the average is over the internal coordinates. The energy $U_{\mathbf{N}}$ is the *potential* of this average force. It is a special case of a potential of average force for specified spatial coordinates.

For a more general definition, we consider a body of fluid whose state is specified by $\mathbf{c},T$: concentration set and temperature. Suppose that some of the molecules, comprising a composition set, $\mathbf{n}$, are fixed to ideal handles. Each handle holds one molecule at a chosen-center-of-mass coordinate within the body of the fluid, but the molecule is free to twist, turn, and stretch about its center of mass and to interact freely with other molecules except for the restriction on the motion of its center of mass. In this case the potential of average force, now written $W(\{\mathbf{n}\},\mathbf{c},T)$ or $W_{\mathbf{n}}(\{\mathbf{n}\},\mathbf{c},T)$ is the reversible, isothermal work which must be performed on the handles to bring the molecules to the spatial coordinate set, $\{\mathbf{n}\}$, from another spatial coordinate set, $\{\mathbf{n}\}_\infty$, in which they are mutually separated to such an extent that the average force on each is negligible. Now if we let $\mathbf{c} \rightarrow 0, 0, \cdots \equiv \mathbf{0}$ we obtain $W(\{\mathbf{n}\},\mathbf{0},T)$, the potential of average force for the set $\mathbf{n}$ at $\{\mathbf{n}\}$ in a vacuum, and this corresponds exactly to $U\{\mathbf{N}\})$. The latter notation is used for this quantity in order to be consistent with the notation of McMillan and Mayer[3] and Hill.[1]

In statistical mechanics there are uses for both kinds of potential of average force, $W(\{\mathbf{n}\},\mathbf{c},T)$ and $W(\{\mathbf{n}\},\mathbf{0},T) = U(\{\mathbf{n}\})$. The latter, which is in general also a function of temperature, is often called a *potential of average force at infinite dilution*, or, more briefly, a *direct potential.* In a later section we shall see that in some problems the role of the direct potential is played by

$$\lim_{c_{\text{solute}}\rightarrow 0} W(\{\mathbf{n}\},\mathbf{c},T)$$

in which the concentration of *solvent* molecules does not vanish. This direct potential depends on the interactions of the solute molecules of the set $\mathbf{n}$ in the presence of the solvent molecules. The discussion in the rest of Section 4 is applicable to potentials of average force and direct potentials of all kinds.

There is a perfectly general expansion of any potential of average force which we now illustrate for $U_{\mathbf{N}}$. This is the expansion in terms of *component potentials* which was introduced by Kahn and Uhlenbeck.[5]

$$\begin{aligned} U_{\mathbf{N}}(\{\mathbf{N}\}) &= \sum_{\text{pairs}} u_{ij}(\{i,j\}) + \sum_{\text{triples}} u_{ijk}(\{i,j,k\}) + \cdots \\ &= \sum_{\{\mathbf{n}\}\subseteq\{\mathbf{N}\}}'' u_{\mathbf{n}}(\{\mathbf{n}\}) \end{aligned} \tag{4.5}$$

The $''$ on the summation indicates that we omit all subsets $\{\mathbf{n}\}$ for which $n < 2$. It is customary also to neglect all u_n for $n > 2$, but for some applications, especially to ionic solutions, the higher component potentials may be important so the use of the complete formalism will be developed here. In general the higher component potentials [i.e., u_n for $n > 2$] have the same origin as the temperature dependence of $U_{\mathbf{N}}$, namely, the averaging over other coordinates than $\{\mathbf{N}\}$. This is illustrated below by calculations for a few simple models.

Since the $U_{\mathbf{N}}$ are more fundamental quantities than the component potentials, the meaning of the latter is easier to see from the inverse of equation (4.5). Using (3.33) we obtain at once,

$$\begin{aligned} u_{\mathbf{n}}(\{\mathbf{n}\}) &= \sum_{\{\mathbf{N}\}\subseteq\{\mathbf{n}\}}'' [-1]^N U_{\mathbf{N}}(\{\mathbf{N}\}) \\ &= U_{\mathbf{n}}(\{\mathbf{n}\}) - \sum_{\substack{\{\mathbf{m}\}\subseteq\{\mathbf{n}\}\\ m=1}} U_{\mathbf{n}-\mathbf{m}}(\{\mathbf{n}-\mathbf{m}\}) \\ &\qquad + \sum_{\substack{\{\mathbf{m}\}\subseteq\{\mathbf{n}\}\\ m=2}} U_{\mathbf{n}-\mathbf{m}}(\{\mathbf{n}-\mathbf{m}\}) - \cdots \end{aligned} \tag{4.6}$$

For example we have, for $n = 2$,

$$u_{i,j}(\{i,j\}) = U_{i,j}(\{i,j\}) \tag{4.7}$$

the pairwise component potential is the same as the potential of average force. As another special case of (4.6) we have

$$\begin{aligned} u_{ijk}(\{i,j,k\}) = U_{i,j,k}(\{i,j,k\}) &- u_{ij}(\{i,j\}) \\ &- u_{ik}(\{i,k\}) - u_{jk}(\{j,k\}) \end{aligned} \tag{4.8}$$

and so this component potential is the difference between $U_{i,j,k}$ and what we should calculate on the basis of pairwise components alone. All of the other $u_{\mathbf{n}}$ for $n > 2$ also have the character of remainders, and indeed of remainders of higher and higher order as n increases. On this basis one may hope that as $N \rightarrow \infty$ equation (4.5) converges in the sense that component potentials, $u_{\mathbf{n}}$, for $n > m$ are negligible, where m is some definite number.

Another basis for expecting the higher component potentials, $u_{\mathbf{n}}$, to be negligible if n is large is the following.[6] The forces which lead to higher component potentials are short-range; they tend to become negligible for distances in excess of several molecular diameters. Let us suppose that in a particular model the forces leading to higher component potentials are negligible for distances larger than two molecular diameters. But only a definite number, M, of molecules can be packed into a sphere whose radius is two molecular diameters, so in this case u_n will be negligible for $n > M$.

Examples of Higher Component Potentials

Example 1. We consider here the potential of average force and the component potentials for a set consisting of two point charges, each of charge z, and a dipole of moment m. Consider first the set consisting of the dipole and one charge at distance r from the center of the dipole. The coordinates for this problem are shown in Fig. 4.1. The potential for a given coordinate set is

$$U(\theta,\varphi,r) = zm \cos \varphi / r^2$$

Now we integrate over the internal coordinates to get (after appropriate normalization)

$$\exp(-U(r)/kT) = [4\pi]^{-1} \int_0^{2\pi} d\theta \int_0^{\pi} e^{-A\cos\varphi} \sin\varphi \, d\varphi$$

where

$$A \equiv zm/kTr^2$$

The integration yields

$$U(r) = -kT \ln (A^{-1} \sinh A)$$

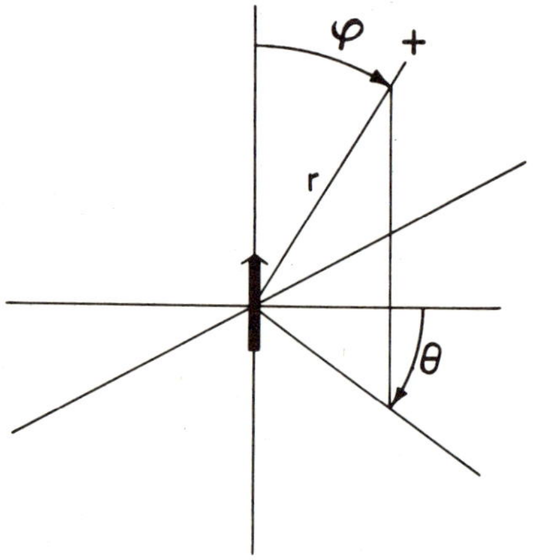

Fig. 4.1. Coordinates for charge-dipole interaction.

We call this u_{cd}, the charge-dipole component potential. The charge-charge component potential, u_{cc}, is z^2/R, where R is the separation of the charges. Now consider the set consisting of the two charges and the dipole, with the dipole halfway between the two charges. For any orientation of the dipole the potential is simply z^2/R ($R = 2r$, where r is the charge-dipole distance). This does not depend on the orientation of the dipole and hence the potential of average force in this case, U_{cdc}, is also z^2/R. Now we apply equation (4.8) to calculate the component potential,

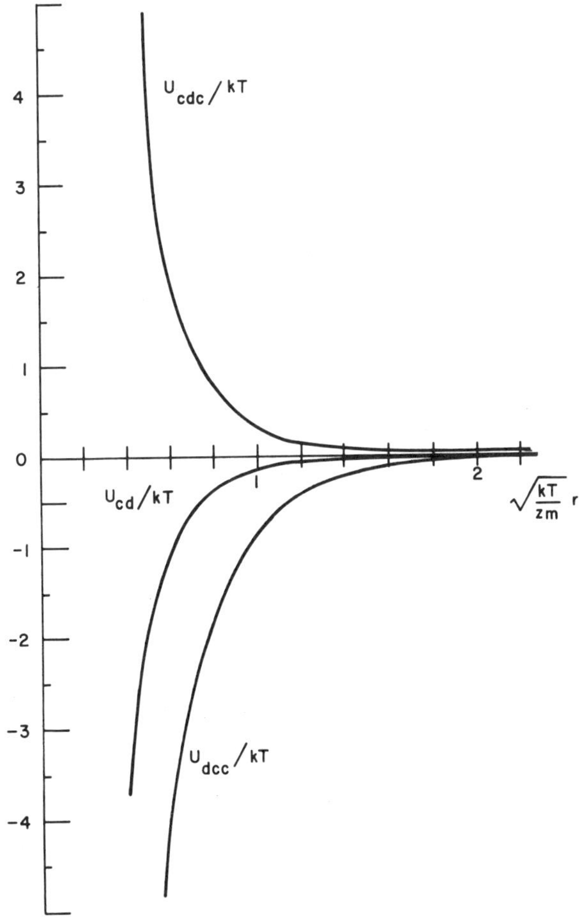

Fig. 4.2. Component potentials for charge-dipole interaction.

$$u_{cdc} = z^2/R - [z^2/R - 2kT \ln(\sinh A/A)]$$
$$= 2kT \ln(\sinh A/A)$$

This is the component potential only for the linear configuration with the dipole in the middle. It is also easy to calculate the component potential for the case in which the two charges are only an infinitesimal distance apart compared to the distance from a charge to the center of the dipole. The charge-charge interaction again cancels in calculating the component potential because this interaction is independent of the orientation of the dipole. The component potential in this case is

$$u_{dcc} = 2kT \ln(\sinh A/A) - kT \ln(\sinh 2A/2A)$$

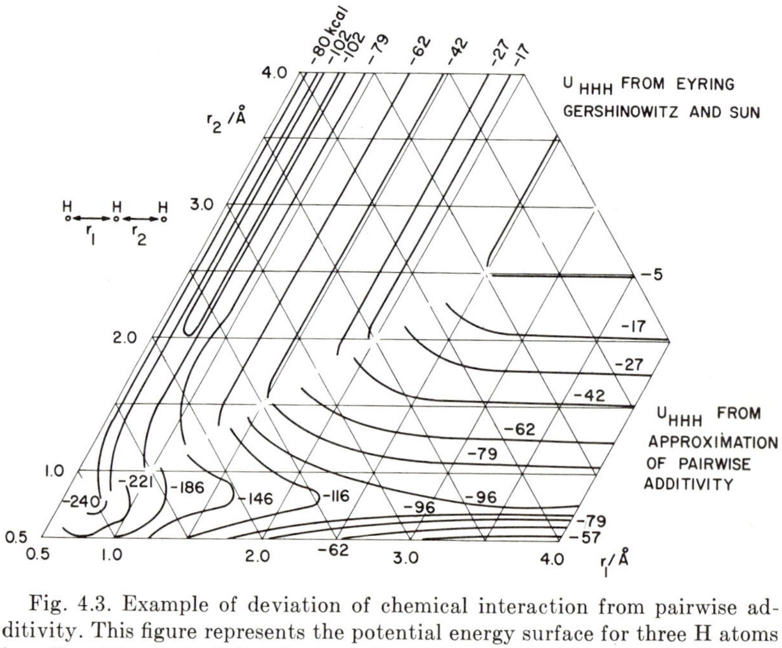

Fig. 4.3. Example of deviation of chemical interaction from pairwise additivity. This figure represents the potential energy surface for three H atoms in a line. The part of the diagram above the median line is U_{HHH} from H. Eyring, H. Gershinowitz, and C. E. Sun, *J. Chem. Phys.* **3**, 786 (1935). The part below represents the potential energy surface in the hypothetical case that the energy is pairwise additive with the pairwise potentials implied by the upper diagram. The deviation of this composite diagram from symmetry about the median line is a direct measure of the third component potential for three *H* atoms in a line.

The charge-dipole component potential and the two higher component potentials are shown as a function of r in Fig. 4.2. Of course u_{cdc} and u_{dcc} are limiting cases of a more general third component potential which is a function of three distances, the charge-charge distance and two charge-dipole distances. The calculation we have outlined illustrates that in general there will be a contribution to this third component potential from an average over internal coordinates.

Example 2. It is a familiar fact that chemical forces are not pairwise additive. A quantitative example is given in Fig. 4.3. A statistical way to explain this deviation from additivity is that it results from averaging over the coordinates of the electrons; only the coordinates of the protons are fixed by choosing a point in the figure.

Example 3. In a similar way van der Waals forces, expressed as a function of molecular center-of-mass coordinates, result from averaging over the coordinates of the electrons. The deviations of van der Waals forces from pairwise additivity have been calculated by Kihara.[7]

Notes and References

1. T. L. Hill, *Statistical Mechanics* (McGraw-Hill Book Company, Inc., New York, 1956).
2. Each element of $\{\mathbf{n}\}$ is the set of spatial coordinates of one molecule, and each element of $\{\mathbf{n}\}_i$ is the set of internal coordinates of one molecule. It is not necessary to treat internal coordinates in a different way from spatial coordinates as we do in this book, but it is convenient. In particular it makes the theory of systems of complex molecules *formally* the same as that of monatomic molecules.
3. W. G. McMillan and J. E. Mayer, *J. Chem. Phys.* **13**, 276 (1945).
4. J. O. Hirschfelder, C. F. Curtiss, and R. B. Bird, *Molecular Theory of Gases and Liquids* (John Wiley & Sons, Inc., New York, 1950).
5. B. Kahn and G. E. Uhlenbeck, *Physica* **5**, 399 (1938).
6. This argument is used in another way by J. G. Kirkwood and Z. W. Salsburg, *Discussions Faraday Soc.* **15**, 28 (1953).
7. T. Kihara, *Advances in Chemical Physics* (Interscience Publishers, Inc., New York, 1958), Vol. I, p. 267.

5. Cluster Expansions

The characteristic mathematical techniques of the cluster theories are developed in this section.

(a) *Cluster Functions and Their Graphical Evaluation*

We define the cluster function for the set $\mathbf{m}$ of molecules at coordinate $\{\mathbf{m}\}$ by the equation

$$\gamma_{\mathbf{m}}(\{\mathbf{m}\}) = \exp\,[-u_{\mathbf{m}}(\{\mathbf{m}\})/kT] - 1 \tag{5.1}$$

where $u_{\mathbf{m}}$ is the component potential introduced in the preceding section. We have

$$U_{\mathbf{N}}(\{\mathbf{N}\}) = \sum_{\{\mathbf{m}\}\subseteq\{\mathbf{N}\}}'' u_{\mathbf{m}}(\{\mathbf{m}\}) \tag{5.2}$$

where $U_{\mathbf{N}}$ is the direct potential for the set $\mathbf{N}$ of molecules at $\{\mathbf{N}\}$. A typical problem involving $U_{\mathbf{N}}$ is the evaluation of the configuration integral

$$Z(\mathbf{N},\mathcal{V},T) \equiv \int_{\mathcal{V}} \exp\,[-U_{\mathbf{N}}/kT]\, d\{\mathbf{N}\} \tag{5.3}$$

but for $N > 10^{20}$ as in ordinary systems the integration is hopelessly difficult to perform by direct methods. However a simplification is achieved by expressing the integrand in terms of cluster functions

$$\begin{aligned}\exp\,[-U_{\mathbf{N}}(\{\mathbf{N}\})/kT] &= \prod_{\{\mathbf{m}\}\subseteq\{\mathbf{N}\}} [1 + \gamma_{\mathbf{m}}(\{\mathbf{m}\})] \\ &= 1 + \sum_{\{i,j\}\subseteq\{\mathbf{N}\}} \gamma_{ij} \\ &\quad + \sum_{\{i,j,k\}\subseteq\{\mathbf{N}\}} [\gamma_{ij}\gamma_{jk} + \gamma_{ij}\gamma_{ik} + \gamma_{ik}\gamma_{jk} \\ &\quad + \gamma_{ij}\gamma_{jk}\gamma_{ik} + \gamma_{ijk} + \gamma_{ijk}\gamma_{ij} + \cdots \\ &\quad + \gamma_{ijk}\gamma_{ij}\gamma_{jk}\gamma_{ik}] + \cdots\end{aligned} \tag{5.4}$$

because if now the order of integration and summation are interchanged, then Z may be expressed as a sum of integrals. The first term is

$$\int_{\mathcal{V}} d\{\mathbf{N}\} = \mathcal{V}^N$$

The next terms are of the form,

$$\int_{\mathcal{V}} \gamma_{ij}\, d\{\mathbf{N}\} = \mathcal{V}^{N-2} \int_{\mathcal{V}} \gamma_{ij}\, d\{ij\}$$

For a given composition set, $\mathbf{n}$, $n = 2$, every integral $\int \gamma_{\mathbf{n}}\, d\{\mathbf{n}\}$ gives the same result, and the number of such terms is simply $\binom{\mathbf{N}}{\mathbf{n}}$. In this way the cluster expansion gives a considerable simplification of the integral. One can, in principle, continue in the way we have begun to describe here, and integrate higher and higher terms of the cluster expansion, but the combinatorial problems in calculating the number of different terms of (5.4) that give the same results on integration become very severe and it appears to be easier to proceed in a less direct manner.

The procedure we shall use is based on grouping the terms of the cluster expansion according to topological criteria. In order to do this we must first establish a set of conventions by which each term of the cluster expansion may be represented as a graph.[1, 2]

Each of the graphs with which we shall be concerned consists of a skeleton and bonds. The skeleton is a set of vertices (points) corresponding to molecules. Each term of the cluster expansion of (5.4) corresponds to a graph with a skeleton of N vertices. The molecules in this equation are distinguishable because each is associated with a definite element of $\{\mathbf{N}\}$, i.e., because the molecules have been assigned to definite spatial coordinates. Therefore we consider the vertices of the graph to be distinguishable from each other, because each corresponds to a definite molecule. The vertices are as distinguishable from each other as though they bore numbers from 1 to N.

The bonds in the graphs represent the cluster functions. A $\gamma_{ij}(\{ij\})$ cluster function is represented by a line connecting vertex i to vertex j of the skeleton. Such a bond will be called a γ_{ij} bond, where i and j represent the *species* of the molecules associated with the two vertices. In certain circumstances in which the species are of no importance, we shall call it a γ_2 bond. The length of a γ_2 bond makes no difference in our representation because in the integration every pair of molecules ij ranges over every separation, $0 < r_{ij} < R$, where R is the diameter of the containing vessel. The important thing about the γ_2 bond is that it connects two particular vertices of the graph: this topological characteristic corresponds to a characteristic of the cluster function γ_{ij} that remains unchanged in the integration.

A $\gamma_{ijk}(\{ijk\})$ cluster function is represented by a triangle connecting vertices i, j, and k of the skeleton. Such a bond will be called a γ_{ijk} bond (where i, j, and k are the species of the three molecules), a γ_m, $m = 3$, bond, or simply a γ_3 bond, depending on the importance of knowing the species at the vertices in a given circumstance. It is least ambiguous if a γ_3 bond is thought of as the surface bounded by the edges of a triangle, rather than as the three edges; otherwise the γ_3 bond may be confused with the product of three γ_2 bonds. Furthermore, just as the length of a γ_2 bond is unimportant, so the shape and area of the γ_3 bond has no significance: The important thing is that it connects three vertices of the skeleton. This is not to imply that every γ_3 bond is the same, rather that if a γ_3 bond connects three particular vertices of the graph then it corresponds to a particular γ_{ijk} cluster function and that is all we need to know about it.

A $\gamma_{hijk}(\{hijk\})$ cluster function is represented by a tetrahedron connecting vertices h, i, j, and k of the skeleton. This bond will be called a γ_{hijk} bond, a γ_{m}, $m = 4$, bond, or simply a γ_4 bond. It would be misleading to represent this bond by a square instead of a tetrahedron, although a square also connects four vertices, because a square connects the vertices in an unsymmetrical manner. The vertices of a square may be classified by pairs as adjacent or opposite, but no such asymmetry is implied by the γ_{hijk} cluster function itself. Therefore if one uses a plane figure to represent a γ_4 bond, he must bear this lack of topological correspondence in mind. It is least ambiguous if a γ_4 bond is thought of as the volume bounded by the four faces of a tetrahedron as this eliminates the possibility of confusing a γ_4 bond with the product of three or four γ_3 bonds or with the product of six γ_2 bonds.

In general a $\gamma_{\mathrm{m}}(\{\mathbf{m}\})$ cluster function is represented in a graph by a γ_{m} bond, which will be called a γ_m bond in cases in which the composition of the vertices (that is, the species of the molecules associated with the vertices) is not important. A γ_m bond may be represented geometrically as a regular polyhedron with m vertices in $m - 1$ dimensions. Such a polyhedron is often called an m-simplex. Although the topology of m-simplexes has been investigated to some extent[3] we have not succeeded in applying any of the results to the problems of the cluster theories.

These conventions are illustrated for some terms of the cluster expansion of (5.4) in Fig. 5.1. Because of typographical difficulties one cannot represent γ_m, $m > 3$, bonds in such figures. This unfortunate circumstance should not obscure the fact that there is a one-to-one correspondence between the terms of the cluster expansion, (5.4), and the graphs of $\gamma_2, \gamma_3, \cdots, \gamma_N$ bonds on a skeleton of N vertices. This correspondence does not at all depend on whether or not we can draw all the graphs.

$1 =$ ⓐ ⓑ ⓒ
    ⓓ ⓔ ⓕ

$\gamma_{ad}\gamma_{de}\gamma_{bf} =$ [diagram]

$\gamma_{ab}\gamma_{bc}\gamma_{ce}\gamma_{ad}\gamma_{de}\gamma_{ef}\gamma_{bf} =$ [diagram]

$\gamma_{de}\gamma_{ae}\gamma_{abd} =$ [diagram]

$\gamma_{cf}\gamma_{abd}\gamma_{bce}\gamma_{ef} =$ [diagram]

$\gamma_{bc}\gamma_{abd}\gamma_{bde}\gamma_{cef} =$ [diagram]

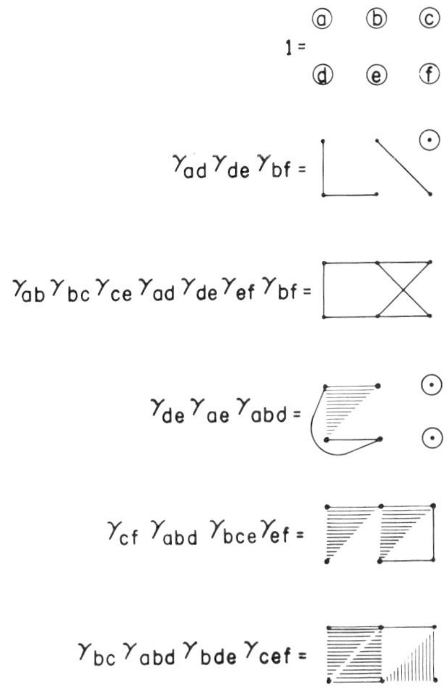

Fig. 5. 1. Some terms of the cluster expansion of $\exp[-U(\{\mathbf{N}\})/kT]$ for $N = 6$. In this case $\mathbf{N} = 1_a, 1_b, 1_c, 1_d, 1_f$. The vertices of the skeleton which do not intersect any bonds are represented as circles. The correspondence of vertices to species is shown only in the first figure. Lines are γ_2 bonds, triangles are γ_3 bonds.

(b) *Connectivity*

The essential feature of the graphs corresponding to the terms of the cluster expansion is their topology: the scheme of connections that each represents. We now proceed to develop the basis for a topological classification of the graphs, which corresponds to a classification of the terms of the cluster expansion according to the way they behave when integrated.

Consider first a graph consisting only of a skeleton and some γ_2 bonds.[1,4] Each γ_2 bond intersects two vertices which are then said to be *directly connected*. A pair of vertices is said to be *indirectly connected* if it is possible to go from one vertex of the pair to the other on a chain of elements of the graph, that is, if there is a sequence: vertex, bond, vertex, $\cdots$, bond, vertex connecting the two vertices of the pair.

These definitions are readily generalized[2] to apply to graphs with higher order bonds. If two vertices are directly connected by any γ_m bond we say that there is an *edge* connecting these vertices. A pair of vertices is said to be indirectly connected if there is a chain: vertex, edge, vertex, $\cdots$, edge, vertex connecting them.

We shall call a graph on a skeleton of n vertices *at least singly connected* (ALSC) if every pair $[n[n - 1]/2$ pairs$]$ of vertices of the skeleton is connected, either directly or indirectly, singly or multiply.

In any graph that is not ALSC the bonds partition the skeleton into subsets that are connected by bonds of the graph. The graph on each of these subsets is called a *reducible cluster*, it is at least singly connected on the subset skeleton. Two special cases are a vertex which is not connected to any other element of a graph and a graph that is already ALSC on its skeleton. It is convenient to also call these reducible clusters. Some examples are given in Fig. 5.2.

Now we are in a position to express the cluster expansion, (5.4), in a form that is amenable to analytical manipulation. First we note that the terms of the cluster expansion are in one-to-one correspondence with the set of all possible, distinguishable graphs on the skeleton of N labeled vertices of composition $\mathbf{N}$. By *possible* we mean to include only graphs of a topology that can be generated by the cluster expansion process. In particular this excludes any graph in which two or more γ_n bonds intersect exactly the same n vertices of the skeleton. By *distinguishable* we mean to include all graphs that can be distinguished from each other by topological criteria, taking into account that the vertices are labeled. (In effect the vertices are numbered from 1 to N.)

Now we define $s_{\mathbf{n}}(\{\mathbf{n}\})$ as the sum of all cluster terms that cor-

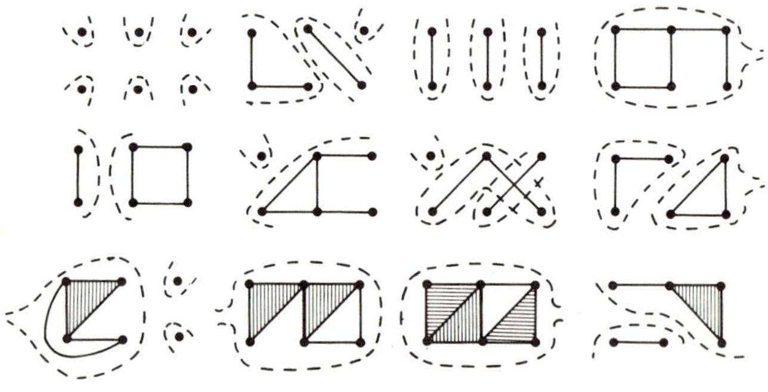

Fig. 5.2. Resolution of a graph into reducible clusters. The dotted lines separate each graph on a skeleton of six vertices into reducible clusters.

respond to ALSC graphs on the skeleton of n labeled vertices of composition $\mathbf{n}$ (but $s_{1_s}(\{1_s\} \equiv 1)$. There will be one term in $s_{\mathbf{n}}$ for every possible, distinguishable ALSC graph on the skeleton of $\mathbf{n}$, where *possible* and *distinguishable* are used in precisely the same sense as above. Then since the cluster expansion of $\exp[-U(\{\mathbf{N}\})/kT]$ corresponds to all possible, distinguishable graphs on the skeleton of $\mathbf{N}$, and since we may classify these graphs according to the partition of the skeleton that each defines, we have the equation

$$\exp[-U_{\mathbf{N}}(\{\mathbf{N}\})/kT] = \sum_{\mathbf{p}]\{\mathbf{N}\}} \prod_{\{\mathbf{k}\}\subseteq\{\mathbf{N}\}} [s_k(\{\mathbf{k}\})]^{p_{\{\mathbf{k}\}}} \tag{5.5}$$

where the elements $p_{\{\mathbf{k}\}}$ of $\mathbf{p}$ are either 1 or 0 depending on whether or not the graph on $\{\mathbf{k}\}$ is a cluster in the partition, $\mathbf{p}$.

When the right side of (5.5) is integrated, $\int_{\mathcal{V}} d\{\mathbf{N}\}$, the integral over the product in each term simply gives the corresponding product of integrals because each of the factors $s_{\mathbf{k}}$ depends on a different, disjoint subset of $\{\mathbf{N}\}$. This is true because each $s_{\mathbf{k}}$ corresponds to a reducible cluster of a graph, and the reducible clusters, by definition, are not connected to each other.

In order to express this observation analytically we define the *reducible cluster integral*,

$$b_{\mathbf{k}} \equiv [1/\mathcal{V}\mathbf{k}!] \int_{\mathcal{V}} s_{\mathbf{k}}(\{\mathbf{k}\})\, d\{\mathbf{k}\}, \qquad \text{with } b_{1_s} \equiv 1 \tag{5.6}$$

and note that every partition, $\mathbf{p}]\{\mathbf{N}\}$, that corresponds to the same partition, $\mathbf{p}]\mathbf{N}$, gives the same result on integration. The number of such partitions is the number of ways of dividing N distinguishable objects, $\mathbf{N}_1$ of species 1, $\mathbf{N}_2$ of species 2, $\cdots$, into piles, $p_{\mathbf{m}}$ piles of composition $\mathbf{m}$, $p_{\mathbf{n}}$ piles of composition $\mathbf{n}$, $\cdots$, with no restriction either on the order of the piles or of the objects within the piles. This number is [1, 5]

$$\mathbf{N}!/\mathbf{p}! \prod_{\mathbf{k}\leq\mathbf{N}} [\mathbf{k}!]^{p_{\mathbf{k}}}$$

Therefore integration of (5.5) yields

$$Z(\mathbf{N},\mathcal{V},T) \equiv \mathbf{N}! \sum_{\mathbf{p}]\mathbf{N}} \prod_{\mathbf{k}\leq\mathbf{N}} [\mathcal{V}b_{\mathbf{k}}]^{p_{\mathbf{k}}}/p_{\mathbf{k}}! \tag{5.7}$$

The last member of this equation is similar to the right side of equation (3.37) and this is the basis of the usefulness of the moment-cumulant relation in the cluster theories. For example, by using the

TABLE 5.1. Analogies

Section 5(b)	Sections 5(c) to 5(f)
reducible integrals	irreducible integrals
reducible clusters	irreducible clusters
partition of a graph	tree of a graph
The moment-cumulant relation generates partitions.	The reversion of a certain series generates trees.

moment-cumulant relation it is easy to revert (5.7) to obtain $b_{\mathbf{k}}$ explicitly as a function of all $Z(\mathbf{N},\mathcal{V},T)$ for $\mathbf{N} \leq \mathbf{k}$. This is a useful way to *define* $b_{\mathbf{k}}$ for quantum mechanical systems.[4, 6, 7]

(c) *Husimi Trees*

Just as in (5.7) the configuration integral is expressed as a sum of products of reducible integrals, so in turn each reducible integral $b_{\mathbf{k}}$ may be written as a sum of products of simpler integrals called irreducible integrals. This is the sense in which the $b_{\mathbf{k}}$ integrals are reducible. In order to describe the reduction of these integrals we need to make some further definitions and classifications of the connectivity of graphs. The procedure is analogous to that of the preceding section, but involves less familiar concepts. As a guide to the reader we express the analogy in a formal way in Table 5.1.

We begin by defining another classification of connectivity. A graph on a skeleton of more than two vertices is *at least doubly connected* (ALDC) if every pair of vertices of the skeleton is either

(a) indirectly connected by at least one chain of edges of bonds and also directly connected, or

(b) indirectly connected by at least two chains that do not intersect at any intermediate vertex.

Some examples are given in Fig. 5.3. In the case of a graph on a skeleton of two vertices, if there is a bond connecting the vertices we call the graph ALDC. It is, of course, not doubly connected in the literal sense, but calling it ALDC makes this classification of the connectivity of graphs more simply related to the reducibility of integrals on the corresponding products of cluster functions.

If a graph is ALSC but not ALDC on its skeleton, then it may be resolved into components, each of which *is* ALDC on its skeleton. This is also illustrated in Fig. 5.3. These components are called *irreducible clusters*. A graph which is ALDC on its skeleton is a single

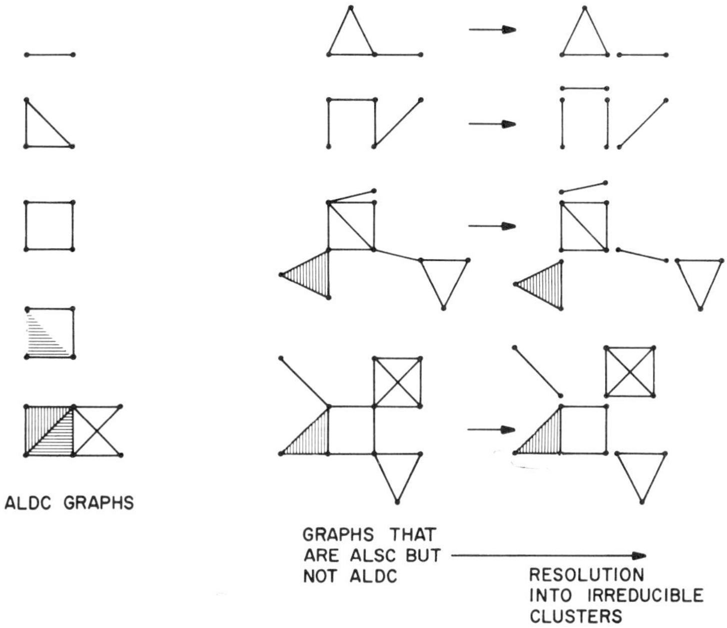

Fig. 5.3. Connectivity of some graphs.

irreducible cluster. Henceforth we shall be concerned only with irreducible clusters and therefore can dispense with the adjective without ambiguity.

Any graph which is ALSC on its skeleton may be formed from a unique set of clusters by allowing the clusters to share vertices, subject to certain restrictions. We note that the reverse process, the resolution of an ALSC graph into clusters, corresponds to a partition of the set of bonds of the graph, but that it does not correspond to a partition of the set of vertices because some of the vertices lie in more than one cluster. However the set of vertices of the clusters is a covering of the set of vertices of the original graph; this kind of a covering we call a *tree*.

We recall that a covering of a set $\mathbf{A}$ is a set of subsets, $\mathbf{B}_1$, $\mathbf{B}_2$, $\cdots$ such that

$$\mathbf{B}_1 \cup \mathbf{B}_2 \cup \cdots = \mathbf{A} \tag{5.8}$$

A partition is a covering in which the subsets are disjoint. A tree is a covering in which the subsets have the following characteristics:

(1) Each subset contains at least two elements of **A**.

(2) The elements of **A** may be collected by the following operations on the subsets of the tree.

(a) Pick arbitrarily any element of any one of the subsets of the tree.

(b) Collect all elements that are in the same subset of the tree as the first element.

(c) Collect all elements that are in the same subset of the tree as any of the elements collected in (b).

(d) Continue in this way until there are no longer any uncollected elements that are in the same subset as any collected element.

This definition of a tree corresponds exactly to what one does when he analyzes a graph into irreducible clusters, provided that we identify each subset of the tree with the *skeleton* of a cluster. To avoid ambiguities we must coin a name for the subsets which comprise a tree. We shall call them *vertex clusters* to remind us that they correspond to the vertices rather than to the bonds of the clusters.

Returning to the set theory aspect, we may easily verify that a tree is a covering with the following conditions on the vertex clusters:

(1) Each vertex cluster contains at least two elements of **A**.

(2) Two vertex clusters may have at most one element in common.

(3) To state this condition we first define a *chain* as a sequence of sets, $\mathbf{B}_1\mathbf{B}_2\mathbf{B}_3 \cdots$, such that each set has an element in common with the set preceding it and each set is different from the one immediately preceding it, but not necessarily different from all of the sets that precede it. Chains of the vertex clusters of a tree have the following properties:

(a) Any pair of vertex clusters may be connected by a chain.

(b) It is not possible to form a chain in which a given vertex cluster appears more than once.

Restriction (3a) assures that the collection procedure by which we define a tree does indeed collect every element of **A**. Restrictions (2) and (3b) assure that every element of **A** appears only once in the collection procedure.

The definition given corresponds to an Husimi tree.[8,9] A Cayley tree implies the additional restriction that there be exactly two elements of **A** in each vertex cluster.

Resolving an ALSC graph into clusters corresponds to making a partition of the set of its bonds and a tree of the set of its vertices. It also corresponds to making a tree of the set of coordinates $\{\mathbf{n}\}$ appearing in the product of cluster functions that corresponds to the graph. For analytical purposes it is convenient to define a *set* corresponding to a tree, but whose elements are non-negative integers. This is similar to the relation between a partition and a partition set.

For a given tree of $\{\mathbf{n}\}$, let the tree set, $\mathbf{t}]\{\mathbf{n}\}$, be the set of numbers

$$\mathbf{t} = t_{\{\mathrm{a}\}}\ , t_{\{\mathrm{b}\}}\ , t_{\{\mathrm{c}\}}\ , \cdots$$

of subsets of $\{\mathbf{n}\}$ that are vertex clusters of the tree. There will be one element of $\mathbf{t}]\{\mathbf{n}\}$ for every subset of $\{\mathbf{n}\}$, and $t_{\{\mathrm{m}\}}$ will be 1 or 0 according to whether $\{\mathbf{m}\} \subseteq \{\mathbf{n}\}$ is or is not a vertex cluster of the tree. Each tree then specifies one tree set, and vice-versa.

Corresponding to the tree of the coordinate set, $\mathbf{t}]\{\mathbf{n}\}$, is the tree of the composition set, $\mathbf{t}]\mathbf{n}$,

$$\mathbf{t} = t_{\mathrm{a}}\ , t_{\mathrm{b}}\ , t_{\mathrm{c}}\ , \cdots$$

whose elements are the numbers of the various subsets of $\mathbf{n}$ that correspond to the compositions of the vertex clusters of the tree. There is one element of $\mathbf{t}]\mathbf{n}$ for every subset of $\mathbf{n}$. The element t_{m} is the number of vertex clusters of composition $\mathbf{m}$ in the corresponding tree of the coordinate set. It is also the number of clusters of composition $\mathbf{m}$ obtained by the resolution of the corresponding graph into clusters. Clearly many distinguishable trees of a given coordinate set $\mathbf{t}]\{\mathbf{n}\}$ correspond to the same tree of a composition set $\mathbf{t}]\mathbf{n}$ unless all of the elements of $\mathbf{n}$ are unity or zero.

(d) *Reducible Cluster Integrals → Irreducible Cluster Integrals*

Now we return to the problem of the cluster integrals. We define $S_{\mathrm{k}}(\{\mathbf{k}\})$ as the sum of all cluster terms (terms generated by the expansion in equation (5.4)) that correspond to ALDC graphs on a skeleton of k vertices of composition set $\mathbf{k}$. Then we have at once the following expansion for the integrand of the reducible cluster integral.

$$s_{\mathrm{n}}(\{\mathbf{n}\}) = \sum_{\mathrm{t}]\{\mathrm{n}\}} \prod_{\{\mathrm{k}\}\subseteq\{\mathrm{n}\}} [S_{\mathrm{k}}(\{\mathbf{k}\})]^{t_{\{\mathrm{k}\}}} \tag{5.9}$$

where the sum is over all distinguishable trees of $\{\mathbf{n}\}$ and the product

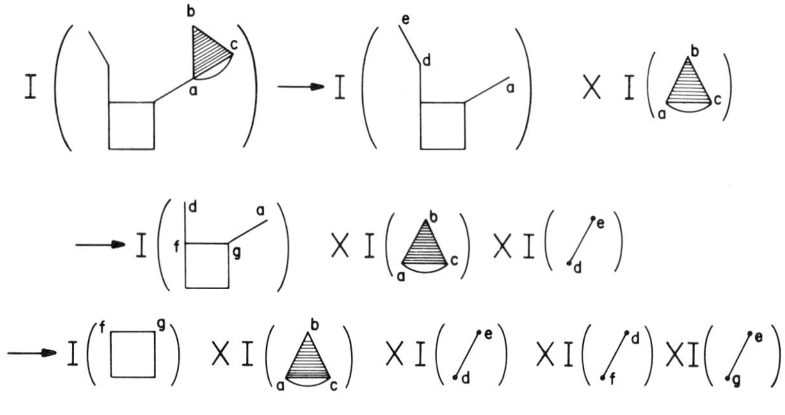

Fig. 5. 4. An integral on a tree may be expressed as a product of integrals on clusters. I( ) is the integral on the indicated graph holding one vertex fixed. First fix a and integrate over b and c. The only bonds which vary in this integration are those among the vertices a, b, c, and these bonds vary over all coordinates consistent with the volume. Therefore the first integration gives a product of integrals, as shown by the arrow. Now integrate over all coordinates of e, holding d fixed. The result is to separate another factor from the first integral, as shown by the next arrow. Finally integrate over all d, with f fixed, and over all a, with g fixed. This factors out two more integrals on single bonds, as shown by the last arrow. Each of the factors after the last arrow is, except for a factor of $1/\mathbf{n}!$, a term in the corresponding $B_{\mathbf{n}}$.

is over all subsets of $\{\mathbf{n}\}$. The following considerations establish the validity of this equation.

Each term of $s_{\mathbf{n}}$ as defined on p. 5.6 corresponds to a definite tree, but the correspondence is not one-to-one. In fact every tree in which at least one vertex cluster has more than three elements corresponds to several terms of $s_{\mathbf{n}}$ which differ among themselves in the number and arrangement of the bonds within each cluster. But one term of (5.9), that is a product of $S_{\mathbf{k}}$ functions corresponding to one tree, does generate all of the terms of $s_{\mathbf{n}}$ that correspond to the same tree $\mathbf{t}]\{\mathbf{n}\}$.

Now we integrate (5.9) over the configuration space that corresponds to all coordinate sets $\{\mathbf{n}\}$ consistent with the volume of the vessel. The left side gives, according to the definition of (5.6), $\mathcal{V}\mathbf{n}!b_{\mathbf{n}}$. To express the right side concisely we define the *irreducible cluster integral*,

$$B_{\mathbf{k}} \equiv [\mathcal{V}\mathbf{k}!]^{-1} \int_{\mathcal{V}} S_{\mathbf{k}}(\{\mathbf{k}\})\, d\{\mathbf{k}\} \qquad (5.10)$$

and note that each term of (5.9) is a product of $S_{\mathbf{k}}$ factors that yields a product of $B_{\mathbf{k}}$ factors on integration. (Here we neglect certain "surface terms" which appear if $\mathcal{V}$ is finite or if γ_{ij} is too long-range. Such terms are discussed in Section (5h)). This is best seen by considering an example, as in Fig. 5.4. We also note that all trees, $\mathbf{t}]\{\mathbf{n}\}$, that correspond to the same tree, $\mathbf{t}]\mathbf{n}$, give the same product of $B_{\mathbf{k}}$ factors on integration. We call the number of such trees of $\{\mathbf{n}\}$ the tree coefficient, $T(\mathbf{t}]\mathbf{n})$. Then the integral of (5.9) is

$$\mathbf{n}!b_{\mathbf{n}} = \sum_{\mathbf{t}]\mathbf{n}} T(\mathbf{t}]\mathbf{n}) \prod_{\mathbf{k}\leq\mathbf{n}}{}'' \, [\mathbf{k}!B_{\mathbf{k}}]^{t_{\mathbf{k}}} \tag{5.11}$$

Some examples of (5.11) for small $\mathbf{n}$ follow.

$$b_{aa} = B_{aa}$$

$$b_{ab} = B_{ab}$$

$$b_{aaa} = B_{aaa} + 2B_{aa}^2$$

$$b_{aab} = B_{aab} + 2B_{ab}B_{aa} + \tfrac{1}{2}B_{ab}^2$$

$$b_{abc} = B_{abc} + B_{ca}B_{ab} + B_{ab}B_{bc} + B_{bc}B_{ac}$$

A difficult and intricate part of the derivations of the cluster theory of non-ionic systems[1, 10, 11, 12] has to do with the determination of the tree coefficient for the general case $\mathbf{t}]\mathbf{n}$, although the problem is not usually described in just this way. Husimi[8] has found a simpler way to determine the tree coefficients in the equations for one-component systems. His procedure is readily generalized to apply to the theory of multicomponent systems but, what is more important, it can be used to avoid altogether the problem of determining the tree coefficients, at least when the surface terms discussed in (5h) are negligible. The reason for this is that Husimi's procedure is via the determination of a function that generates trees just as the moment-cumulant relation generates partitions and, if surface terms are neglected, it is only this generating function that is needed for the derivation of the cluster theory. It may be helpful to note that a complete discussion of the use of generating functions in combinatorial problems has been given by Riordan.[5]

(e) *The Recursion Formula for Tree Coefficients*

The first step in the generalized Husimi procedure is to find a recursion formula for the tree coefficients. The formula is based on the fact that if we delete one vertex cluster from a tree of $\{\mathbf{n}\}$ we get a number of trees on disjoint subsets of $\{\mathbf{n}\}$. Indeed if we delete a vertex cluster of composition $\mathbf{q}$ we get q smaller trees. The preceding statement is obviously true if every vertex of the deleted cluster is also a vertex of at least one other cluster. We make it true in general by defining the empty tree as a tree containing no vertex clusters, and by beginning an empty tree at each vertex of the deleted cluster that is not also the vertex of another cluster. The tree coefficient for the empty tree beginning at a vertex of composition s is $T(\mathbf{0}]1_s)$ and we define it as equal to unity for convenience in writing the equations that follow. These relations are illustrated in Fig. 5.5. The essential idea is that if we delete a vertex cluster from a tree, then a new tree, either empty or non-empty, "begins" at each vertex of the deleted vertex cluster.

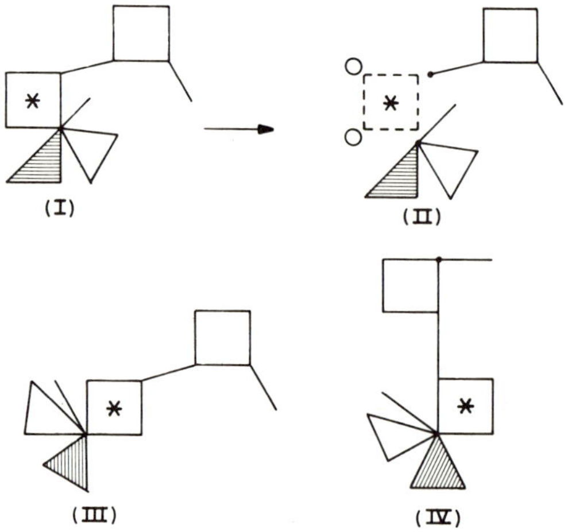

Fig. 5.5.‡ This illustrates the basis of the recursion formula for the tree coefficients. In (I) the cluster which we arbitrarily decide to delete is marked with an asterisk. The effect of deleting this cluster is shown in (II). Four new trees are formed, two of which are empty, as indicated by the 0's.

In calculating the tree coefficients we regard (I), (III), and (IV) as distinguishable trees on the coordinate set. They differ only in their connections to the cluster to be deleted. There are 4!/2! trees which differ in this way, because the vertices of the coordinate set are distinguishable, but interchanging two identical trees (the empty trees in this case) does not give a new distinguishable graph. (I), (III), and (IV) are just three examples of these twelve trees.

‡ See Fig. 5.1.

The recursion formula we derive from this is

$$t_{\mathbf{q}} T(\mathbf{t}]\mathbf{n}) = \frac{\mathbf{n}!}{\mathbf{q}!} \sum_{\substack{\mathbf{t}_1,\mathbf{t}_2,\cdots,\mathbf{t}_q]\mathbf{t}-\mathbf{1}_\mathbf{q} \\ \mathbf{n}_1,\mathbf{n}_2,\cdots,\mathbf{n}_q]\mathbf{n}}} \prod_{i=1}^{q} \frac{T(\mathbf{t}_i]\mathbf{n}_i)}{[\mathbf{n}_i - \mathbf{1}_i]!} \tag{5.12}$$

Here the sum is over all partitions (not trees) of $\mathbf{t} - \mathbf{1}_\mathbf{q}$ into q subsets. If the original tree, $\mathbf{t}]\mathbf{n}$, is trivial in the sense that it consists of a single cluster then all of these subsets must be empty, but if the original tree consists of more than one vertex cluster than at least one of the subset trees must be non-empty. The partitions of $\mathbf{t} - \mathbf{1}_\mathbf{q}$ are partitions in the sense of equation (3.34) because $\mathbf{t}]\mathbf{n}$, like the composition set, is a set of non-negative integers.

In order to define the product in (5.12) we number the vertices of the deleted cluster from 1 to q in an arbitrary way. The factors of the product correspond, one-to-one, to a tree on each of these vertices and the ith subset tree is assigned to the ith vertex. The set 1_i is the composition set of the ith vertex.

We may partition the n distinguishable vertices of the composition set $\mathbf{n}$ into subsets whose compositions are

$$\mathbf{q},\ \mathbf{n}_1 - 1_1\,,\ \mathbf{n}_2 - 1_2\,,\ \cdots,\ \mathbf{n}_\mathbf{q} - 1_\mathbf{q}$$

in the following number of ways:

$$\mathbf{n}!/\mathbf{q}! \prod [\mathbf{n}_i - 1_i]!$$

and this is a factor in $T(\mathbf{t}]\mathbf{n})$, just as the product of subset T's is.

In any tree on the coordinate set, $\mathbf{t}]\{\mathbf{n}\}$, the vertices (elements of the subsets of the tree) are all distinguishable and hence trees which differ only in the interchange of the subset trees which begin at two different vertices of an arbitrary cluster are distinguishable. (See Fig. 5.5 for an example.) Rather than calculating another combinatorial factor to account for this we choose to retain this distinguishability in the recursion formula. Therefore we sum over all *ordered* partitions of $\mathbf{t} - 1_\mathbf{q}$: not only over all partitions but over all permutations in the order in which we write the subsets, $\mathbf{t}_1$., $\mathbf{t}_2$, $\cdots$, $\mathbf{t}_\mathbf{q}$, of each partition. We note that if $\mathbf{t}_i]\mathbf{n}_i$ contains no vertex of the same species as vertex i of the deleted cluster, then the corresponding term must vanish. This is taken care of in an elementary way if we agree to calculate $1/[\mathbf{n}_i - 1_i]!$ by the expression

$$1/[\mathbf{n}_i - 1_i]! = n_{ii}/\mathbf{n}_i! \tag{5.13}$$

where n_{ii} is the element of $\mathbf{n}_i$ corresponding to the same species as the ith vertex of the deleted cluster.

The origin of the factor t_{q} remains to be explained. Summing over all ordered partitions of $\mathbf{t} - 1_{\mathrm{q}}$ gives us t_{q} times too many terms because in these partitions any particular cluster of composition $\mathbf{q}$ may appear either in one of the subsets, $\mathbf{t}_i$, or as the deleted cluster. On the other hand the principle of the recursion formula is the calculation of the relation of T to the product of subset T's corresponding to the deletion of a *particular* vertex cluster.

(f) *The Generating Function for the Tree Coefficients*

Now we shall use the recursion formula to specify a generating function for the tree coefficients. We define the generating function by the equation

$$C(\mathbf{z},\mathbf{y}) \equiv \sum_{n}[\mathbf{z}^{\mathbf{n}}/\mathbf{n}!] \sum_{\mathbf{t}]\mathbf{n}} T(\mathbf{t}]\mathbf{n})\mathbf{y}^{\mathbf{t}} \qquad (5.14)$$

where T is the tree coefficient and where $\mathbf{z}$ and $\mathbf{y}$ are sets of variables:

$$\mathbf{z} \equiv z_1, z_2, \cdots, z_\sigma \qquad (5.15)$$

$$\mathbf{y} = y_{\mathrm{a}}, y_{\mathrm{b}}, \cdots \qquad (5.16)$$

The number of components is σ (cf. equation (3.1)). The set $\mathbf{y}$ has an infinite number of elements, and in order to have $\mathbf{y}^{\mathbf{t}}$ always defined we set zero equal to the exponent of every y_{m} for which there is no t_{m} (cf. equation (3.10)). Note that the second sum in (5.14) corresponds to (5.11) when one defines $y_{\mathrm{k}} = \mathbf{k}!B_{\mathrm{k}}$.

We also define two kinds of derivatives of C.

$$c_s \equiv z_s[\partial C/\partial z_s] = \sum n_s[\mathbf{z}^{\mathbf{n}}/\mathbf{n}!] \sum T(\mathbf{t}]\mathbf{n})\mathbf{y}^{\mathbf{t}} \qquad (5.17)$$

$$C_{\mathrm{a}} \equiv \partial C/\partial y_{\mathrm{a}} = (1/y_{\mathrm{a}}) \sum [\mathbf{z}^{\mathbf{n}}/\mathbf{n}!] \sum t_{\mathrm{a}}T(\mathbf{t}]\mathbf{n})\mathbf{y}^{\mathbf{t}} \qquad (5.18)$$

That the following relation is satisfied by the derivatives is readily verified by direct substitution. (Note that $\mathbf{t}]\mathbf{n}$ implies that $\sum_a''[a - 1]\mathbf{t}_{\mathrm{a}} = n - 1$.)

$$c \equiv \sum_s c_s = C + \sum_a'' [a - 1]y_{\mathrm{a}}C_{\mathrm{a}} \qquad (5.19)$$

By inspection of equation (5.18) we see that the coefficient of $\mathbf{z}^{\mathbf{n}}\mathbf{y}^{\mathbf{t}}/y_{\mathrm{q}}$ in $\mathbf{n}!C_{\mathrm{q}}$ is the left side of (5.12),

$$t_{\mathrm{q}}T(\mathbf{t}]\mathbf{n})$$

We may obtain the right side of (5.12) in terms of (5.17) as follows.

First we use (5.17) to obtain the expansion of $\mathbf{c}^{\mathbf{q}}$:

$$c^{\mathbf{q}} = \prod_{i=1}^{q} \sum \frac{\mathbf{z}^{\mathbf{n}_i}}{[\mathbf{n}_i - 1_i]!} \sum T(\mathbf{t}]\mathbf{n}_i)\mathbf{y}^{\mathbf{t}} \tag{5.20}$$

where the first sum is over all $\mathbf{n}_i$ for which $1_i \in \mathbf{n}_i$ and the second is over all trees of $\mathbf{n}_i$. The product is over q factors, q_1 with i = species 1, q_2 with i = species 2, $\cdots$, q_σ with i = species σ; where

$$\mathbf{q} = q_1, q_2, \cdots, q_\sigma$$

The coefficient of $\mathbf{z}^{\mathbf{n}}\mathbf{y}^{\mathbf{t}}$ in (5.20) is

$$\sum_{\substack{\mathbf{t}_1,\mathbf{t}_2,\cdots,\mathbf{t}_\sigma]\mathbf{t} \\ \mathbf{n}_1,\mathbf{n}_2,\cdots,\mathbf{n}_\sigma]\mathbf{n}}} \prod_{i=1}^{q} T(\mathbf{t}_i]\mathbf{n}_i)/[\mathbf{n}_i - 1_i]! \tag{5.21}$$

that is, we simply sum over all partitions of $\mathbf{t}$ and $\mathbf{n}$. However this expression represents each product only once whereas, unless the elements of q are each either 0 or 1, each algebraic product is obtained several times corresponding to different orders of picking the terms in the sums to make the product. As an elementary example consider the operations in forming the product

$$[a + b][a + b] = a^2 + ab + ba + b^2 = a^2 + 2ab + b^2$$

The second equality follows because $ab = ba$. The multiplication of tree coefficients is also commutative, but to avoid having to calculate the multinomial coefficients in this problem (compare with equation (3.19)) we *define* the partition of $\mathbf{t}$ that appears in expression (5.21) as an *ordered* partition and specify that we sum over all partitions of $\mathbf{t}$ that are distinguishable when the subsets, $\mathbf{t}_1, \mathbf{t}_2, \cdots, \mathbf{t}_q$, are written out in a line. This corresponds to agreeing to write

$$[a + b][a + b] = a^2 + ab + ba + b^2$$

where again all coefficients are unity.

Now compare expression (5.21) with the right side of equation (5.12). Apparently the latter is the coefficient of $\mathbf{z}^{\mathbf{n}}\mathbf{y}^{\mathbf{t}}/y_{\mathbf{q}}$ in $[\mathbf{n}!/\mathbf{q}!]\mathbf{c}^{\mathbf{q}}$. Combining this with the result we obtained for the left side of (5.12) we find that

$$\mathbf{n}!C_{\mathbf{q}} = [\mathbf{n}!/\mathbf{q}!]\mathbf{c}^{\mathbf{q}}$$

or

$$\mathbf{q}!C_{\mathbf{q}} = \mathbf{c}^{\mathbf{q}} \tag{5.22}$$

We may combine this with (5.19) to get

$$c = C + \sum'' [a - 1]y_{\mathbf{a}}\mathbf{c}^{\mathbf{a}}/\mathbf{a}! \tag{5.23}$$

Equations (5.22) and (5.23) are equivalent statements of the condition imposed on the generating function C by the recursion relation among the tree coefficients.

A particularly useful result is obtained by eliminating C from equation (5.23) in the following way. We define the functions

$$\mathfrak{S}(\mathbf{c},\mathbf{y}) \equiv \sum'' \mathbf{c}^{\mathbf{a}}y_{\mathbf{a}}/\mathbf{a}! \tag{5.24}$$

$$\mathfrak{S}_s \equiv \partial\mathfrak{S}/\partial c_s = [1/c_s] \sum'' a_s\mathbf{c}^{\mathbf{a}}y_{\mathbf{a}}/\mathbf{a}! \tag{5.25}$$

$$\mathfrak{S}_{si} \equiv \partial^2\mathfrak{S}/\partial c_s\partial c_i \tag{5.26}$$

and substitute these in (5.23) to get

$$c = C + \sum c_s\mathfrak{S}_s - \mathfrak{S} \tag{5.27}$$

This is differentiated with respect to c_i to obtain

$$1 = \sum_s [\partial C/\partial z_s][\partial z_s/\partial c_i] + \mathfrak{S}_i + \sum_s c_s\mathfrak{S}_{si} - \mathfrak{S}_i \tag{5.28}$$

or

$$1 = \sum_s c_s[\partial/\partial c_i][\ln z_s + \mathfrak{S}_s] \tag{5.29}$$

This equation is a relation between C and $\mathfrak{S}$ that is consistent with the condition that C is a generating function for the tree coefficients. Any function, $\mathfrak{S}$, that satisfies this equation will also be, at least implicitly, a generating function for the tree coefficients. It is readily verified that if

$$\ln c_s = \ln z_s + \mathfrak{S}_s + I_s \tag{5.30}$$

then $\mathfrak{S}$ satisfies (5.29). The constant of integration is found to be zero by comparing (5.30) with (5.17) and (5.25) in the limit $\mathbf{z} \to \mathbf{0}$. Then we have

$$z_s = c_s \exp(-\mathfrak{S}_s) \tag{5.31}$$

This equation is the reversion of equation (5.17). The tree coefficients are by definition the coefficients of (5.17) and so we have found that the reversion of (5.31) generates trees. When we apply this result in Section 8 we are able to identify z_s as the fugacity and c_s as the concentration of species s and then to show that $\mathfrak{S}$ is directly related to the excess Helmholz free energy. As noted, following (5.16) we have $y_{\mathbf{k}} = \mathbf{k}!B_{\mathbf{k}}$ so (5.24) is the series expansion, with irreducible cluster integrals as coefficients, of the thermodynamic function $\mathfrak{S}$.

If we neglect the surface terms (discussed below in Section 5(h)) this is all we need to know about the reversion of (5.17) to derive the cluster theory, but in order to provide a comparison with the work of Fuchs[12] we shall now outline the method by which one may use this result to evaluate the tree coefficients.

We see from (5.17) that $n_s T(\mathbf{t}]\mathbf{n})/\mathbf{n}!$ is the coefficient of $\mathbf{z}^{\mathbf{n}}\mathbf{y}^{\mathbf{t}}$ in c_s. On the other hand the coefficient of $\mathbf{z}^{\mathbf{n}}$ in c_s may be obtained from (5.31) by differentiation or by contour integration. Using the latter method we have (see Smirnow[13] for a discussion of the extension of Cauchy's residue theorem to many variables)

$$\sum_{\mathbf{t}]\mathbf{n}} \frac{n_s T(\mathbf{t}]\mathbf{n})}{\mathbf{n}!} \mathbf{y}^{\mathbf{t}} = \left[\frac{1}{2\pi i}\right]^{\sigma} \oint \frac{c_s}{\mathbf{z}^{\mathbf{n}}} \prod_{i=1}^{\sigma} \frac{dz_i}{z_i} \tag{5.32}$$

where the contour is a small circle around the origin in the complex z_s plane for each s. One next substitutes (5.31) in the integrand, which then depends explicitly on $\mathbf{c}$ but not on $\mathbf{z}$. After integration one equates coefficients of $\mathbf{y}^{\mathbf{t}}$ on both sides. Although the evaluation of this contour integral is not easy, all of the necessary steps are given in Fuchs's paper.[12] The method for one component is given by Husimi[8] and a similar contour integral for the one-component case is evaluated below (Section 5(h)).

(g) *Comparison with Fuchs's Method*

The following is an outline in our notation of the part of Fuchs's paper[12] that is equivalent to the solution of the Husimi tree problem. His first step is to calculate the tree coefficients by a direct solution of the very difficult combinatorial problem (Fuchs (3.21)). Now if we define

$$y_{\mathbf{k}} \equiv \mathbf{k}!B_{\mathbf{k}} \tag{5.33}$$

then the left side of (5.32) is the same as the right side of (5.11). This leads to

$$n_s b_{\mathbf{n}} = \left[\frac{1}{2\pi i}\right]^{\sigma} \oint \frac{c_s}{\mathbf{z}^{\mathbf{n}}} \prod_{i=1}^{\sigma} \frac{dz_i}{z_i} \tag{5.34}$$

The second step in Fuchs's procedure is to define a function c_s in such a way that this contour integral yields the tree coefficients which he has derived in the first step (Fuchs (4.5)). The coefficients in the resulting equation for c_s (i.e., those variables which are constant in the contour integration) may then be identified with the irreducible cluster integrals because they have the same relation to $b_{\mathbf{n}}$. The last step is to show that the function c_s is the concentration of species s. (Fuchs (8.19)). Interspersed with these are several other steps which have to do with the part of the problem which we solve by the moment-cumulant relation and which Fuchs, following Born and Fuchs,[11] solves by applying the method of steepest descents to a certain contour integral, different from (5.34). This outline does not do justice to the great mathematical virtuosity exhibited in Fuchs's paper, which is not a trivial generalization of the one-component case. Our procedure seems to be less demanding in this respect.

In closing it may be pointed out that our result on the reversion of (5.31), which is equivalent to the result of Fuchs, is a partial generalization of Lagrange's theorem on the reversion of series.[14] It is a generalization in the sense that it is a reversion in many variables instead of only one. It is less general in that it applies to a particular function of $\mathbf{c}$, namely

$$\exp\left[-[\partial/\partial c_s] \sum{}'' \mathbf{c}^{\mathbf{a}} y_{\mathbf{a}}/\mathbf{a}!\right]$$

while a complete generalization of Lagrange's theorem would apply to any function of $\mathbf{c}$.

(h) *The Surface Terms*

Equation (5.11) is not exact. In order to illustrate this clearly we now introduce a quantity $\beta_{\mathbf{k}}$ which is *defined* by the equation

$$\mathbf{n}!b_{\mathbf{n}} = \sum_{\mathbf{t}]\mathbf{n}} T(\mathbf{t}]\mathbf{n}) \prod_{\mathbf{k}\leq\mathbf{n}} [\mathbf{k}!\beta_{\mathbf{k}}]^{t_{\mathbf{k}}} \tag{5.35}$$

the reverse of which is

$$\mathbf{n}!\beta_{\mathbf{n}} = \sum_{\mathbf{t}]\mathbf{n}} \theta(\mathbf{t}]\mathbf{n}) \prod_{\mathbf{k}\leq\mathbf{n}} [\mathbf{k}!b_{\mathbf{k}}]^{t_{\mathbf{k}}}, \tag{5.35r}$$

where θ, the appropriate combinational factor, is related to the tree coefficient by reversion. Here we only indicate how θ may be calculated for a multicomponent system and then we simplify the problem by passing to a one-component system. This is adequate for our purpose which is only to illustrate that $\beta_{\mathbf{k}} - B_{\mathbf{k}}$ is not generally zero.

We begin with

$$\mathfrak{S} = \sum{}'' \mathbf{c}^{\mathbf{n}}\beta_{\mathbf{n}} \tag{5.36}$$

which, with (5.31) gives

$$\ln [c_s/z_s] = \sum{}'' n_s \mathbf{c}^{\mathbf{n}-\mathbf{1}_s}\beta_{\mathbf{n}} \tag{5.37}$$

therefore we have

$$\begin{aligned} n_s\beta_{\mathbf{n}} &= [2\pi i]^{-\sigma} \oint c_s \ln (c_s/z_s) \prod_{i=1}^{\sigma} dc_i/c_i^{\,n_i+1} \\ &= [[\mathbf{n} - \mathbf{1}_s]!]^{-1} \sum_{\mathbf{t}]\mathbf{n}} \theta(\mathbf{t}]\mathbf{n}) \prod_{\mathbf{k}\le\mathbf{n}} [\mathbf{k}!b_{\mathbf{k}}]^{t_{\mathbf{k}}} \end{aligned} \tag{5.38}$$

Now if we substituted (5.17) in the form

$$c_s = z_s \sum n_s \mathbf{z}^{\mathbf{n}-\mathbf{1}_s} b_{\mathbf{n}} \tag{5.39}$$

in the integrand, performed the integration, and equated coefficients of $\prod [b_{\mathbf{k}}]^{t_{\mathbf{k}}}$ then we could determine $\theta(\mathbf{t}]\mathbf{n})$. This calculation is difficult and has not yet been carried through.

Next we pass to a one-component system. Then (5.38) becomes

$$[[n - 1]!]^{-1} \sum_{\mathbf{t}]n} \theta(\mathbf{t}]n) \prod_{k\le n} [k!b_k]^{t_k} = [2\pi i]^{-1} \oint \ln (c/z) c^{-n}\, dc \tag{5.40}$$

In this case it is easy to simplify the integral by successive partial integration since

$$\oint u\, dv = -\oint v\, du$$

if the path of integration is the same in both cases.[8] So we have

$$\begin{aligned} \oint \ln (c/z) c^{-n}\, dc &= [n - 1]^{-1} \oint c^{-[n-1]}\, d \ln (c/z) \\ &= -[n - 1]^{-1} \oint [z/c]^{n-1} z^{-n}\, dz \end{aligned} \tag{5.41}$$

The path of integration in the first integral is a circle around the origin, and, since we are determining concentration-independent coefficients, the circle may be chosen arbitrarily small. Then on this circle $z \sim c$ so the path of integration in the last integral is again a small circle around the origin, but this time in the complex z plane. Therefore the partial integration is justified.

Next we substitute (5.17) in the form

$$c/z = \sum z^n [n + 1]b_{n+1} \tag{5.42}$$

Using a method like that used in deducing the moment-cumulant relation we find

$$[c/z]^{-[n-1]} = \sum z^m A_m^{(n-1)} \tag{5.43}$$

$$A_m^{(n-1)} = \sum_{\mathbf{p}]m} \frac{[p + n - 2]!}{[n - 2]!} [-1]^p \prod_{h \leq m} [[h + 1]b_{h+1}]^{p_h}/p_h! \tag{5.44}$$

Hence (5.41) is

$$-2\pi i[n - 1]^{-1} A_{n-1}^{(n-1)}$$

and

$$\sum_{\mathbf{t}]n} \theta(\mathbf{t}]n) \prod [k!b_k]^{t_k}$$

$$= \sum_{\mathbf{p}]n-1} [-1]^{p+1}[p + n - 2]! \prod_{h \leq n-1} [[h + 1]b_{h+1}]^{p_h}/p_h! \tag{5.45}$$

$$= \sum_{\mathbf{t}]n} [-1]^{t+1}[t + n - 2]! \prod_{k \leq n} [kb_k]^{t_k}/t_k!$$

where the last member follows because

$$\mathbf{p}]n - 1 \qquad \text{implies} \qquad \sum ip_i = n - 1$$

while

$$\mathbf{t}]n \qquad \text{implies} \qquad \sum [i - 1]t_i = n - 1$$

Therefore

$$\theta(\mathbf{t}]n) = [-1]^{t+1}[t + n - 2]! \prod_{k \leq n} [[k - 1]!]^{-t_k}/t_k! \tag{5.46}$$

and (5.35r) becomes (for one component)

$$n!\beta_n = \sum_{\mathbf{t}]n} [-1]^{t+1}[t + n - 2] \prod [kb_k]^{t_k}/t_k! \tag{5.47}$$

This result was first obtained by Mayer[15] and then by Kilpatrick,[7] both in notation different from that employed here

From (5.47) we deduce

$$\beta_2 \equiv b_2 = B_2 \tag{5.48}$$

$$\beta_3 = b_3 - 2b_2^2 \tag{5.49}$$

$$\beta_4 \equiv b_4 - 6b_2b_3 + \tfrac{20}{3}b_2^3 \tag{5.50}$$

Now in general we define the surface term as

$$\sigma_n \equiv \beta_n - B_n \tag{5.51}$$

where B_n is defined by (5.10). The simplest such term is

$$\sigma_3 \equiv \beta_3 - B_3 = [3!\mathcal{V}]^{-1} \int_{\mathcal{V}} [S_{abc} + 3\gamma_{ab}\gamma_{bc}]\, d\{abc\} - 2B_2^2 - B_3$$

or

$$2\mathcal{V}^2\sigma_3 = \mathcal{V} \int \gamma_{ab}\gamma_{bc}\, d\{abc\} - \left[\int \gamma_{ab}\, d\{ab\}\right]^2 \tag{5.52}$$

It is convenient to define the function

$$\Psi(\{b\}) = \int_{\mathcal{V}} \gamma_{ab}\, d\{a\} = \int_{\mathcal{V}} \gamma_{bc}\, d\{c\} \tag{5.53}$$

Then (5.52) becomes

$$2\mathcal{V}^2\sigma_3 = \left[\int_{\mathcal{V}} d\{b\}\right] \int_{\mathcal{V}} \Psi^2\, d\{b\} - \left[\int_{\mathcal{V}} \Psi\, d\{b\}\right]^2 \tag{5.54}$$

According to the Schwartz inequality[16] $|\sigma_3| > 0$ unless Ψ is independent of $\{b\}$ within $\mathcal{V}$. This is never the case, but if it can be shown that the right side of (5.54) increases less rapidly than $\mathcal{V}^2$ as $\mathcal{V} \to \infty$ then we may be sure that σ_3 is negligible. This will be the case if the intermolecular forces are negligible beyond a certain range.

We now consider some examples of the behavior of σ_3 at large $\mathcal{V}$. Consider first a one-component gas of hard spheres of radius A. Writing x for the distance between two molecules, we have

$$\begin{aligned} \gamma(x) = \gamma_2(x) &= -1 \quad \text{if} \quad x < A \\ &= 0 \quad \text{if} \quad x > A \end{aligned} \tag{5.55}$$

If this is substituted in (5.53) and if we suppose that the system is spherical with radius R and with y the distance from the center to molecule b, then we obtain for $R > A$

$$\Psi(y) = \frac{\pi\gamma(R - y)}{y}\left[\frac{A^4 - [R - y]^4}{4} + \frac{y}{3}[A^3 - [R - y]^3]\right] + \frac{2\pi A^3}{3} \tag{5.56}$$

Combining this with (5.54) leads to

$$\sigma_3 = \pi^2 A^7/6R + \text{terms of order } R^{-2}, R^{-3}, \cdots \tag{5.57}$$

For another example suppose that the cluster function is

$$\gamma(x) = \exp[Ae^{-\alpha x}/x] - 1 \tag{5.58}$$

in which case the intermolecular potential has the same distance dependence as the Debye potential. To investigate the behavior of σ_3 at large $\mathcal{V}$ we may replace γ by the largest term of its series expansion.

$$\gamma(x) \simeq Ae^{-\alpha x}/x + \cdots \tag{5.59}$$

Then we find

$$\Psi(y) = [4\pi A/\alpha^2][1 - [1 + \alpha R]^{-\alpha R} \sinh(\alpha y)/\alpha y] \tag{5.60}$$

and

$$\sigma_3 = 12\pi^2 A^2/\alpha^5 R + \text{terms of order } R^{-2}, R^{-3}, \cdots \tag{5.61}$$

Finally suppose that

$$\gamma(x) = e^{A/x} - 1 \tag{5.62}$$

as in a system of ionic species. Then we have

$$\gamma(x) = A/x + \cdots \tag{5.63}$$

and corresponding to this

$$\Psi(y) = 2\pi[R^2 - y^2/3] \tag{5.64}$$

which leads to

$$\sigma_3 = R^4[1/21 - 1/25]/8 \tag{5.65}$$

It is plausible that the higher σ_n behave in the same way as σ_3 as $\mathcal{V} \to \infty$. If this is the case then one can show that surface terms are negligible in this limit for all systems in which the pairwise potential at large distances decreases at least as fast as r^{-4} with increasing r. This is not true of ionic systems: These are discussed separately in Section 11.

In summary we have shown that in general (5.11) is not exact if B_k is defined by (5.10) as is customary. However if we introduce

β_k *defined* by (5.10) (cf. (5.35)) then we have in complete generality

$$\mathfrak{S} = \sum'' \mathbf{c}^{\mathbf{n}}\beta_{\mathbf{n}}$$

where $\mathfrak{S}$ is the thermodynamic function discussed above and, in greater detail, in Sections 6 and 8. To recover the irreducible cluster integrals B_k defined by (5.10) we introduce the coefficients of the surface terms (5.51) and then we have

$$\mathfrak{S} = \sum'' \mathbf{c}^{\mathbf{n}}B_{\mathbf{n}} + \sum'' \mathbf{c}^{\mathbf{n}}\sigma_{\mathbf{n}}\,. \tag{5.66}$$

The surface terms vanish in the limit $\mathcal{V} \to \infty$ except for systems with Coulombic interactions. Therefore we may neglect the surface terms in various equations based on (5.66) until we come to ionic systems.

Notes and References

1. J. E. Mayer and M. G. Mayer, *Statistical Mechanics* (John Wiley & Sons, Inc., New York, 1940). J. E. Mayer, *Handbuch der Physik* (Springer-Verlag, Berlin, 1958), B. XII, p. 73.
2. H. L. Friedman, *Molecular Phys.* **2,** 23 (1959).
3. P. S. Aleksandrov, *Combinatorial Topology* (Graylock Press, Rochester, N. Y., 1956), Vol. 1.
4. T. L. Hill, *Statistical Mechanics* (McGraw-Hill Book Company, Inc., New York, 1956), Chapter 5.
5. J. Riordan, *Introduction to Combinatorial Analysis* (John Wiley & Sons, Inc., New York, 1958).
6. S. Ono, *J. Chem. Phys.* **19,** 504 (1941).
7. J. E. Kilpatrick, *J. Chem. Phys.* **21,** 274 (1953).
8. K. Husimi, *J. Chem. Phys.* **18,** 682 (1950).
9. R. J. Riddell and G. E. Uhlenbeck, *J. Chem. Phys.* **21,** 2056 (1953).
10. J. E. Mayer and S. F. Harrison, *J. Chem. Phys.* **6,** 87 (1938).
11. M. Born and K. Fuchs, *Proc. Roy. Soc. (London)* **A166,** 391 (1938).
12. K. Fuchs, *Proc. Roy. Soc. (London)* **A179,** 408 (1942).
13. W. I. Smirnow, *Lehrgang der Hoheren Mathematik* (Veb Deutscher Verlag der Wissenschaften, Berlin, 1955), Vol. III. 2.
14. E. T. Whittaker and G. N. Watson, *A Course of Modern Analysis* (Cambridge University Press, New York, 1935) p. 133.
15. J. E. Mayer, *J. Chem. Phys.* **10,** 629 (1942).
16. H. Margenau and G. Murphy, *The Mathematics of Physics and Chemistry* (D. Van Nostrand Company, Inc., Princeton, N. J., 1943).

CHAPTER 2

The Cluster Theory of Non-Ionic Systems

6. Thermodynamics. Volume as an Independent Variable

In this section we derive some specialized thermodynamic relations that are useful in the cluster theories. In a later section we shall obtain some additional thermodynamic relations, in which pressure rather than volume is the independent variable, which are more convenient for comparison with experiment.

It is to be noted that script capital letters will be used to represent *extensive* thermodynamic variables.

For a reversible isothermal expansion of a closed system from state 1 to state 2 the change in Helmholz free energy is

$$\mathfrak{F}(2) - \mathfrak{F}(1) = -\int_1^2 P\, d\mathcal{V} \tag{6.1}$$

where P is the pressure and $\mathcal{V}$ the volume. It is convenient to introduce N, the total number of molecules in the system, and $c = N/\mathcal{V}$, the total concentration, and to write this equation in the form

$$\mathfrak{F}(2) - \mathfrak{F}(1) = -N \int_1^2 P\, d(1/c) \tag{6.2}$$

$$= -NkT \int_1^2 c\, d(1/c) - N \int_1^2 [P - kTc]\, d(1/c) \tag{6.3}$$

The first integral in (6.3) is the change in free energy for the expansion of an ideal gas over the same states of c,T as those covered in the expansion of the real system. Therefore the second integral corresponds to the part of the free energy change in the real system that arises from interactions among the molecules. We shall call this the change in *excess free energy*. With this in mind the excess free energy of a system of concentration c and temperature T is defined by the equation,[1]

$$\mathfrak{F}^{\text{ex}} \equiv -N \int_{c=0}^{c} [P - kTc]\, d(1/c) \tag{6.4}$$

The ideal free energy of the same system, defined to correspond to the first integral in (6.3), may be written[2]

$$\mathfrak{F}^{\text{id}} \equiv \sum N_s \mu_s^{\dagger} - NkT + kT \sum N_s \ln c_s \tag{6.5}$$

where N_s is the number of molecules of species s in the system and $c_s \equiv N_s/\mathcal{V}$.

Now we note the general relation for the chemical potential,

$$(\partial \mathfrak{F}/\partial N_s)_{T,\mathcal{V},\mathbf{N}-N_s} = \mu_s$$

corresponding to which we have

$$(\partial \mathfrak{F}^{\text{id}}/\partial N_s)_{T,\mathcal{V},\mathbf{N}-N_s} \equiv \mu_s^{\text{id}} = \mu_s^{\dagger} + kT \ln c_s \tag{6.6}$$

$$(\partial \mathfrak{F}^{\text{ex}}/\partial N_s)_{T,\mathcal{V},\mathbf{N}-N_s} \equiv \mu_s^{\text{ex}} = kT \ln (z_s/c_s) \tag{6.7}$$

where z_s, the fugacity of component s, is defined by the equation

$$z_s \equiv c_s \exp(\mu_s^{\text{ex}}/kT) \tag{6.8}$$

From (6.6) we see that $\mu_s^{\dagger}$ is the chemical potential of pure component s in the hypothetical ideal gas state of $c_s = 1$. It depends on the choice of this state and on the temperature, but not on the set $\mathbf{c}$ that characterizes the mixture. The definition of the fugacity assures that $z_s \rightarrow c_s$ as $\mathbf{c} \rightarrow \mathbf{0}$ and that z_s corresponds to the quantity represented by the same symbol by McMillan and Mayer[3] and by Hill[4] although the name varies. The relation of z_s to f_s, the fugacity of Lewis and Randall[5] is

$$z_s = f_s/kT.$$

Finally we note that we have

$$\begin{aligned} \mu_s &= \mu_s^{\text{id}} + \mu_s^{\text{ex}} \\ &= \mu_s^{\dagger} + kT \ln z_s \end{aligned} \tag{6.9}$$

Another thermodynamic function which is very closely related to $\mathfrak{F}^{\text{ex}}$ and is particularly convenient in the cluster theories[6,7] is

$$\mathfrak{S} \equiv -\, \mathfrak{F}^{\text{ex}}/\mathcal{V}kT \tag{6.10}$$

This form of the excess free energy is useful because it has especially simple expansions in all three branches of the cluster theory: imperfect gases, non-ionic solutions, and ionic solutions. The definition implies that $\mathfrak{S}$ has the units of concentration.

An important thermodynamic relation for $\mathfrak{S}$, equivalent to (6.4), is derived by dividing (6.10) by c and then differentiating by $1/c$ to obtain the excess pressure:

$$kT\left(\frac{\partial(\mathfrak{S}/c)}{\partial(1/c)}\right)_{T,\mathbf{x}} = P - ckT \qquad (6.11)$$

where $\mathbf{x}$ is the set of mole fractions,

$$\mathbf{x} \equiv c_1/c,\ c_2/c,\ \cdots,\ c_\sigma/c \qquad (6.12)$$

An equation for $\mathfrak{S}$ that is equivalent to (6.8) is

$$z_s = c_s \exp[-(\partial\mathfrak{S}/\partial c_s)_{\mathbf{c}-c_s,T}]. \qquad (6.13)$$

Another equation for z_s that we shall need may be derived from the properties of $P\mathcal{V}$ as a thermodynamic potential.[8] We begin with the Gibbs-Duhem equation,

$$-\mathcal{V}\,dP + \mathcal{S}\,dT + \sum N_s\,d\mu_s = 0, \qquad (6.14)$$

which may be rearranged to read

$$d(P\mathcal{V}) = \mathcal{S}\,dT + P\,d\mathcal{V} + \sum N_s\,d\mu_s\,. \qquad (6.15)$$

In order to change from the set $\boldsymbol{\mu}$ to the set $\mathbf{z}$ as independent variables we substitute (6.9) in (6.15) to get

$$d(P\mathcal{V}) = [\mathcal{E}^{\text{ex}} + P\mathcal{V}](dT/T) + P\,d\mathcal{V} + kT\sum[N_s/z_s]dz_s \qquad (6.16)$$

where the excess internal energy[9]

$$\mathcal{E}^{\text{ex}} = T\mathcal{S} + T\sum N_s(\partial\mu_s^\dagger/\partial T)_{\mathcal{V}} + kT\sum N_s \ln z_s - P\mathcal{V} \qquad (6.17)$$

has the same relation to $\mathcal{F}^{\text{ex}}$ as $\mathcal{E}$ has to $\mathcal{F}$. From (6.16) we deduce at once

$$\left(\frac{\partial(P\mathcal{V})}{\partial z_s}\right)_{T,\mathcal{V},\mathbf{z}-z_s} = kTN_s/z_s \qquad (6.18)$$

or

$$z_s(\partial P/\partial z_s)_{T,\mathcal{V},\mathbf{z}-z_s} = kTc_s \qquad (6.19)$$

which is the desired equation. In Section 8 we shall see that in an important sense (6.19) is the inverse of (6.13). Equation (6.13) gives the fugacity from an equation of state (an expression for $\mathfrak{S}$) in which the concentration set $\mathbf{c}$ is the independent variable. On the other hand (6.19) gives the concentration from an equation of state (an expression for P) in which the fugacity set $\mathbf{z}$ is the independent variable. The same inverse relation connects (5.17) and (5.31).

As is well known the natural independent variables for the thermodynamic potential $\mathfrak{F}$ are $\mathbf{N},\mathcal{V},T$ or, what is equivalent $\mathbf{c},\mathcal{V},T$. In a similar way the natural variables for the thermodynamic potential $P\mathcal{V}$ are $\boldsymbol{\mu},\mathcal{V},T$ or $\mathbf{z},\mathcal{V},T$. A corollary is that for an independent variable x we have

$$\left(\frac{\partial(P\mathcal{V})}{\partial x}\right)_{\boldsymbol{\mu},\mathcal{V},T} = -\left(\frac{\partial\mathfrak{F}}{\partial x}\right)_{\mathbf{c},\mathcal{V},T} \tag{6.20}$$

For proof we begin with

$$P\mathcal{V} = \mathcal{G} - \mathfrak{F} = \sum_s \mu_s N_s - \mathfrak{F} \tag{6.21}$$

and differentiate to get

$$\left(\frac{\partial(P\mathcal{V})}{\partial x}\right)_{\boldsymbol{\mu},\mathcal{V},T} = \sum_s \mu_s \left(\frac{\partial N_s}{\partial x}\right)_{\boldsymbol{\mu},\mathcal{V},T} - \left(\frac{\partial\mathfrak{F}}{\partial x}\right)_{\boldsymbol{\mu},\mathcal{V},T} \tag{6.22}$$

The identity

$$\left(\frac{\partial\mathfrak{F}}{\partial x}\right)_{\boldsymbol{\mu},\mathcal{V},T} = \left(\frac{\partial\mathfrak{F}}{\partial x}\right)_{\mathbf{c},\mathcal{V},T} + \sum_s \left(\frac{\partial\mathfrak{F}}{\partial c_s}\right)_{x,\mathcal{V},T}\left(\frac{\partial c_s}{\partial x}\right)_{\boldsymbol{\mu},\mathcal{V},T} \tag{6.23}$$

is a readily derived generalization of the familiar identity

$$(\partial w/\partial x)_y = (\partial w/\partial x)_z + (\partial w/\partial z)_x(\partial z/\partial x)_y \,.$$

We note that in (6.23)

$$(\partial\mathfrak{F}/\partial c_s)(\partial c_s/\partial x) = \mu_s\,(\partial N_s/\partial x)$$

and combine (6.22) and (6.23) to get (6.20).

If we define the standard states of $\mathbf{z}$ and $\mathfrak{F}$ to be at the same value of x as the real state to which $P\mathcal{V}$ and $\mathfrak{F}$ pertain, then (6.20) is equivalent to

$$\left(\frac{\partial(P\mathcal{V})}{\partial x}\right)_{\mathbf{z},\mathcal{V},T} = -\left(\frac{\partial\mathfrak{F}^{\text{ex}}}{\partial x}\right)_{\mathbf{c},\mathcal{V},T} \tag{6.24}$$

For proof we begin with an identity of the same form as (6.23)

$$\begin{aligned}\left(\frac{\partial(P\mathcal{V})}{\partial x}\right)_{z,\mathcal{V},T} &= \left(\frac{\partial(P\mathcal{V})}{\partial x}\right)_{\mathbf{u},\mathcal{V},T} + \sum_s \left(\frac{\partial(P\mathcal{V})}{\partial \mu_s}\right)_{x,\mathcal{V},T} \left(\frac{\partial \mu_s}{\partial x}\right)_{z,\mathcal{V},T} \\ &= \left(\frac{\partial(P\mathcal{V})}{\partial x}\right)_{\mathbf{u},\mathcal{V},T} + \sum N_s(\partial\mu_s^\dagger/\partial x)_{\mathcal{V},T}\end{aligned} \qquad (6.25)$$

In the last member we have used (6.9) and (6.15). The sum is $\partial\mathcal{G}^\dagger/\partial x = \partial\mathfrak{F}^\dagger/\partial x$. Combining (6.20) and (6.25) we have

$$\left(\frac{\partial(P\mathcal{V})}{\partial x}\right)_{z,\mathcal{V},T} = -\left(\frac{\partial(\mathfrak{F} - \mathfrak{F}^\dagger)}{\partial x}\right)_{c,\mathcal{V},T} \qquad (6.26)$$

This is equivalent to (6.24) because the ideal part of $\mathfrak{F} - \mathfrak{F}^\dagger$ is independent of x.

Slight variations in the procedure are required[9] if x is T or $\mathcal{V}$. Equation (6.24) is used in Section 9.

Notes and References

1. We note that for an ionic system the exponent of c in the expression

$$\lim_{c\to 0} (P - kTc) = Ac^n$$

is 3/2 and in other cases it is 2, so there is no difficulty with the integral in (6.4) at the lower limit. Of course this is not true of the first integral in (6.3) if we allow the concentration in state 1 to vanish.

The formulation of excess functions is also described by Rowlinson (*Handbuch der Physik*) (Springer-Verlag, Berlin, 1958) B. 12, p. 1), but he calls them residual functions. Michels, Geldemann, and de Groat (*Physica* **12,** 105 (1946), Eq. (3)) call them internal functions.

2. Equations of this form, but containing an extra term, are often seen. Examples are E. A. Guggenheim, *Thermodynamics* (Interscience, New York, 1950), second ed., Equation 6.10.1 and T. L. Hill, *Statistical Mechanics*, (McGraw-Hill Book Company, Inc., New York, 1956), Equation A.2.3. The differences imply corresponding differences in the definition of $\mu\dagger$ (or μ^0).

3. W. G. McMillan and J. E. Mayer, *J. Chem. Phys.* **13,** 276 (1945).
4. T. L. Hill, *Statistical Mechanics* (McGraw-Hill Book Company, Inc., New York, 1956).
5. G. N. Lewis and M. Randall, *Thermodynamics* (McGraw-Hill Book Company, Inc., New York, 1923).
6. H. L. Friedman, *Molecular Phys.* **2,** 23 (1959).
7. H. L. Friedman, *J. Chem. Phys.* **32,** 135 (1960).
8. R. Fowler and E. A. Guggenheim, *Statistical Thermodynamics* (Cambridge University Press, New York, 1952).
9. H. L. Friedman, *J. Chem. Phys.* **34,** 73 (1961).

7. The Theory of the Grand Partition Function

Although the procedures in Sections 2 and 5 indicate that one may get useful expansions of the configuration integral (or canonical partition function) by cluster methods, it is in some ways easier to apply these methods to the grand partition function. In this section we give the necessary formulation of this and of several other useful statistical functions. This treatment closely follows the classic paper of McMillan and Mayer.[1]

At first we consider systems that obey the laws of quantum mechanics, as this is simpler than beginning with classical systems.

Consider a grand canonical ensemble[2,3,4,5] of total volume $M\mathcal{V}$ divided into M regions, each of volume $\mathcal{V}$. The boundaries of the regions are permeable to all of the molecules that we shall put into the system, but are "thick" enough so that the interactions among molecules in different regions are negligible. However we shall also assume that we may consider every molecule in the system to be in one region or another; the numbers in the boundaries are negligible.

The ensemble is an isolated system, hence the energy of the ensemble and the composition set of the molecules in it are constant. Therefore the entropy of the ensemble is a maximum when it is at equilibrium with respect to the transfer of molecules within it from one region to another. We shall use this maximum property to derive the distribution law for the grand canonical ensemble.

Within a given region the possible quantum states depend on the composition set, $\mathbf{N}$, of the molecules in the region. We suppose that for a particular composition set ω is the number of quantum states having a particular energy E. (Note that two different regions, both of which happen to contain molecules of the composition set $\mathbf{N}$, may both be in one of these ω quantum states without violating the Pauli exclusion principle because the *complete* specification of a quantum state cannot be the same in two spatially distinct regions.)

If a region contains molecules of the composition set $\mathbf{N}$ and they are collectively in one of the ω quantum states of energy E we shall say that the region is $\mathbf{N},\omega,E$. Let $\Omega(m)$ be the total number of quantum states accessible to the ensemble, subject to the condition that exactly m of the regions are $\mathbf{N},\omega,E$. This number may be decomposed into factors in the following way.

$$\Omega(m) = \binom{M}{m} \omega^m \Omega_r(M - m) \tag{7.1}$$

The binomial coefficient is the number of ways of dividing M distinguishable regions into two sets, one of m and the other of $M - m$ regions. Any one of the m regions may be in any one of ω quantum states, giving rise to the factor ω^m. The number of quantum states accessible to the part of the ensemble consisting of the other $M - m$ regions is defined as $\Omega_r(M - m)$.

We identify the entropy of the ensemble with $k \ln \Omega$. Then we have

$$\mathcal{S}(m) = k\left[\ln\binom{M}{m} + m \ln \omega\right] + \mathcal{S}_r(M - m) \tag{7.2}$$

where the last term is the entropy of the part of the ensemble that consists of $M - m$ regions that are not $\mathbf{N},\omega,E$. For another state of the ensemble, in which only $m - 1$ regions are $\mathbf{N},\omega,E$ we have

$$\mathcal{S}(m - 1) = k \ln\binom{M}{m - 1} + [m - 1]k \ln \omega + \mathcal{S}_r(M - m + 1) \tag{7.3}$$

The entropy change for the process in which one of the regions of the isolated system changes from $\mathbf{N},\omega,E$ to another set of states is obtained by subtracting.

$$\Delta\mathcal{S}(m \rightarrow m - 1) = k \ln\left[\frac{m}{\omega[M - m + 1]}\right] + \Delta\mathcal{S}_r(M - m \rightarrow M - m + 1) \tag{7.4}$$

We assume that both M and m are very large compared to unity. Then if m is the number of regions of the ensemble that are $\mathbf{N},\omega,E$ at equilibrium, $\mathcal{S}(m)$ must be a maximum of $\mathcal{S}$ for a variation of m and we have

$$\Delta\mathcal{S}(m \rightarrow m - 1) = 0 \qquad m = m_{\text{eq}} \tag{7.5}$$

We also have

$$\Delta\mathcal{S}_r = \Delta\mathcal{E}_r/T + \Delta(P\mathcal{V})_r/T - \Delta(\boldsymbol{\mu}\cdot\mathbf{N})_r/T \tag{7.6}$$

where the Δ quantities on the right pertain to the changes in the remainder of the ensemble when the number of regions in it is increased by one. Now for the entire ensemble to be at equilibrium P, T, and $\boldsymbol{\mu}$ must be the same throughout. The energy and composition set of the ensemble do not change with variation of m and therefore $\Delta\mathcal{E}_r$ and $\Delta\mathbf{N}_r$ of (7.6) may be identified with E and $\mathbf{N}$, respectively, of one of the regions that is $\mathbf{N},\omega,E$. We combine this interpretation of (7.6) with the equilibrium condition,

$$0 = k \ln \left[\frac{m}{\omega[M - m + 1]} \right] + \Delta S_r$$

to obtain

$$kT \ln[m/\omega[M - m + 1]] = \boldsymbol{\mu} \cdot \mathbf{N} - P\mathcal{V} - E \tag{7.7}$$

where all of the quantities pertain to one of the regions that is $\mathbf{N},\omega,E$ in the ensemble at equilibrium.

Finally we may consider both M and $\mathcal{V}$ to be very large so that $M \gg m \gg 1$ and hence with negligible error $M - m + 1 \simeq M$. This changes (7.7) to the desired distribution law.

$$\begin{aligned} m/M\omega &= \exp([\boldsymbol{\mu} \cdot \mathbf{N} - P\mathcal{V} - E_l]/kT) \\ &\equiv \mathsf{P}(\mathbf{N},\mathcal{V},l) \end{aligned} \tag{7.8}$$

The ratio m/M is the probability that at equilibrium any particular region of the ensemble is $\mathbf{N},\omega,E$. Dividing this by ω gives us the probability, $\mathsf{P}(\mathbf{N},\mathcal{V},l)$, that the region be in a particular one, say l, of the ω quantum states of energy $E = E_l$. (E_l is the energy of the particular quantum state, l.) Of course this probability depends not only on $\mathbf{N},\mathcal{V}$, and l as explicitly noted, but also on the equilibrium state of the ensemble as specified by $\boldsymbol{\mu},P$, and T.

The normalization condition for P is clearly

$$\sum_{\mathbf{N}} \sum_{l} \mathsf{P}(\mathbf{N},\mathcal{V},l) = 1 \tag{7.9}$$

With this equation and the usual definition of the petit canonical partition function,

$$Q(\mathbf{N},\mathcal{V},T) \equiv \sum_{l} e^{-E_l/kT}, \tag{7.10}$$

the distribution law for the grand canonical ensemble becomes

$$1 = \exp(-P\mathcal{V}/kT) \sum_{N} \exp(\boldsymbol{\mu} \cdot \mathbf{N}/kT)\, Q(\mathbf{N}, \mathcal{V}, T) \tag{7.11}$$

It is convenient to make two further substitutions. We define the function

$$Z(\mathbf{N},\mathcal{V},T) \equiv \exp(\boldsymbol{\mu}^{\dagger} \cdot \mathbf{N}/kT)\, \mathbf{N}!Q(\mathbf{N},\mathcal{V},T) \tag{7.12}$$

which is the configuration integral if the system is classical. Here $\mu_s\dagger$

is the chemical potential of species s in the standard state (cf. Section 6). In general $Z(\mathbf{N},\mathcal{V},T)$ reduces to $\mathcal{V}^N$ for a gas of non-interacting particles and its deviation from this value is a measure of the deviation of the real system from a perfect gas in the same state, $\mathbf{N},\mathcal{V},T$. We also introduce the *grand partition function*,

$$\Xi \equiv \exp\,[P\mathcal{V}/kT] \tag{7.13}$$

We substitute these definitions of Z and Ξ into (7.11) and simplify the resulting equation with the help of (6.9) in order to obtain the more useful form of the distribution law for the grand ensemble:

$$\Xi(\mathbf{z},\mathcal{V},T) = \sum\,[\mathbf{z}^{\mathbf{N}}/\mathbf{N}!]\;Z(\mathbf{N},\mathcal{V},T) \tag{7.14}$$

Both Q and Z are thermodynamic functions which may in principle be calculated from a molecular model for the system. The methods of the cluster theories may be applied to these calculations and indeed this procedure is used in the early versions of the cluster theories as well as in a recent contribution by Brout.[6] However a less direct method introduced by Fowler and Guggenheim[2] in which we work through the grand partition function enables us to decompose a difficult mathematical problem into two parts, each of which is somewhat simpler than the original problem. This seems to be especially relevant for the derivation of the theory for multicomponent systems. On the other hand the non-physical nature of the boundaries of the regions of the grand canonical ensemble[5] leads us to suspect that this is not a satisfactory basis for investigating the surface properties of real systems.

The Generalization of the Grand Partition Function

If our objective were only a theory of multicomponent imperfect gases equation (7.14) could be used as a starting point for the application of cluster methods. However, as shown by McMillan and Mayer[1] we may generalize (7.14) in such a way that the multicomponent gas theory is applicable at once to solutions. This generalization of the grand partition function also serves to clarify the definition of Z, which we have introduced in a completely arbitrary way.

Up to this point the derivation has been based on the most general quantum mechanical systems. A characteristic difference between quantum mechanical and classical systems is that only in the latter are the coordinates of the molecules satisfactory variables for the description of a microscopic state of the system. We now wish to make the transition from a quantum mechanical system to one in which the coordinates of the molecules are good variables for describing a state. As we shall see this can be done in a formal way without abandoning the quantum mechanical laws; however this approach does not seem to be convenient for applications in which quantum effects are very important.[7]

We begin with the general result from quantum mechanics that if $\Psi_l(\{\mathbf{N}\})$ is the wave function for quantum state l of a system of molecules of composition set $\mathbf{N}$ in volume $\mathcal{V}$, then the probability density for observing the molecules at the coordinates $\{\mathbf{N}\}$ in this quantum state is given by

$$|\Psi_l(\{\mathbf{N}\})|^2$$

where $\{\mathbf{N}\}$ specifies internal as well as spatial coordinates. This probability density is *specific* rather than *generic*.[3,4]

In order to explain this important distinction we first consider the partition of the set $\{\mathbf{N}\}$ according to molecular species:

$$\{N_1\} \cup \{N_2\} \cup \cdots \cup \{N_s\} \cup \cdots \cup \{N_\sigma\} = \{\mathbf{N}\}$$

and note that the N_s elements of $\{N_s\}$ are distinguishable from each other except in the negligible case that $\{N_s\}$ corresponds to having more than one molecule at a given set of coordinates. A generic probability density pertains to the observation of $\{\mathbf{N}\}$ with no restriction as to *which* of the N_s molecules of species s is at any particular one of the N_s distinguishable elements of $\{N_s\}$. To define the specific probability density we consider the molecules of each species to be distinguishable from each other, as though they were numbered from 1 to N_s for every s. The specific probability density pertains to the observation of molecule 1 of species 1 at a specified element of $\{N_1\}$, molecule 2 of species 1 at a different specified element of $\{N_1\}$, $\cdots$, and molecule N_σ of species σ at a specified element of $\{N_\sigma\}$.

The integral of the specific probability density over the whole range of coordinates accessible to the molecules must be unity: Each molecule must be found somewhere in the vessel. Therefore we have

$$\int_{\mathcal{V}} |\Psi_l(\{\mathbf{N}\})|^2 \, d\{\mathbf{N}\} = 1. \tag{7.15}$$

Next we calculate the specific probability density for observing molecules of the set $\mathbf{N}$ in the configuration $\{\mathbf{N}\}$ in one of the regions of the grand ensemble characterized by $\boldsymbol{\mu},T$. This probability density is obtained by multiplying Ψ^2 by the probability that the region be occupied by molecules of the set $\mathbf{N}$, collectively in the state l, and summing over l:

$$\begin{aligned} \mathsf{P}(\{\mathbf{N}\},\mathcal{V}) &\equiv \sum_l |\Psi_l(\{\mathbf{N}\})|^2 \mathsf{P}(\mathbf{N},\mathcal{V},l) \\ &= \Xi^{-1} \exp(\boldsymbol{\mu}\cdot\mathbf{N}/kT) \sum_l |\Psi_l(\{\mathbf{N}\})|^2 e^{-E_l/kT} \end{aligned} \tag{7.16}$$

If we integrate this over the volume of the region we obtain

$$\int_{\mathcal{V}} \mathsf{P}(\{\mathbf{N}\},\mathcal{V})\, d\{\mathbf{N}\} = \Xi^{-1} \exp(\boldsymbol{\mu}\cdot\mathbf{N}/kT) \sum_l e^{-E_l/kT} \tag{7.17}$$

where we have used (7.15). This integral is simply the probability that the region of the ensemble be occupied by molecules corresponding to the composition set $\mathbf{N}$.

A more interesting integral is

$$\int_{\mathcal{V}} \mathsf{P}(\{\mathbf{N}+\mathbf{n}\},\mathcal{V})\, d\{\mathbf{N}\}$$

This is the probability density in the grand ensemble for observing molecules of composition set $\mathbf{N}+\mathbf{n}$ in a particular region, with molecules of composition set $\mathbf{N}$ at unspecified coordinates, but with molecules of $\mathbf{n}$ at particular coordinates $\{\mathbf{n}\}$. This probability density is specific as far as the coordinates, $\{\mathbf{n}\}$, are concerned. The corresponding generic probability density is obtained by multiplying by $[\mathbf{N}+\mathbf{n}]!$ and dividing by $\mathbf{N}!$. There are $[\mathbf{N}+\mathbf{n}]!$ ways of assigning the molecules of composition set $\mathbf{N}+\mathbf{n}$ to the coordinates $\{\mathbf{N}+\mathbf{n}\}$, but $\mathbf{N}!$ of these give the same integral even if the molecules are all distinguishable. The generic probability density is thus

$$([\mathbf{N}+\mathbf{n}]!/\mathbf{N}!) \int_{\mathcal{V}} \mathsf{P}(\{\mathbf{N}+\mathbf{n}\},\mathcal{V})\, d\{\mathbf{N}\}$$

This expression is now summed over $\mathbf{N}$ to obtain the generic probability density in the grand ensemble that a region have some molecules of composition set $\mathbf{n}$ at coordinates $\{\mathbf{n}\}$, regardless of the composition set and coordinates of the other molecules in the region:

$$\mathbf{c}^{\mathbf{n}} g_{\mathbf{n}}(\{\mathbf{n}\},\boldsymbol{\mu}) \equiv \sum_N \frac{[\mathbf{N}+\mathbf{n}]!}{\mathbf{N}!} \int_{\mathcal{V}} \mathsf{P}(\{\mathbf{N}+\mathbf{n}\},\mathcal{V})\, d\{\mathbf{N}\} \tag{7.18}$$

This equation serves as the definition of the generic correlation function, $g_{\mathbf{n}}(\{\mathbf{n}\},\boldsymbol{\mu},\mathcal{V},T) = g_{\mathbf{n}}$ which plays an important role in the following development.[8] The definition implies that if $n \ll N$ then $\int_{\mathcal{V}} g_{\mathbf{n}}\, d\,\{\mathbf{n}\} = \mathcal{V}^n$ as one may verify directly.

We now combine (7.18) with some previously derived equations. We introduce the expression (7.16) for $\mathsf{P}(\{\mathbf{N}+\mathbf{n}\},\mathcal{V})$ to obtain

$$g_{\mathbf{n}} \equiv \frac{\exp\,(\boldsymbol{\mu}\cdot\mathbf{n}/kT)}{\Xi\mathbf{c}^{\mathbf{n}}} \sum \frac{\exp\,(\boldsymbol{\mu}\cdot\mathbf{N}/kT)}{\mathbf{N}!}\,[\mathbf{N}+\mathbf{n}]! \cdot \sum_l e^{-E_l/kT} \int_{\mathcal{V}} |\Psi_l(\{\mathbf{N}+\mathbf{n}\})|^2\, d\{\mathbf{N}\} \tag{7.19}$$

Then we use equation (6.9) to replace the chemical potential set, $\boldsymbol{\mu}$, by the fugacity set, $\mathbf{z}$, and the set $\boldsymbol{\mu}^\dagger$.

$$\Xi[\mathbf{c}^{\mathbf{n}}/\mathbf{z}^{\mathbf{n}}]g_{\mathbf{n}}(\{\mathbf{n}\},\mathbf{z}) = \sum_N [\mathbf{z}^{\mathbf{N}}/\mathbf{N}!]\exp\,([\mathbf{n}+\mathbf{N}]\cdot\boldsymbol{\mu}^\dagger/kT) \times [\mathbf{n}+\mathbf{N}]! \sum_l e^{-E_l/kT} \int_{\mathcal{V}} |\Psi_l(\{\mathbf{n}+\mathbf{N}\})|^2\, d\{\mathbf{N}\} \tag{7.20}$$

Now taking the limit of both sides of (7.20) as $\mathbf{z} \to \mathbf{0}$, noting that

$$\Xi \equiv \exp\,(P\mathcal{V}/kT) \to 1, \cdot$$

$$c_s/z_s \to 1,$$

we find that the only non-vanishing term on the right is that for $N = 0$ and so we have

$$g_{\mathbf{n}}(\{\mathbf{n}\},\mathbf{0}) = \mathbf{n}!\exp\,(\mathbf{n}\cdot\boldsymbol{\mu}^\dagger/kT)\sum_l e^{-E_l/kT}\,|\Psi_l(\{\mathbf{n}\})|^2 \tag{7.21}$$

Comparison of this result with (7.12), the definition of Z, shows at once that we have

$$Z(\mathbf{n},\mathcal{V},T) = \int_{\mathcal{V}} g_{\mathbf{n}}(\{\mathbf{n}\},\mathbf{0})\, d\{\mathbf{n}\} \tag{7.22}$$

In equations (7.15) to (7.22) We have used $\{\mathbf{n}\}$ to represent a complete set of coordinates, i.e. including internal coordinates. Now we return to the notation used in the rest of this book, writing $\{\mathbf{n}\}$ for *spatial* coordinates and $\{\mathbf{n}\}_i$ for internal coordinates. Then (7.22) becomes

$$Z(\mathbf{n},\mathcal{V},T) = \int_{\mathcal{V}} g_{\mathbf{n}}(\{\mathbf{n}\},\{\mathbf{n}\}_i, \mathbf{0})\, d\{\mathbf{n}\}\, d\{\mathbf{n}\}_i \tag{7.23}$$

Now comparing (7.23) with the more familiar form for the configuration integral (5.3),

$$Z(\mathbf{n},\mathcal{V},T) = \int_{\mathcal{V}} \exp\left(-U_{\mathbf{n}}/kT\right) d\{\mathbf{n}\}$$

we find that these are equivalent provided that we have

$$\int g_{\mathbf{n}}(\{\mathbf{n}\},\{\mathbf{n}\}_i, \mathbf{0})\, d\{\mathbf{n}\}_i = \exp\left(-U_{\mathbf{n}}(\{\mathbf{n}\})/kT\right) \tag{7.24}$$

In all of these equations $U_{\mathbf{n}}$ is the potential of average force in a vacuum as discussed in Section 5. Equation (7.24) is readily derived in an independent way.[1, 4] Thus it is apparent that defining Z by equation (7.12) is consistent with the usual definition.

Now returning to (7.21) and substituting it in (7.20), we get

$$\Xi[\mathbf{c}^{\mathbf{n}}/\mathbf{z}^{\mathbf{n}}] g_{\mathbf{n}}(\{\mathbf{n}\},\{\mathbf{n}\}_i, \mathbf{z}) = \sum [\mathbf{z}^{\mathbf{N}}/\mathbf{N}!] \int_{\mathcal{V}} g_{\mathbf{n}+\mathbf{N}}(\{\mathbf{n}+\mathbf{N}\},\{\mathbf{n}+\mathbf{N}\}_i, \mathbf{0})\, d\{\mathbf{N}\}\, d\{\mathbf{N}\}_i \tag{7.25}$$

The last step is to integrate both sides of this equation over the entire range of $\{\mathbf{n}\}_i$ and to introduce the *spatial generic correlation function*,

$$g_{\mathbf{n}}(\{\mathbf{n}\},\mathbf{z}) \equiv \int_{\mathcal{V}} g_{\mathbf{n}}(\{\mathbf{n}\},\{\mathbf{n}\}_i, \mathbf{z})\, d\{\mathbf{n}\}_i \tag{7.26}$$

which is also a function of T although this is not specified. It is also a function of $\mathcal{V}$ except that as $\mathcal{V}$ goes to infinity at constant $\mathbf{z}$ this dependence vanishes.[9] We repeat that henceforth we shall use $\{\mathbf{n}\}$ only to represent a set of *spatial* coordinates.

The important equations which follow become difficult to read unless we simplify the notation as far as possible. Therefore instead of $g_{\mathbf{n}}(\{\mathbf{n}\},\mathbf{z})$, which specifies a correlation function for molecules of composition set $\mathbf{n}$ at $\{\mathbf{n}\}$ in a system of fugacity set $\mathbf{z}$, we shall write only $g_{\mathbf{n}}$ or $g_{\mathbf{n}}(\mathbf{z})$, depending on whether or not the fugacity set is obvious from the context.

With these changes (7.25) becomes

$$\Xi[\mathbf{c}^{\mathbf{n}}/\mathbf{z}^{\mathbf{n}}]g_{\mathbf{n}}(\mathbf{z}) = \sum_{N} [\mathbf{z}^{\mathbf{N}}/\mathbf{N}!] \int_{\mathcal{V}} g_{\mathbf{n}+\mathbf{N}}(\mathbf{0})\, d\{\mathbf{N}\} \tag{7.27}$$

If $n = 0$ this equation reduces to

$$\begin{aligned} \Xi &= \sum_{N} [\mathbf{z}^{\mathbf{N}}/\mathbf{N}!] \int_{\mathcal{V}} g_{\mathbf{N}}(\mathbf{0})\, d\{\mathbf{N}\} \\ &= \sum_{N} \mathbf{z}^{\mathbf{N}} Z(\mathbf{N},\mathcal{V},T)/\mathbf{N}! \end{aligned} \tag{7.28}$$

and in this sense (7.27) is a generalization of the grand partition function.

A further generalization of the grand partition function can be deduced by noting that (7.27) represents the general member of a sequence of series to which the methods of Section 3 apply. In order to see this more clearly we introduce the abbreviation

$$G(\{\mathbf{n}\},\mathbf{z}) \equiv \Xi[\mathbf{c}^{\mathbf{n}}/\mathbf{z}^{\mathbf{n}}]g_{\mathbf{n}}(\mathbf{z}) \tag{7.29}$$

and note that

$$G(\{\mathbf{n}\},\mathbf{0}) = g_{\mathbf{n}}(\mathbf{0}) \tag{7.30}$$

Now equation (7.27) may be written in the form

$$G(\{\mathbf{n}\},\mathbf{z}) = \sum [\mathbf{z}^{\mathbf{N}}/\mathbf{N}!] \int_{\mathcal{V}} G(\{\mathbf{n}+\mathbf{N}\},\mathbf{0})\, d\{\mathbf{N}\} \tag{7.31}$$

Comparison with (3.25) shows that the inverse of (7.31) is

$$G(\{\mathbf{n}\},\mathbf{0}) = \sum [[-\mathbf{z}]^{\mathbf{N}}/\mathbf{N}!] \int_{\mathcal{V}} G(\{\mathbf{n}+\mathbf{N}\},\mathbf{z})\, d\{\mathbf{N}\} \tag{7.31$'$}$$

which would give us a way to determine the potentials of average force in a vacuum from the spatial correlation functions in a gas at finite pressure, if only the latter were experimentally accessible. A more interesting result is obtained by comparison of (7.31) with (3.25)″ which leads to the generalization of (7.31)

$$G(\{\mathbf{n}\},\mathbf{z}+\mathbf{y}) = \sum [\mathbf{z}^{\mathbf{N}}/\mathbf{N}!] \int_{\mathcal{V}} G(\{\mathbf{n}+\mathbf{N}\},\mathbf{y})\, d\{\mathbf{N}\} \tag{7.32}$$

where $\mathbf{z}$ and $\mathbf{y}$ are two fugacity sets. This equation gives a general (although implicit) relation between the spatial correlation functions in the state characterized by $\mathbf{z} + \mathbf{y}, T$ in terms of the state characterized by $\mathbf{y}, T$. The corresponding generalization of (7.27) is obtained simply by substituting (7.29) in (7.32) but it is rather more difficult to read:

$$\Xi(\mathbf{z} + \mathbf{y}) \frac{[\mathbf{c}(\mathbf{z} + \mathbf{y})]^{\mathbf{n}}}{[\mathbf{z} + \mathbf{y}]^{\mathbf{n}}} g_{\mathbf{n}}(\mathbf{z} + \mathbf{y}) = \frac{\Xi(\mathbf{y})[\mathbf{c}(\mathbf{y})]^{\mathbf{n}}}{\mathbf{y}^{\mathbf{n}}} \sum \frac{\mathbf{z}^{\mathbf{N}}[\mathbf{c}(\mathbf{y})]^{\mathbf{N}}}{\mathbf{N}!\,\mathbf{y}^{\mathbf{N}}} \int_{\mathcal{V}} g_{\mathbf{n}+\mathbf{N}}(\mathbf{y})\, d\{\mathbf{N}\} \tag{7.33}$$

Osmotic Equilibrium

An application of the equations for two fugacity sets is readily made to a system consisting of a solution in osmotic equilibrium with its solvent. We let $\mathbf{y}$ be the fugacity set of solvent species, which in this system is the same for the solution and for the pure solvent. Then $\mathbf{z}$ is the fugacity set of the solute species in the solution and $\mathbf{z} = \mathbf{0}$ is the fugacity set of the solute species in the pure solvent. We get especially simple equations if we restrict $\mathbf{n}$, appearing on the left of (7.32) and (7.33), to be a composition set of solute species only. Then we have

$$G(\{\mathbf{n}\}, \mathbf{z} + \mathbf{y}) = \Xi\,(\mathbf{z} + \mathbf{y}) \frac{[\mathbf{c}(\mathbf{z} + \mathbf{y})]^{\mathbf{n}}}{\mathbf{z}^{\mathbf{n}}} g_{\mathbf{n}}(\mathbf{z} + \mathbf{y}) \tag{7.34}$$

where the concentrations are all solute concentrations in the solution and where

$$\Xi(\mathbf{z} + \mathbf{y}) = \exp\,([P_0 + \Pi]\mathcal{V}/kT)$$

with P_0 the pressure on the pure solvent and Π the osmotic pressure.

We also have

$$G(\{\mathbf{N} + \mathbf{n}\}, \mathbf{y}) = \Xi(\mathbf{y}) g_{\mathbf{N}+\mathbf{n}}(\mathbf{y}) \lim_{\mathbf{z}\to 0} \left(\frac{[\mathbf{c}(\mathbf{z} + \mathbf{y})]^{\mathbf{N}+\mathbf{n}}}{[\mathbf{z} + \mathbf{y}]^{\mathbf{N}+\mathbf{n}}} \right) \tag{7.35}$$

The limit in this equation is a finite quantity which depends on the interaction of solute with the solvent when the latter is in the state $\mathbf{y}, T$. The Ξ appearing here is $\exp\,(P_0\mathcal{V}/kT)$. Now we combine these equations with (7.33) to obtain

$$\Xi_{\text{osm}}[\mathbf{c}^{\mathbf{n}}/\mathbf{a}^{\mathbf{n}}] g_{\mathbf{n}}(\mathbf{z} + \mathbf{y}) = \sum\, [\mathbf{a}^{\mathbf{N}}/\mathbf{N}!] \int_{\mathcal{V}} g_{\mathbf{n}+\mathbf{N}}(\mathbf{y})\, d\{\mathbf{N}\} \tag{7.36}$$

where we have introduced the abbreviations

$$\Xi_{\text{osm}} \equiv \exp\,(\Pi\mathcal{V}/kT) \tag{7.37}$$

and

$$a_s \equiv z_s \lim_{\mathbf{z}\to 0} \frac{c_s(\mathbf{z}+\mathbf{y})}{z_s} \tag{7.38}$$

The function a_s is called the *activity* of solute species s in the solution. It plays the same role in solutions as the corresponding fugacity, z_s, does in the gas phase. In particular note that as the fugacity of solute fugacities vanishes in the solution ($\mathbf{z} \to \mathbf{0}$), $a_s \to c_s$ for every solute species s.

We also remark that the summation in (7.36) is over composition sets $\mathbf{N}$ of solute particles alone. We derive this by noting in (7.32) that if $\mathbf{z}$ is a set of solute fugacities, then the exponent in any non-zero $\mathbf{z}^{\mathbf{N}}$ must be a solute composition set. Furthermore only solute concentrations appear in (7.36) because of the restriction that $\mathbf{n}$ be a solute composition set. Finally we note that the state $\mathbf{z} + \mathbf{y}$ corresponds to a state specified by solute activity $\mathbf{a}$ while the state $\mathbf{y}$ corresponds to $\mathbf{a} = \mathbf{0}$. Therefore we may write (7.36) in a form in which only solute activities appear:

$$\Xi_{\text{osm}}[\mathbf{c}^{\mathbf{n}}/\mathbf{a}^{\mathbf{n}}]g_{\mathbf{n}}(\mathbf{a}) = \sum_{N} [\mathbf{a}^{\mathbf{N}}/\mathbf{N}!] \int_{\mathcal{V}} g_{\mathbf{n}+\mathbf{N}}(\mathbf{0})\, d\{\mathbf{N}\} \tag{7.39}$$

and for the special case $n = 0$,

$$\Xi_{\text{osm}} = \sum_{N} [\mathbf{a}^{\mathbf{N}}/\mathbf{N}!] \int_{\mathcal{V}} g_{\mathbf{N}}(\mathbf{0})\, d\{\mathbf{N}\} \tag{7.40}$$

Discussion

The most important result of this section is the complete analogy of (7.39) to (7.27) and, as a special case, (7.40) to (7.28). This is illustrated in Fig. 7.1 from which we see that each equation relates the correlation function of a "gas" under pressure to the correlation functions in the infinitely dilute "gas," but that in (7.39) and (7.40) the gas is only such in the van't Hoff sense. In fact in Section 8 we shall derive a generalization of the van't Hoff osmotic pressure equation from (7.40) so it is apparent that the van't Hoff model of a solution in osmotic equilibrium with its solvent is the physical basis of the observed analogy of (7.39) to (7.27).

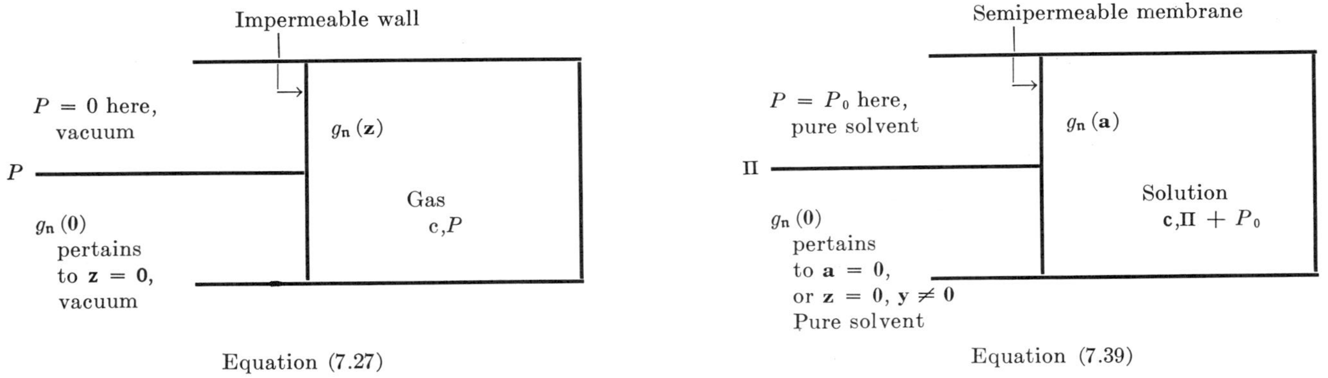

Fig. 7.1. Illustrating the analogy of (7.27) to (7.39).

We now describe this analogy in greater detail. For the imperfect gas the functions assumed to be given, namely the set of $g_{\mathbf{n}}(\mathbf{0})$, are each determined by the forces among a set, **n**, of molecules in a vacuum, while for a solution the corresponding functions, again the set of $g_{\mathbf{n}}(\mathbf{0})$, are each determined by the forces among a set, **n**, of *solute* molecules in the solvent in the state P_0. The correspondence between the fugacity set **z** of the gas and the activity set **a** of the solute in the solution is obvious. The correlation functions $g_{\mathbf{n}}(\mathbf{z})$ provide a detailed description of the equilibrium state of the gas at a pressure P. (Consider that the pressure is a function of **z** and appears implicitly in Ξ in (7.27).) The correlation functions $g_{\mathbf{n}}(\mathbf{a})$ provide a detailed description of the equilibrium state of the solution at pressure $P_0 + \Pi$. (P_0 is the pressure on the pure solvent in osmotic equilibrium and Π, a function of the set **a**, implicitly appears in $\Xi_{\text{osm.}}$.) However the analogy of (7.39) to (7.27) is incomplete in one respect: The detailed description of the equilibrium state of the *solution* is incomplete because we have restricted **n** to be a composition set of solute species, hence (7.39) can not be used to calculate correlation functions for any set $\{\mathbf{n}\}$ which specifies the coordinates of one or more solvent molecules.

The spatial correlation functions correspond to what one could determine experimentally if he could determine, as by photography, the positions of the centers of mass of the molecules in a vessel at a particular instant. By observing the coordinates of every set of composition **n** in such photographs and making suitable averages one could get an experimental value of $g_{\mathbf{n}}(\mathbf{z})$. On the other hand $g_{\mathbf{n}}(\mathbf{a})$ of equation (7.39) corresponds to what one would obtain in this way from photographs of solutions if only the solute molecules registered in the photographic process.

The general philosophical import of these observations has been expressed in the following way by McMillan and Mayer[1]:

"It is thus to be noted that, for a world of constant temperature at least, the presence of intangible molecules which act with forces on ordinary molecules but which could pass through them, would not alter the applicability of the Gibbs statistics to equilibrium systems. Since the intangible molecules pass through all barriers their fugacity would everywhere be constant, though not their concentration, and all measured pressures would be osmotic pressures. The measured forces between ordinary molecules would all be average forces, influenced by the all-pervading fluid molecules, but the potentials calculated from them, the potentials of average force, would be strictly applicable to the calculation of all observed thermodynamic properties."

What we have quoted may be deduced by comparing equations (7.28) and (7.40) which lead, respectively, to Ξ and Ξ_{osm} and hence to all other *thermodynamic* properties. If the molecules of the "all-pervading fluid" were too subtle for direct observation by *any* means then it would follow that one would be unable to tell whether, in a given case, he were applying (7.27) or (7.39) and then the conclusion of McMillan and Mayer could be generalized by changing the last words of their statement from "thermodynamic properties" to "equilibrium properties."

Perhaps the best example of an observable equilibrium property that is more detailed than a thermodynamic property is the radial distribution function for electrons, which is the spatial correlation function, $g_{ee}(\mathbf{z})$, for pairs of electrons as determined by x-ray or electron diffraction experiments.[10] The electrons in ordinary chemical systems are of course associated with atomic or molecular centers but, except for systems containing only monatomic molecules of a single species, $g_{ee}(\mathbf{z})$ represents a much less detailed description of the system than $g_{\mathbf{n}}(\mathbf{z})$ or even $g_{\mathbf{n}}(\mathbf{a})$. A similar discussion applies to the radial distribution function that one may determine by neutron diffraction. Therefore it seems that the spatial correlation functions are in fact, if not in principle, experimentally inaccessible and that the most we can do by experimental observations on equilibrium systems without ancillary assumptions is to determine thermodynamic functions. Some applications of correlation functions are described in Sections 9 and 10.

Our main interest in the analogy of (7.28) to (7.40) lies in the fact that the mathematical manipulation of (7.28) by cluster theory methods to obtain a more useful form for calculating experimental quantities for imperfect gases applies at once to (7.40) and solutions. However since the form of (7.40) forces us to make a distinction between solute and solvent molecules in the solutions the resulting theory of solutions is most useful for cases in which such a distinction is natural. This is indeed the case for systems which have an intrinsic asymmetry among the components, such as polystyrene–benzene or sodium chloride–water. At the other extreme we have mixtures such as H_2–D_2, benzene–carbon tetrachloride, or NaCl–KCl, in which an asymmetry can only be imposed by artificial means such as restricting the composition range so that the chosen "solvent" species is always in excess.

We remark that equation (7.39) gives a complete description of the equilibrium state of a solution at finite **a**, and hence at finite **c**, in terms of the correlation functions of the infinitely dilute solution ($\mathbf{a} = \mathbf{c} = \mathbf{0}$). This is perhaps especially startling to one who is accustomed to think about electrolyte solutions and all of their inherent complexity. This aspect may be clarified by considering the analogy of the most general of the equations in this section, (7.32), to a Taylor series expansion (in one variable for simplicity),

$$f(x + y) = \sum [y^n/n!]f^{(n)}(x)$$

where

$$f^{(n)}(x) \equiv (\partial^n f(z)/\partial z^n) \qquad \text{at} \qquad z = x$$

Just as $f(x + y)$ is determined at any y by $f(x)$ and all of the derivatives of f at x, so in (7.32) $G(\{\mathbf{n}\})$ at the fugacity set $\mathbf{z} + \mathbf{y}$ is determined by $G(\{\mathbf{n}\})$ and all of the $G(\{\mathbf{n} + \mathbf{N}\})$ at the fugacity set $\mathbf{y}$.

Indeed as Ono[11] has shown one may derive (7.32) from (7.31) by a Taylor's series method, rather than by making use of the properties of sequences of series. However the analogy of (7.33) to a Taylor's series is not complete for the Taylor series expansion of a function of a real variable is only *necessarily* valid in the range of the independent variable in which the function and all of its derivatives are continuous. This restriction does not pertain to the theorems on sequences of series and hence the method which we have used, which is due to Mayer,[12] is preferable to the Taylor series method for the most general considerations and for application to the problem of phase transitions. It must be pointed out, however, that there are other characteristic and very serious difficulties which impede the application of the cluster theories to the investigation of phase equilibria. The difficulties, which were first pointed out by Kahn and Uhlenbeck[13] and have not yet been resolved,[5,14] do not appear in the equations developed in this section, but only in certain equations obtained by further manipulation.

The final topic to be considered here is the convergence of the series expansions we have obtained for Ξ and its generalizations. It is sufficient to discuss only (7.32), the most general equation.

The function $G(\{\mathbf{N}\},\mathbf{z})$ is now more appropriately written

$$G(\{\mathbf{N}\},\mathbf{z},\mathcal{V},T)$$

for, as the chain of definitions leading to this function shows, it depends on the temperature and the volume $\mathcal{V}$ of each region of the grand ensemble. The integrals in (7.32) are each over all spatial coordinate sets $\{\mathbf{N}\}$ that are consistent with the molecules whose coordinates are $\{\mathbf{N}\}$ being within this volume. Now we recognize that as a result of the repulsive forces among molecules at small mutual separations, only some definite maximum number of molecules may be found within a given region of the grand ensemble with non-vanishing probability. For molecules that are hard spheres this is a perfectly sharp result; more molecules than some definite N^* are simply unable to squeeze in and $G(\{\mathbf{N}\},\mathbf{z},\mathcal{V},T) = 0$ for $N > N^*$ for some finite N^*, depending on $\mathcal{V}$. This assures the convergence of the series because the number of non-zero terms is finite. For real molecules the result is qualitatively the same but depends in detail on the form of the repulsive potentials.

Although the mathematical convergence of these series is readily established in this way, these series are poorly suited for the calculation of observable quantities from a molecular model because the largest term in the series is found at quite large N: For ordinary systems this is for N of the order of Avogadro's number. A further defect in these series from this point of view is the appearance of the fugacities as independent variables; for experimental work the concentrations are generally much more convenient. In the following section we employ the methods of the cluster theory to rearrange the terms of these series to obtain much more rapidly convergent expansions for Ξ and Ξ_{osm} in which the independent variables are concentrations rather than activities. In Section 9 we shall see how to obtain the corresponding expansions of the spatial correlation functions.

Notes and References

1. W. G. McMillan and J. E. Mayer, *J. Chem. Phys.* **13**, 276 (1945).
2. R. Fowler and E. A. Guggenheim, *Statistical Thermodynamics* (Cambridge University Press, New York, 1952).
3. D. ter Haar, *Elements of Statistical Mechanics*, (Holt, Rinehart, and Winston, Inc., New York, 1954).
4. T. L. Hill, *Statistical Mechanics*, (McGraw-Hill Book Company, Inc., New York, 1956).
5. J. E. Mayer, *Handbuch der Physik* (Springer-Verlag, Berlin, 1958) B, XII, p. 73.

6. R. Brout, *Phys. Rev.* **115,** 824 (1959).
7. Evidence for this is the large number of papers treating quantum mechanical systems by cluster methods ab initio, rather than by applying the final results of the McMillan-Mayer theory. For example see: T. Kihara, *Advances in Chemical Physics*, I. Prigogine, ed. (Interscience Publishers, Inc., New York, 1958), Vol. I, p. 267; E. W. Montroll and J. C. Ward, *Phys. Fluids* **1,** 55 (1958); C. D. Hartogh and H. A. Tolhoek, *Physica* **24,** 896 (1958); S. Baldursson, J. E. Mayer, and H. Aroeste, *J. Chem. Phys.* **31,** 814 (1959). On the other hand the approach outlined here has been used in a quantum mechanical problem by T. Morita, Prog. Theor. Phys., **22,** 757 (1959).
8. McMillan and Mayer use the symbol F_n instead of g_n.
9. This is true for a system in which surface effects are negligible. For a discussion see: T. L. Hill, *Statistical Mechanics* (McGraw-Hill Book Company, Inc., New York, 1956), p. 234.
10. The statistical mechanical aspects of such measurements on one-component systems of monatomic molecules are discussed by T. L. Hill, *Statistical Mechanics* (McGraw-Hill Book Company, Inc., New York, 1956), p. 185. For a more complete discussion see: G. H. Vineyard, *Liquid Metals and Solidification* (American Society for Metals, Cleveland, Ohio, 1958), pp. 34–46; J. Waser and V. Schomaker, *Revs. Modern Phys.* **25,** 671 (1953).
11. S. Ono, *Progr. Theoret. Phys.* (Kyoto) **6,** 447 (1951). T. L. Hill, *Statistical Mechanics* (McGraw-Hill Book Company, Inc., New York, 1956), p. 246.
12. J. E. Mayer, *J. Chem. Phys.* **10,** 629 (1942).
13. B. Kahn and G. E. Uhlenbeck *Physcia* **5,** 399 (1938).
14. For recent articles on the subject see R. Brout, *Phys. Rev.* **115,** 824 (1959) and S. Katsura, *Prog. Theoret. Phys. (Kyoto)* **13,** 571 (1955). T. L. Hill, *Statistical Mechanics* (McGraw-Hill Book Company, Inc., New York, (1956), p. 264 and Appendix G.

8. The Density Expansion of the Excess Free Energy

Imperfect Gases

In this section we derive the basic equation for the density or concentration expansions for thermodynamic properties in equilibrium systems.

We begin with the fugacity expansion for Ξ (equations (7.14), (7.28))

$$\Xi(\mathbf{z},\mathcal{V},T) = \sum [\mathbf{z}^{\mathbf{N}}/\mathbf{N}!] \int_{\mathcal{V}} g_{\mathbf{N}}(\mathbf{0})\, d\{\mathbf{N}\} \tag{8.1}$$

where $\{\mathbf{N}\}$ is a set of spatial coordinates; prior integration over internal coordinates is assumed. We also have

$$g_{\mathbf{N}}(\mathbf{0}) = \exp\left[-U_{\mathbf{N}}/kT\right] \tag{8.2}$$

where $U_{\mathbf{N}}$, the "direct potential," is the potential of average force for the set $\mathbf{N}$ at $\{\mathbf{N}\}$ in a vacuum.

We introduce the expansion of $U_{\mathbf{N}}$ in component potentials (equation (4.5))

$$U_{\mathbf{N}} = \sum_{\{\mathbf{n}\}\subseteq\{\mathbf{N}\}}'' u_{\mathbf{n}} \tag{8.3}$$

and the definition of the cluster functions

$$\gamma_{\mathbf{n}} \equiv \exp\left[-u_{\mathbf{n}}/kT\right] - 1 \tag{8.4}$$

and so obtain the cluster expansion of $g_{\mathbf{N}}(\mathbf{0})$. (equation (5.5))

$$g_{\mathbf{N}}(\mathbf{0}) = \sum_{\mathbf{p}]\{\mathbf{N}\}} \prod_{\{\mathbf{k}\}\subseteq\{\mathbf{N}\}} [s_{\mathbf{k}}(\{\mathbf{k}\})]^{p\{\mathbf{k}\}} \tag{8.5}$$

where the sum is over partitions of $\{\mathbf{N}\}$ and where $s_{\mathbf{k}}$ is the sum of all products of cluster functions that correspond to at-least-singly-connected graphs on the skeleton of $\mathbf{k}$.

The integration of (8.5) depends on two observations.

(1) In a given term of (8.5) each factor of the product depends on a distinct $\{\mathbf{k}\}$, no element of which appears in any other factor. Therefore, the integral of a given term is a product of integrals of the factors $s_{\mathbf{k}}$ of the term.

(2) There will be many terms of (8.5) which correspond to different partitions of $\{\mathbf{N}\}$ but to the same partition of $\mathbf{N}$. Each such term gives the same result when integrated. The number of such terms is

$$\mathbf{N}!/\prod_{\mathbf{k}\leq\mathbf{N}} [[\mathbf{k}!]^{p_{\mathbf{k}}}/p_{\mathbf{k}}!] \tag{8.6}$$

We define the reducible cluster integral by the equation

$$b_{\mathbf{k}} = (1/\mathcal{V}\mathbf{k}!) \int_{\mathcal{V}} s_{\mathbf{k}}\, d\{\mathbf{k}\} \tag{8.7}$$

and then the above observations lead to the result (equation (5.7))

$$\int_{\mathcal{V}} g_{\mathbf{N}}(\mathbf{0})\, d\{\mathbf{N}\} = \mathbf{N}! \sum_{\mathbf{p}]\mathbf{N}} \prod_{\mathbf{k}\leq\mathbf{N}} [\mathcal{V} b_{\mathbf{k}}]^{p_{\mathbf{k}}}/p_{\mathbf{k}}! \tag{8.8}$$

where the sum is over partitions of $\mathbf{N}$. Comparing this result with the cumulant-moment relation ((3.35) and (3.38)) we find at once

$$\Xi(\mathbf{z},\mathcal{V},T) = \exp\left(\sum{}' \mathbf{z}^{\mathbf{n}} b_{\mathbf{n}} \mathcal{V}\right) \tag{8.9}$$

or, taking logarithms,

$$P/kT = \sum{}' \mathbf{z}^{\mathbf{n}} b_{\mathbf{n}} \tag{8.10}$$

Now we use the thermodynamic relation (equation (6.19))

$$z_s(\partial P/\partial z_s)_{T,\mathcal{V},\mathbf{z}-z_s} = kTc_s \tag{8.11}$$

to convert (8.10) to the form

$$c_s = \sum{}' n_s \mathbf{z}^{\mathbf{n}} b_{\mathbf{n}} \tag{8.12}$$

The next step is to define the irreducible cluster integral (equation (5.10)),

$$B_{\mathbf{k}} = [\mathcal{V}\mathbf{k}!]^{-1} \int_{\mathcal{V}} S_{\mathbf{k}}\, d\{\mathbf{k}\} \tag{8.13}$$

where $S_{\mathbf{k}}$ is the sum of all products of cluster functions that correspond to at-least-doubly-connected graphs on the skeleton of **k**. The reducible cluster integrals may be expressed as sums of products of irreducible cluster integrals in the following way (equation (5.11)),

$$\mathbf{n}!b_{\mathbf{n}} = \sum_{\mathbf{t}]\mathbf{n}} T(\mathbf{t}]\mathbf{n}) \prod_{\mathbf{k}\leq\mathbf{n}}{}'' [\mathbf{k}!B_{\mathbf{k}}]^{t_{\mathbf{k}}} \tag{8.14}$$

where the sum is over all Husimi trees of **n** and where T is a combinatorial factor, the tree coefficient.

Now we define

$$\mathfrak{S}^* \equiv \sum{}'' \mathbf{c}^{\mathbf{n}} B_{\mathbf{n}} \tag{8.15}$$

From this equation and (8.12) and (8.14) we deduce that

$$c_s = z_s \exp\,(\partial\mathfrak{S}^*/\partial c_s) \tag{8.16}$$

because reversion‡ of this equation gives an equation of the form of (8.12) in which the coefficients are trees of the $B_{\mathbf{n}}$ and, as (8.14) shows, the $b_{\mathbf{n}}$ are indeed trees of the $B_{\mathbf{n}}$. (Cf. the relation of (5.31) to (5.24).)

Comparing (8.16) with the thermodynamic equation (6.13),

$$c_s = z_s \exp\,(\partial\mathfrak{S}/\partial c_s) \tag{8.17}$$

‡ *Reversion* here refers to the fact that equation (8.12) expresses the concentration of any component in terms of the fugacities of all the components while (8.16), after a minor rearrangement, expresses the fugacity of any component in terms of the concentrations of all of the components.

we see that

$$\mathfrak{S}^* = \mathfrak{S} = -\mathfrak{F}^{ex}/\mathcal{V}kT \tag{8.18}$$

By these mathematical procedures we have converted (8.1), an expansion of Ξ in powers of fugacity, to

$$\mathfrak{F}^{ex} = -\mathcal{V}kT \sum{}'' \mathbf{c}^{\mathbf{n}} B_{\mathbf{n}} \tag{8.19}$$

an expansion of the excess free energy in powers of concentration. In both cases the coefficients of the expansion are determined by the interactions of sets of molecules in the vacuum, but the change from $Z(\mathbf{n},\mathcal{V},T)$ to $B_{\mathbf{n}}$ as coefficients gives a tremendous improvement in convergence. The physical reason for this is that the Z's have appreciable contributions from almost all regions of the configuration space while the $B_{\mathbf{n}}$ have negligible contributions from almost all regions of the configuration space. The terms of $g_{\mathbf{n}}(\mathbf{0})$ include many disconnected graphs on the skeleton of $\mathbf{n}$, these correspond to functions which approach unity as the distance between the disconnected parts becomes larger. The terms of $S_{\mathbf{n}}$ includes only ALDC graphs on the skeleton of $\mathbf{n}$; these terms rapidly approach zero as the distances between molecules increase because of the short range of the bonds.

If we compare $b_{\mathbf{n}}$ and $B_{\mathbf{n}}$ in the same way we see that, for the same $\mathbf{n}$, $b_{\mathbf{n}}$ has appreciable contributions from a larger region of the configuration space than $B_{\mathbf{n}}$ because some of the terms of $b_{\mathbf{n}}$ are more loosely connected than any terms of $B_{\mathbf{n}}$. The bonds all correspond to factors in the integrand that are less than unity for large separations so a more tightly connected graph has appreciable contributions from a smaller part of the configuration space.

On the basis of such considerations one is at once led to speculate on the possibility of further manipulations to gain an additional increase in the connectivity of the graphs and an additional improvement in the convergence of the series. There has been some progress in this direction, all based on the convolution method introduced by Mayer in his adaptation of the cluster theory to ionic solutions. These developments are referred to again at the end of Section 14.

Equation (8.19) is based on the assumption that the surface terms are negligible. The more general result is

$$-\mathfrak{F}^{ex}/\mathcal{V}kT = \mathfrak{S} = \sum{}'' \mathbf{c}^{\mathbf{n}} B_{\mathbf{n}} + \sum{}'' \mathbf{c}^{\mathbf{n}} \sigma_{\mathbf{n}} \tag{8.19*}$$

where the general surface term, $\sigma_{\mathbf{n}}$, is defined by equation (5.51). It is customary to neglect these terms. This may be justified for terms of finite n, at least for non-ionic systems, if one considers (8.19*) in the limit as $\mathcal{V} \to \infty$ (Section 5(h)).

Solutions

The procedures of equations (8.1) to (8.19) can be repeated for solutions, beginning with (7.40) instead of (7.28). It is apparent at once that the result is

$$-\mathfrak{F}^{\text{ex}}/\mathcal{V}kT = \mathfrak{S} = \sum{}'' \mathbf{c}^{\mathbf{n}} B_{\mathbf{n}} \tag{8.20}$$

where now (cf. Fig. 7.1) $\mathfrak{F}^{\text{ex}}$ is the excess free energy of the solution in equilibrium with the solvent in its reference state, P_0. Hence $\mathfrak{F}^{\text{ex}}$ pertains to the solution at pressure $P_0 + \Pi$. $\mathbf{c}$ is the set of solute concentrations in the solution at $P_0 + \Pi$. The integrand of $B_{\mathbf{n}}$ has the same structure as in the imperfect gas case except that now the cluster functions are given by

$$\gamma_{\mathbf{m}} = \exp\left[-w(\{\mathbf{m}\},\mathbf{0})/kT\right] - 1 \tag{8.21}$$

where $w(\{\mathbf{m}\},\mathbf{0})$ is a component of the potential of average force of the solute set $\mathbf{m}$ at $\{\mathbf{m}\}$ in the pure solvent in its reference state. In other words instead of having as direct potentials the potentials of average force $U_{\mathbf{m}}(\{\mathbf{m}\})$ for set $\mathbf{m}$ at $\{\mathbf{m}\}$ in a vacuum, we have the potentials of average force $W_{\mathbf{m}}(\{\mathbf{m}\},\mathbf{z},\mathbf{a} = \mathbf{0})$ for the set $\mathbf{m}$ of solute particles in the solvent (fugacity of solvent species $\mathbf{z}$, activities of solute species $\mathbf{a} = \mathbf{0}$).

Explicit Internal Coordinates

In some cases the procedure of integrating over internal coordinates at the outset may result in the effects of greatest interest being carried in the higher component potentials. Then it may be preferable to carry the internal coordinates explicitly through the calculation.

If this is done all of the equations in this section are valid provided that the following changes are made:

(1) Interpret $\{\mathbf{N}\}$ as the set of coordinates, including internal coordinates carried in the calculation.

(2) In equation (8.13) $\mathcal{V} \equiv \int d\{1\}$ where the integration is over the complete configuration space for a particle.

(3) In equation (8.12) and later equations

$$c_s = N_s / \int d\{1\}$$

For example, if orientational coordinates are carried then $\int d\{1\}_i = 8\pi^2$ and

$$c_s = N_s/8\pi^2 \times \text{geometric volume of system}$$

The same modifications serve to allow Section 9 to be applied to calculations in which some internal coordinates are carried explicitly.

9. Spatial Correlation Functions

Introduction

The spatial generic correlation functions (equation (7.26)) provide a more detailed description of the properties of systems at equilibrium than do the thermodynamic functions. The correlation functions have mostly been used in equilibrium theories as tools for deriving expressions for thermodynamic functions from molecular models or their equivalent, but they are also useful in interpreting certain experimental observations, particularly those relating to association equilibria.[1] There is a strong and obvious connection between the tendency of some molecules to associate and the behavior of the correlation functions which describe the equilibrium spatial distribution of these molecules in a system. On this basis it seems that correlation functions are of interest even when exact and useful expressions for the thermodynamic functions in terms of a model are already available.

The calculation of thermodynamic functions from expressions for correlation functions is a familiar procedure.[2,3] However, here we employ the reverse procedure, namely the calculation of density expansions for the correlation functions from such expansions for thermodynamic functions. This is of interest in itself as an extension of the known relations among these functions and it also seems to be useful as a practical method to calculate density expansions of correlation functions in certain cases. As an example in Section 14 the method is used to obtain the correlation functions of ionic solutions from the density expansion for the free energy of these solutions.

The following procedure, which was suggested by Lebowitz and Percus,[4] is both simpler and more general than that originally proposed.[5] It is based on the concept of the *functional derivative* from the calculus of variations.

The configuration integral of a system $Z(\mathbf{N},\mathcal{V},T)$ depends not only on $\mathbf{N},\mathcal{V}$, and T but also on the direct potential.

$$U_{\mathbf{N}}(\{\mathbf{N}\}) = \sum_{\{\mathbf{n}\}\subseteq\{\mathbf{N}\}} u_{\mathbf{n}}(\{\mathbf{n}\}) \tag{9.1}$$

For our purpose we consider $\mathbf{N},\mathcal{V},T$, and the $u_{\mathbf{n}}$ to be the independent variables and examine the change in Z caused by a change in some particular component potential $u_{\mathbf{r}}$ at some particular coordinates $\{\mathbf{r}\}$. This is done formally by differentiating both sides of

$$Z(\mathbf{N},\mathcal{V},T) = \int_{\mathcal{V}} \exp(-\beta U_{\mathbf{N}})\, d\{\mathbf{N}\} \tag{9.2}$$

with respect to the function $u_{\mathbf{r}}(\{\mathbf{r}\})$ to obtain

$$\frac{\partial Z}{\partial u_{\mathbf{r}}} = -\beta \binom{\mathbf{N}}{\mathbf{r}} \int_{\mathcal{V}} \exp(-\beta U_{\mathbf{N}})\, d\{\mathbf{N}-\mathbf{r}\} \tag{9.3}$$

The range of the integral is reduced from all $\{\mathbf{N}\}$ in $\mathcal{V}$ to all $\{\mathbf{N}-\mathbf{r}\}$ in $\mathcal{V}$ because a subset $\{\mathbf{r}\}$ of $\{\mathbf{N}\}$ must correspond to $\{\mathbf{r}\}$ in the independent variable on the left. The combinatorial factor is the number of different subsets of $\{\mathbf{N}\}$ that may satisfy this condition.

Now we have, from Section 7,

$$\Xi = \sum [\mathbf{z}^{\mathbf{N}}/\mathbf{N}!]Z(\mathbf{N},\mathcal{V},T) \tag{9.4}$$

$$g_{\mathbf{r}} = [\mathbf{z}^{\mathbf{r}}/\mathbf{c}^{\mathbf{r}}\Xi] \sum [\mathbf{z}^{\mathbf{M}}/\mathbf{M}!] \int_{\mathcal{V}} \exp(-\beta U_{\mathbf{M}+\mathbf{r}})\, d\{\mathbf{M}\} \tag{9.5}$$

We differentiate Ξ with respect to $u_{\mathbf{r}}$ and then combine with (9.3) and (9.5) to get

$$\left(\frac{\partial \ln \Xi}{\partial u_{\mathbf{r}}(\{\mathbf{r}\})}\right)_{\mathbf{z},\mathcal{V},T} = -\beta[\mathbf{c}^{\mathbf{r}}/\mathbf{r}!]g_{\mathbf{r}}(\{\mathbf{r}\},\mathbf{z}) \tag{9.6}$$

This is to be compared to the thermodynamic relation (6.24)

$$\left(\frac{\partial \ln \Xi}{\partial x}\right)_{\mathbf{z},\mathcal{V},T} = \mathcal{V}\left(\frac{\partial \mathfrak{S}}{\partial x}\right)_{\mathbf{c},\mathcal{V},T} \tag{9.7}$$

to get

$$\mathcal{V}\left(\frac{\partial \mathfrak{S}}{\partial u_{\mathbf{r}}}\right)_{\mathbf{c}} = -\beta[\mathbf{c}^{\mathbf{r}}/\mathbf{r}!]g_{\mathbf{r}} \tag{9.8}$$

It is clear on physical grounds that $\partial\mathfrak{F}^{\text{ex}}/\partial u_{\mathbf{r}}$ is independent of $\mathcal{V}$ as long as $\{\mathbf{r}\}$ is within $\mathcal{V}$. As usual, to calculate bulk properties we consider the limit as $\mathcal{V} \to \infty$. It is convenient to proceed to this limit only after the factor $\mathcal{V}$ appearing here cancels another factor $\mathcal{V}$ that eventually appears (equation (9.16)). A result equivalent to (9.8) is more readily derived from (9.3) in the canonical ensemble, but the longer derivation in the grand ensemble is given here for consistency.

The Cluster Expansion of $g_{\mathbf{r}}$

We apply (9.8) to the cluster expansion of $\mathfrak{S}$.

$$\mathfrak{S} = \sum{}'' \mathbf{c}^{\mathbf{n}} B_{\mathbf{n}} \tag{9.9}$$

$$\left(\frac{\partial \mathfrak{S}}{\partial u_{\mathbf{r}}}\right)_{\mathbf{c}} = \sum{}'' \mathbf{c}^{\mathbf{n}} \frac{\partial B_{\mathbf{n}}}{\partial u_{\mathbf{r}}} \tag{9.10}$$

The functional derivative of $B_{\mathbf{n}}$ is obtained in the same way as that of Z, but is more complicated because of the form of $S_{\mathbf{n}}(\{\mathbf{n}\})$, the integrand of $B_{\mathbf{n}}$. It is convenient to write

$$S_{\mathbf{n}}(\{\mathbf{n}\}) = \Big[\prod_{\{\mathbf{m}\}\subseteq\{\mathbf{n}\}}{}'' [1 + \gamma_{\mathbf{m}}]\Big]{:}\mathfrak{S} \tag{9.11}$$

where the notation $:\mathfrak{S}$ instructs us to retain only those terms of the product that correspond to ALDC graphs on the skeleton of $\mathbf{n}$. Next we assume $\{\mathbf{r}\} \subseteq \{\mathbf{n}\}$ and then we have

$$\frac{\partial S_{\mathbf{n}}}{\partial u_{\mathbf{r}}} = \frac{\partial \gamma_{\mathbf{r}}}{\partial u_{\mathbf{r}}}\Bigg[\prod_{\substack{\{\mathbf{m}\}\subseteq\{\mathbf{n}\}\\ \{\mathbf{m}\}\neq\{\mathbf{r}\}}}{}'' [1 + \gamma_{\mathbf{m}}]\Bigg]{:}\mathfrak{S} - \gamma_{\mathbf{r}} \tag{9.12}$$

where the notation $:\mathfrak{S} - \gamma_{\mathbf{r}}$ instructs us to retain only those terms of the product that correspond to ALDC graphs on the skeleton of $\mathbf{n}$ from which a $\gamma_{\mathbf{r}}$ bond has been removed. We note that for every $\{\mathbf{m}\}$ that is a subset of $\{\mathbf{r}\}$ we may factor $[1 + \gamma_{\mathbf{m}}]$ from this product because this $\gamma_{\mathbf{m}}$ bond cannot determine whether or not a graph satisfies the connectivity condition. Furthermore the derivative on the right of (9.12) is $-\beta[1 + \gamma_{\mathbf{r}}]$. So we have

$$\frac{\partial S_{\mathbf{n}}}{\partial u_{\mathbf{r}}} = -\beta \exp(-\beta U_{\mathbf{r}}) S(\mathbf{r}{:}\mathbf{n} - \mathbf{r}) \tag{9.13}$$

where $S(\mathbf{r}:\mathbf{m})$, a function of $\{\mathbf{r}+\mathbf{m}\}$, is the sum of all products of cluster functions corresponding to graphs that satisfy the following conditions:

(1) There are no bonds intersecting only the skeleton of $\mathbf{r}$.

(2) Every vertex of $\mathbf{m}$ is connected to at least two vertices of $\mathbf{r}$ by independent paths.

Furthermore in order to simplify the notation we specify

$$S(\mathbf{r}:\mathbf{n}-\mathbf{r}) = 0 \qquad \text{unless} \qquad \mathbf{r} \subseteq \mathbf{n}$$

$$S(\mathbf{r}:\mathbf{0}) = 1.$$

Next we define the cluster integral

$$P(\mathbf{r}:\mathbf{m}) \equiv [\mathbf{m}!]^{-1} \int S(\mathbf{r}:\mathbf{m})\, d\{\mathbf{m}\} \tag{9.14}$$

which is a function of $\{\mathbf{r}\}$, and then we have

$$\begin{aligned}\frac{\partial B_{\mathbf{n}}}{\partial u_{\mathbf{r}}} &= \frac{-\beta \exp(-\beta U_{\mathbf{r}})}{\mathbf{n}!\mathcal{V}} \binom{\mathbf{n}}{\mathbf{r}} [\mathbf{n}-\mathbf{r}]! P(\mathbf{r}:\mathbf{n}-\mathbf{r}) \\ &= -\beta \exp(-\beta U_r)[\mathbf{r}!\mathcal{V}]^{-1} P(\mathbf{r}:\mathbf{n}-\mathbf{r})\end{aligned} \tag{9.15}$$

The multinomial coefficient in the second member of (9.15) arises in the same way as that in (9.3). The other combinatorial factors in (9.15) come from the definitions of $B_{\mathbf{n}}$ and $P(\mathbf{r}:\mathbf{n}-\mathbf{r})$. The change in range of the integral in going from $B_{\mathbf{n}}$ to $P(\mathbf{r}:\mathbf{n}-\mathbf{r})$ is again analogous to the change from (9.2) to (9.3).

Finally we combine (9.15) with (9.10) to get

$$\left(\frac{\partial \mathfrak{S}}{\partial u_{\mathbf{r}}}\right)_{\mathbf{c}} = \frac{-\beta\, \mathbf{c}^{\mathbf{r}} \exp(-\beta U_{\mathbf{r}})}{\mathcal{V}\mathbf{r}!} \sum \mathbf{c}^{\mathbf{m}} P(\mathbf{r}:\mathbf{m}) \tag{9.16}$$

which may be compared with (9.8) to obtain the desired cluster expansion of the correlation function

$$g_{\mathbf{r}}(\{\mathbf{r}\},\mathbf{c}) = g_{\mathbf{r}}(\{\mathbf{r}\},\mathbf{0}) \sum_{m} \mathbf{c}^{\mathbf{m}} P(\mathbf{r}:\mathbf{m}) \tag{9.17}$$

The density expansion of the potential of average force is readily obtained by taking the logarithm of (9.17) with the help of the cumulant-moment relation. The result is

$$\begin{aligned}W(\{\mathbf{n}\},\mathbf{c}) &= -kT \ln g_{\mathbf{n}}(\{\mathbf{n}\},\mathbf{c}) \\ &= U_{\mathbf{n}}(\{\mathbf{n}\}) - kT \sum_{m}{}' \mathbf{c}^{\mathbf{m}} Q(\mathbf{n}:\mathbf{m})\end{aligned} \tag{9.18}$$

where $Q(\mathbf{n}:\mathbf{m})$ is the same as $P(\mathbf{n}:\mathbf{m})$ except for the following additional restriction on the graphs that correspond to the integrand of $Q(\mathbf{n}:\mathbf{m})$:

(3) Every pair of vertices of the skeleton of $\mathbf{m}$ is connected by at least one path that does not pass through any vertices of $\mathbf{n}$.

The coefficients $Q(\mathbf{n}:\mathbf{m})$ of the expansion of the potential of average force contain only the more highly connected terms of the coefficients $P(\mathbf{n}:\mathbf{m})$ of the expansion of the correlation function. In a similar way the coefficients of the expansion of the highest component, $w_{\mathbf{n}}(\{\mathbf{n}\})$, of the potential of average force, $W_{\mathbf{n}}(\{\mathbf{n}\})$, contain only the more highly connected terms of $Q(\mathbf{n}:\mathbf{m})$. This is illustrated in Section 10.

Discussion

The graphs of some of the lower $P(\mathbf{n}:\mathbf{m})$ and $Q(\mathbf{n}:\mathbf{m})$ are given in Fig. 9.1. Note that when sets $\mathbf{n}$ and $\mathbf{m}$ consist, respectively, of n and m different species, then the terms of the integrand of $P(\mathbf{n}:\mathbf{m})$

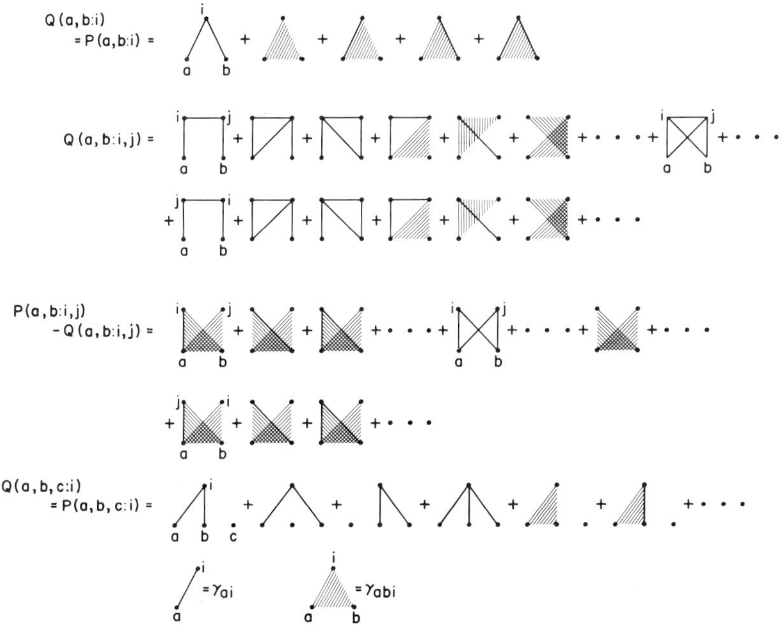

Fig. 9.1. Diagrams of the terms of the integrands of several coefficients of the cluster expansions of g_{ab}, W_{ab}, g_{abc}, and W_{abc}.

correspond, one-to-one, to all of the distinguishable graphs that satisfy the connectivity criteria. For smaller numbers of species, the number of terms of the integrand of $P(\mathbf{n}:\mathbf{m})$ remains the same, but some of them now correspond to indistinguishable graphs. This is the same principle that determines the number of terms of the integrand of the irreducible cluster integrals, and it is the same for $Q(\mathbf{n}:\mathbf{m})$.

Cluster expansions of the correlation functions and potentials of average force were first given for one-component systems by Mayer and Montroll[6] and de Boer,[7] and for multicomponent systems by Meeron.[3] The expansions given in (9.17) and (9.18) are of the same form as the earlier results but differ in being of the same high generality as the McMillan-Mayer theory. Thus they are valid for multicomponent systems in which the direct potentials are not necessarily pairwise and which are just sufficiently classical so that it makes sense to specify the center-of-mass coordinates of the molecules. There is of course still the condition that the validity of the expansions is only assured if the composition set $\mathbf{c}$ is such that there is no phase transition as the composition is changed from $\mathbf{c} = \mathbf{0}$ to $\mathbf{c}$.[8]

It follows from the discussion in Section 7 that (9.17) and (9.18) may be used to calculate correlation functions and potentials of average force of sets of solute particles in solution provided that the appropriate direct potentials are used, namely, the potentials of average force for sets of solute ions in the pure solvent.

A typical use for correlation functions is in the interpretation of the non-equilibrium properties of a system. An example is the spectroscopic work cited above.[1] Another use is in the partial summation of cluster series expansions to obtain either more rapidly converging series or integral equations that represent the sum of the most important terms of the series. These developments lie outside the scope of this book although a recent contribution by Meeron[9] promises to be applicable to ionic solutions.

Sometimes it is advantageous to formulate thermodynamic functions explicitly in terms of correlation functions. The general expressions, which are readily derived,[5] are listed below. They show in a particularly obvious way the conjugate character of the pair $g_{\mathbf{n}}, u_{\mathbf{n}}$ in determining the excess functions.

The excess energy is

$$\mathcal{E}^{\text{ex}} = \sum_{n}'' [\mathbf{c}^{\mathbf{n}}/\mathbf{n}!] \int_{\mathcal{V}} e_{\mathbf{n}} g_{\mathbf{n}} \, d\{\mathbf{n}\} \tag{9.19}$$

where

$$e_{\mathbf{n}}(\{\mathbf{n}\}) = \partial(u_{\mathbf{n}}/kT)/\partial(1/kT) \tag{9.20}$$

is the component energy corresponding to $u_{\mathbf{n}}$ as a component free energy.

The excess pressure is

$$P - ckT = -[3\mathcal{V}]^{-1} \sum_{n}'' [\mathbf{c}^{\mathbf{n}}/\mathbf{n}!] \int_{\mathcal{V}} v_{\mathbf{n}} g_{\mathbf{n}} \, d\{\mathbf{n}\} \tag{9.21}$$

where

$$v_{\mathbf{n}}(\{\mathbf{n}\}) = \sum_{\{p\} \in \{\mathbf{n}\}} R_{\{p\}} \nabla_{\{p\}} u_{\mathbf{n}} \tag{9.22}$$

is a generalization of the virial. The sum is over each set of spatial coordinates $\{p\}$ that is an element of $\{\mathbf{n}\}$. That is, in each term of the sum $\{p\}$ represents the spatial coordinates of one molecule of $\mathbf{n}$. The radius vector from the arbitrary origin of coordinates to the molecule at $\{p\}$ is $R_{\{p\}}$ and $\nabla_{\{p\}}$ is the gradient operator for a change in $\{p\}$.

For a general variable x, defined in Section 6, the excess function is

$$\mathcal{V} \left(\frac{\partial \mathfrak{S}}{\partial x}\right)_{\mathbf{c}, \mathcal{V}, T} = \sum_{n}'' [\mathbf{c}^{\mathbf{n}}/\mathbf{n}!] \int_{\mathcal{V}} x_{\mathbf{n}} g_{\mathbf{n}} \, d\{\mathbf{n}\} \tag{9.23}$$

where

$$x_{\mathbf{n}}(\{\mathbf{n}\}) = \partial u_{\mathbf{n}}/\partial x \tag{9.24}$$

In order to express the excess free energy in terms of correlation functions we must employ a "charging" process. For instance consider a hypothetical system in which each component potential is $\xi u_{\mathbf{n}}$, where $u_{\mathbf{n}}$ is the function in the real system. A value of the charging parameter $\xi = 0$ corresponds to an ideal gas or solution while $\xi = 1$ corresponds to the real system. All changes in ξ are supposed to be at constant $\mathcal{V}$ and T. For any ξ the correlation function $g_{\mathbf{n}}(\xi)$ may be calculated from (9.17) by substituting $\xi u_{\mathbf{n}}$ for every $u_{\mathbf{n}}$. We define

$$\bar{g}_{\mathbf{n}} \equiv \int_0^1 g_{\mathbf{n}}(\xi) \, d\xi \tag{9.25}$$

and then the excess free energy is given by

$$\mathfrak{F}^{\text{ex}} = \sum_{n}{}'' [\mathbf{c}^{\mathbf{n}}/\mathbf{n}!] \int_{\mathcal{V}} u_{\mathbf{n}}\bar{g}_{\mathbf{n}}\, d\{\mathbf{n}\}. \tag{9.26}$$

The expressions for the bulk properties are obtained by taking the limit as $\mathcal{V} \to \infty$, after first dividing by $\mathcal{V}$ in equations (9.19), (9.23), and (9.26). As far as the operation of integrating is concerned this limit is equivalent to integrating $d\{\mathbf{n} - 1_s\}$ over infinite volume, i.e., to holding one particle fixed in the integration.

Notes and References

1. See for example J. van Kranendonk, *Physica* **23**, 825 (1957); *Physica* **25**, 337 (1957).
2. T. L. Hill, *Statistical Mechanics* (McGraw-Hill Book Co., Inc., New York, 1956).
3. E. Meeron, *J. Chem. Phys.* **27**, 1238, (1957).
4. J. L. Lebowitz and J. K. Percus *Phys. Rev.* **122**, 1675 (1961).
5. H. L. Friedman, *J. Chem. Phys.* **34**, 73 (1961).
6. J. E. Mayer and E. Montroll, *J. Chem. Phys.* **9**, 2 (1941).
7. J. de Boer, *Repts. Progr. in Phys.* **12**, 305 (1949).
8. J. E. Mayer, "Theory of Real Gases," in *Handbuch der Physik* (Springer Verlag, Berlin, 1958), B. XII, p. 73.
9. E. Meeron, *J. Math. Phys.* **1**, 192 (1960).

10. Some Applications of the Cluster Theory

In this section we consider several applications of the equations derived in Sections 8 and 9 before going on to recast these equations in a form that is suitable to ionic systems. These examples of the application of the cluster theory have been chosen to bring out various interesting features.

Symmetrical Ideal Solutions

From the point of view of the McMillan-Mayer theory an ideal solution is one for which the cluster integral sum $\mathfrak{S}$ vanishes. The osmotic pressure of such a solution is given by

$$\Pi = kTc$$

This may be termed an *unsymmetrical* ideal solution for in it, just as in the McMillan-Mayer theory, solute and solvent do not play equivalent roles. The usual *symmetrical* ideal mixture is defined by the equation for the chemical potential of each component i

$$\mu_i = \mu_i^p + kT \ln x_i \tag{10.1}$$

where μ_i^p corresponds to the pure component in the same state of aggregation as the mixture and where x_i is the mole fraction of the component in the mixture. For only two components, 1 and 2, with a membrane permeable only to 1, equation (10.1) is equivalent to the equation for the osmotic pressure

$$\Pi = [-kT/v_1] \ln x_1 \tag{10.2}$$

where v_1 is the molecular volume of component 1 (cm^3 of bulk component 1 per molecule). To find what model is needed to get an equation of this form from the cluster theory it is convenient to begin with the equations for a one-component gas of non-attracting hard spheres. For this model the pairwise cluster function is

$$\begin{aligned} \gamma_{ij}(r) = \exp(-u_{ij}(r)/kT) - 1 &= -1 \quad \text{if} \quad r < a \\ &= 0 \quad \text{if} \quad r > a \end{aligned} \tag{10.3}$$

where a is the diameter of a sphere. The higher cluster functions all vanish for this model.

We have at once

$$B_2 = [2!\mathcal{V}]^{-1} \int_{\mathcal{V}} \gamma_{ij}\, d\{ij\} = -v/2 \tag{10.4}$$

where $v = 4\pi a^3/3$ is the volume that the presence of one molecule in $\mathcal{V}$ excludes to the center of a second molecule. The computation of the higher cluster integrals is complicated by the fact that the volume that two molecules exclude to the center of a third is $2v$ only if the first two molecules are far enough apart (Fig. 10.1). An account of the exact calculation of the first few cluster integrals for the hard sphere model is given by Hirschfelder et al.[1] For our purpose we shall neglect the geometrical factors. This corresponds to a lattice model[2] for the gas in which the volume is divided into cells of volume v, each of which may be occupied by at most one molecule. For this model

$$B_3 = [3!\mathcal{V}]^{-1} \int_{\mathcal{V}} \gamma_{ab}\gamma_{bc}\gamma_{ca}\, d\{abc\} = -v^2/6$$

$$\begin{aligned} B_4 &= [4!\mathcal{V}]^{-1} \int_{\mathcal{V}} \gamma_{ab}\gamma_{bc}\gamma_{cd}\gamma_{da}[3 + 6\gamma_{ac} + \gamma_{ac}\gamma_{bd}]\, d\{abcd\} \\ &= [3 - 6 + 1]v^3/4! = -v^3/12 \end{aligned}$$

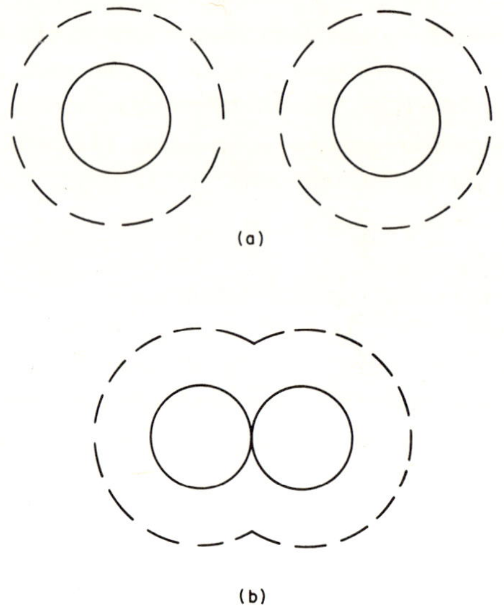

Fig. 10.1. Calculation of the third virial coefficient of a hard sphere gas. The solid circles represent the molecules, the broken circles represent the regions that the presence of the molecule within excludes to the center of another molecule. For the configuration of two molecules shown in (a) the volume from which the center of a third molecule is excluded is $2v$. For the configuration of two molecules shown in (b) the volume from which the center of a third molecule is excluded is less than $2v$.

It is clear that in general for this model

$$B_n = -v^{n-1}d_n/n! \tag{10.5}$$

where d_n is the *disparity*, defined as the excess of terms of the integrand of B_n with odd numbers of γ_2 bonds over those with even numbers of bonds. It has been shown that[2]

$$d_n = [n-2]! \tag{10.6}$$

Thus the model we treat here leads to the equations

$$B_n = -v^{n-1}/n[n-1] \tag{10.7}$$

$$\mathfrak{S} = -c - v^{-1}V_f \ln V_f \tag{10.8}$$

$$P = -kTv^{-1} \ln V_f \tag{10.9}$$

where $V_f \equiv 1 - cv$ is the fraction of the volume that is not occupied by molecules.

It is easily seen that these equations reduce to the equations for an ideal gas as $v \to 0$. With suitable interpretation and with $v > 0$ they also become the equations for a symmetrical ideal mixture obtained from the unsymmetrical (McMillan-Mayer) point of view. In this case u_{ij} in (10.3), the direct potential, is the potential for two solute molecules in the pure solvent and (10.9) is the expression for the osmotic pressure for a membrane permeable only to solvent molecules. Then (10.3) no longer implies that there are no forces in addition to the hard-sphere repulsion between the solvent molecules, but only that solute-solute, solute-solvent, and solvent-solvent intermolecular forces are all the same. Furthermore if the cells occupied by solute and solvent species in the lattice model have the same size then V_f is x_1, the mole fraction of solvent, and (10.9) becomes[2]

$$\Pi = -kTv^{-1} \ln x_1$$

the same as (10.2).

Note that for this model the cluster theory yields the equation for a symmetrical ideal mixture only if the infinite series of cluster integrals is summed. An approximation valid only at low c ($\sim$low x_2) cannot be used because by definition equation (10.2) is valid for an ideal solution over the entire composition range. Unfortunately it is not known how to generalize the above calculations to apply to a fluid system.

Kirkwood Superposition Approximation

The procedures discussed in Section 4 for defining component potentials are not limited to direct potentials but may be applied to other potentials of average force as well. Thus in a gas of composition **c** we may write

$$W_{\mathbf{n}}(\{\mathbf{n}\},\mathbf{c}) = \sum_{\{\mathbf{m}\}\subseteq\{\mathbf{n}\}}'' w_{\mathbf{m}}(\{\mathbf{m}\},\mathbf{c}) \qquad (10.10)$$

with $w_{ij} \equiv W_{ij}$. The case in which $n = 3$ is of particular interest. Then we have

$$W_{ijk} = W_{ij} + W_{jk} + W_{ki} + w_{ijk} \qquad (10.11)$$

The Kirkwood superposition approximation is $w_{ijk} = 0$. This is a widely used[3] device to achieve an approximate solution of the sequences of integral equations that relate the various correlation functions $g_{\mathbf{n}}$. The component potential w_{ijk}, sometimes called the superposition defect, can be obtained as a cluster expansion in the following way.

We begin with the expansions (Section 9) for W_{ijk} and W_{ij}.

$$W_{ijk} = U_{ijk} - kT \sum' \mathbf{c}^{\mathbf{n}} Q(ijk:\mathbf{n}) \tag{10.12}$$

$$W_{ij} = U_{ij} - kT \sum' \mathbf{c}^{\mathbf{n}} Q(ij:\mathbf{n}) \tag{10.13}$$

These expansions are substituted in (10.11) to obtain

$$w_{ijk} = u_{ijk} - kT \sum' \mathbf{c}^{\mathbf{n}} q(ijk:\mathbf{n}) \tag{10.14}$$

where we define

$$q(ijk:\mathbf{n}) = Q(ijk:\mathbf{n}) - Q(ij:\mathbf{n}) - Q(jk:\mathbf{n}) - Q(ik:\mathbf{n}) \tag{10.15}$$

After recalling the definition of $Q(\mathbf{m}:\mathbf{n})$, following equation (9.18), we can easily construct a similar graphical description of $q(\mathbf{m}:\mathbf{n})$, the generalization of $q(ijk:\mathbf{n})$. Thus $\mathbf{n}!q(\mathbf{m}:\mathbf{n})$ is the integral over a sum of terms, each of which corresponds to a graph of γ_2, γ_3, $\cdots$ bonds that satisfies the following connectivity conditions.

(1) There are no bonds intersecting only the skeleton of $\mathbf{m}$.

(2) Every element of $\mathbf{n}$ is connected to at least two elements of $\mathbf{m}$ by independent paths.

(3) Every pair of elements of $\mathbf{n}$ is connected by at least one path that does not intersect $\mathbf{m}$.

(4) Every element of $\mathbf{m}$ is connected to some element of $\mathbf{n}$. The definition of $q(\mathbf{m}:\mathbf{n})$ is thus just the same as that of $Q(\mathbf{m}:\mathbf{n})$ except for the inclusion of the fourth restriction on the connectivity of the graphs.

For the case most commonly treated, namely a one-component system with pairwise-additive direct potentials, the superposition defect is‡

$$w_{ijk} = -kTc \int \gamma_{is}\gamma_{js}\gamma_{ks}\, d\{s\} + O(c^2) \tag{10.16}$$

We shall only attempt to estimate the validity of the superposition approximation in the limit of low c. Consider first the case in which $\{ijk\}$ represents a set of spatial coordinates that are mutually related as the vertices of an equilateral triangle. Then to the first order in c equation (10.11) becomes

‡ This $O(x)$ notation is explained in Section 15.

$$W_{ijk} = u_{ij} + u_{jk} + u_{ik} - kTc\left[3\int \gamma_{is}\gamma_{sj}\, d\{s\} + \int \gamma_{is}\gamma_{js}\gamma_{ks}\, d\{s\}\right] \quad (10.17)$$

Suppose now we have a gas of non-attracting hard spheres and that the edge of the equilateral triangle is slightly longer than a, the diameter of a spherical molecule. In this case $u_{ij} + u_{jk} + u_{ik}$ vanishes, the first integral in brackets is roughly v, and the second is roughly $-v$. In this particular case we then have

$$W_{ijk}(\text{exact}) \simeq -2cvkT$$

$$W_{ijk}(\text{superposition}) \simeq -3cvkT$$

which implies a quantitative inaccuracy but not a qualitatively serious error in this procedure. Moreover this seems to be the worst case for the hard sphere gas: For other configurations $\{ijk\}$ the superposition defect is smaller relative to the superposition W_{ijk}. Finally this conclusion about the validity of the superposition approximation at low c seems to carry over to other models in which the intermolecular forces are short range.

The Donnan Equilibrium

The Donnan Equilibrium in ionic systems is readily treated by specializing the results of the non-ionic cluster theory of Sections 7 and 8. This was first pointed out by Hill, from whose papers[4,5,6] the present discussion derives. A particularly interesting aspect of these results is their comparison with the result of applying the Debye-Hückel theory to the Donnan equilibrium, so this calculation is included here as well.

Although it does seem appropriate to consider the Donnan equilibrium here, it will be necessary to make many references to the results of later sections of this book.

Donnan-Debye-Hückel theory. We consider two subsystems, "out" and "in," separated by a membrane that is permeable to all but one of the molecular species in the system. This non-diffusing species has charge $z\,|\,\epsilon\,|$ per molecule and is present at concentration ρ in the "in" region (the charge on an electron is ϵ). For the rest we have the solvent and ions of various species s of charge $z_s\,|\,\epsilon\,|$, present at concentration c_s^* in the "in" region and c_s in the "out" region. The solvent and these ions are at equilibrium with respect to diffusion through the membrane.

We list here various functions of the charges and concentrations that are useful in the calculation.

$\mu_n \equiv \sum_s c_s z_s^n$ is the nth moment of the concentration of charge types in the "out" region.

$\lambda = 4\pi\epsilon^2/DkT$ where D is the dielectric constant of the pure solvent.

$\kappa = [\lambda\mu_2]^{1/2}$ is the Debye parameter in the "out" region.

$y = \sum_s [c_s^* - c_s] z_s^2$.

$\kappa^* = [\lambda[\mu_2 + y + \rho z^2]]^{1/2}$ is the Debye parameter in the "in" region.

$\delta = [\kappa^*/\kappa] - 1$.

$\alpha = \kappa\lambda/8\pi$.

$x/|\epsilon|$ = the membrane potential in units of kT/ϵ.

The Debye-Hückel activity coefficient for species s in the "out" region is γ_s, given by

$$\ln \gamma_s = -z_s^2\alpha \tag{10.18}$$

The corresponding activity coefficient in the "in" region is $-z_s^2\alpha[1 + \delta]$. The basic notion is that the diffusing species are at electrochemical equilibrium, for which the condition is

$$\ln (c_s^*/c_s) = -z_s x + z_s^2\alpha\delta \tag{10.19}$$

The osmotic pressure is given by

$$\Pi/kT = \rho + \sum [c_s^* - c_s] + [\kappa^3 - \kappa^{*3}]/24\pi \tag{10.20}$$

where the first terms express the ideal contribution to Π and the κ terms the ion-atmosphere contribution. This equation is exact for a system in which (10.18) is valid for every ion (with the additional factor $(1 + \delta)$ in (10.18) for ions in the "in" region). When (10.19) is substituted in (10.20), we obtain, after a little manipulation

$$\begin{aligned}\Pi/kT - \rho &= \mu_2[x^2/2 - \alpha\delta^2] - \mu_3\alpha\delta x + \mu_4\alpha^2\delta^2/2 + O(\rho^3) \\ &\equiv -B_2\rho^2 + \cdots\end{aligned} \tag{10.21}$$

where $-B_2$ is the second virial coefficient of the osmotic pressure. At low ρ the functions x and δ are proportional to ρ. The terms in the second member of (10.21) are then each proportional to ρ^2 while the omitted terms are of higher order in ρ.

To find the second virial coefficient we must express x and δ as functions of ρ. In addition to the equations given we have the electroneutrality conditions in each phase:

$$\mu_1 = 0$$
$$z\rho + \sum c_s^* z_s = 0 \qquad (10.22)$$

which lead to

$$\rho z = \mu_2 x + \mu_3 \alpha\delta + \cdots \qquad (10.23)$$

$$y = \mu_3 x + \mu_4 \alpha\delta + \cdots \qquad (10.24)$$

$$\delta = z^2\rho/\mu_2 + y/2\mu_2 + \cdots \qquad (10.25)$$

where terms of order ρ^2 and higher are omitted. We eliminate y and δ from these equations to get

$$x = \rho\left[\frac{z}{\mu_2} + \frac{\alpha\mu_3}{2\mu_2^2}\left[z - \frac{\mu_3}{\mu_2}\right]\right] + O(\alpha^2) \qquad (10.26)$$

$$\delta = \frac{z\rho}{\mu_2}\left[z - \frac{\mu_3}{\mu_2}\right] + O(\alpha) \qquad (10.27)$$

When these are substituted in (10.27) we find

$$B_2 = -[z^2/2\mu_2][1 - [\alpha/2][z - \mu_3/\mu_2]^2] + O(\alpha^2) \qquad (10.28)$$

This is the asymptotic form of B_2 in the limit of low ionic strength in the "out" region. Higher terms in α would be without significance because the calculation is based on (10.18) which is valid only in the limit of small α.

Next we see how the cluster theory can lead not only to B_2, but also to expressions for c_s^*/c_s and for the membrane potential.

McMillan-Mayer theory. We have at once

$$B_2 = \frac{1}{2}\int_0^\infty [\exp(-u(r)/kT) - 1]4\pi r^2\, dr \qquad (10.29)$$

where $u(r)$ is the direct potential for a pair of non-diffusing ions in the "out" region. We shall assume that the direct potential for this problem (i.e., the potential of average force for two non-diffusing ions of charge z in the "out" solution) is given by the equation (Section 14)

$$u(r_{ij})/kT = z^2\lambda q(r_{ij}) + z^2\mu_3\lambda^3\left[z\int q(r_{ip})[q(r_{pj})]^2\, d\{p\} - \frac{1}{2}\mu_3\lambda\int q(r_{ip})[q(r_{pq})]^2 q(r_{qj})\, d\{p,q\}\right] \qquad (10.30)$$

where $q(r) = e^{-\kappa r}/4\pi r$ and where the graphs corresponding to the integrals are chains of alternating q and q^2 bonds. The first term in (10.30) is the Debye-Hückel potential of average force while the other terms are the leading corrections to this potential for point ions in an ideal dielectric.

Now we integrate the first few terms of

$$B_2 = \frac{1}{2} \int \sum_n{}' \frac{[-u/kT]^n}{n!} 4\pi r^2 \, dr. \tag{10.31}$$

We first have

$$\begin{aligned} B_{2,1} &= -\frac{1}{2} z^2 \lambda \int_0^\infty q(r) 4\pi r^2 \, dr \\ &= -z^2/2\mu_2 \end{aligned} \tag{10.32}$$

which corresponds to the leading term of (10.28). Hill has pointed out[4] that this term is obtained for any $u(r)$ that is consistent with electrical neutrality. The next term is

$$B_{2,2} = -\frac{1}{2} z^3 \mu_3 \lambda^3 \int q(r_{ip})[q(r_{pj})]^2 \, d\{pj\} \tag{10.33}$$

This is a convolution integral (see Section 12) in which one bond is a q bond, the next is a q^2 bond, and the third bond (from i to j) is unity. Applying the convolution theorem, noting that the Fourier transform of unity is the Dirac delta function, and referring to Table 12.1 for the other transforms, we find

$$\begin{aligned} B_{2,2} &= -z^3 \mu_3 \lambda^3 \int \frac{1}{\kappa^2 + t^2} \frac{\tan^{-1}(t/2\kappa)}{8\pi t} \delta(t) \, dt \\ &= -z^3 \mu_3 \alpha / 2\mu_2^2 \end{aligned} \tag{10.34}$$

By using the same technique we find

$$\begin{aligned} B_{2,3} &= \frac{1}{4} z^2 \mu_3^2 \lambda^4 \int q(r_{ip})[q(r_{pq})]^2 q(r_{qj}) \, d\{p,q,j\} \\ &= z^2 \mu_3^2 \alpha / 4\mu_2^3 \end{aligned} \tag{10.35}$$

In terms of the total concentration of the "out" solution,

$$c = \sum_s c_s$$

$B_{2,1}$ shows a c^{-1} dependence and both $B_{2,2}$ and $B_{2,3}$ show a $c^{-1/2}$ dependence. There is just one other term in B_2 with a $c^{-1/2}$ dependence and it comes from the $n = 2$ term of (10.31). It is

$$\begin{aligned} B_{2,4} &= [\lambda^2 z^4/4] \int [q(r)]^2 4\pi r^2 \, dr \\ &= z^4 \alpha / 4\mu_2 \end{aligned} \tag{10.36}$$

The sum of these contributions to B_2 is

$$\sum_{j=1}^{4} B_{2,j} = \frac{-z^2}{2\mu_2}\left[1 - \frac{\alpha}{2}[z - \mu_3/\mu_2]^2\right] \tag{10.36}$$

This agrees with (10.28) but it is remarkable that the μ_3 term in (10.36) comes from the μ_3 term in the direct potential, (10.30), while the μ_3 term in (10.28) comes not from the direct potential through the thermodynamic expression (10.18), but comes from the electroneutrality conditions. This new result suggests that the electroneutrality condition also puts restrictions on the higher terms of $u(r)$.

The inclusion of the higher terms in (10.30) corresponds to an equation for $\ln \gamma_s$ with more terms than (10.18). This may be established by applying (9.26). If the new activity coefficients and the corresponding osmotic coefficients are used in the Donnan method, equation (10.28) is again obtained. It is probably in general true that for any model and for any degree of accuracy the Donnan and McMillan-Mayer methods give the same result, but that more terms of the direct potential must be carried in the latter method to achieve this. On the other hand the equilibrium conditions (such as (10.19)) and the electroneutrality condition need not be explicitly invoked in the McMillan-Mayer method.

Distribution of diffusing species.‡ In a Donnan equilibrium system one may measure the concentration c_s^* of a diffusing species in the "in" region. (It is assumed that the concentrations c_s of the diffusing species in the "out" region are independent variables that are fixed by the conditions of the experiment.) Therefore it is desirable to have a cluster theory expression for c_s^* or, what is equivalent, c_s^*/c_s.

We begin with the equations for an osmotic equilibrium system in which some of the non-diffusing species are present on both sides of

‡ In this subsection z_s is fugacity of species s rather than ionic charge.

the membrane. We write (7.33) for the case in which the solute fugacity set $\mathbf{z}$ is replaced by $\mathbf{z}^*$, while $\mathbf{y}$ is replaced by $\mathbf{z} + \mathbf{y}$. The solvent fugacity set $\mathbf{y}$ is the same in both compartments but the non-diffusing solutes have fugacity set $\mathbf{z}^*$ in the "in" compartment and $\mathbf{z}$ in the "out" compartment. It is sufficient to consider only the $n = 0$ case of (7.33). Then we get

$$\Xi(\mathbf{z}^* + \mathbf{y}) = \Xi(\mathbf{z} + \mathbf{y}) \sum [\zeta^{\mathbf{N}}/\mathbf{N}!] \int g_{\mathbf{N}}\, d\{\mathbf{N}\} \quad (10.37)$$

where

$$\zeta_s \equiv [z_s^* - z_s]c_s/z_s \quad (10.38)$$

In this equation $c_s \equiv c_s(\mathbf{z} + \mathbf{y})$ pertains to the "out" region. We abbreviate the two grand partition functions as Ξ^* and Ξ, respectively, and note that

$$\exp\,(\Pi \mathcal{V}/kT) = \Xi^*/\Xi = \sum [\zeta^{\mathbf{N}}/\mathbf{N}!] \int g_{\mathbf{N}}\, d\{\mathbf{N}\} \quad (10.39)$$

where Π is the osmotic pressure and $\mathcal{V}$ is the volume of each region. This equation is formally identical to (7.40) and the calculus of Section 8 can now be applied to obtain

$$\mathfrak{S} = \sum_n{}'' \mathbf{x}^{\mathbf{n}} B_{\mathbf{n}} \quad (10.40)$$

where the direct potentials for the integrands of the $B_{\mathbf{n}}$ are the potentials of average force for sets $\mathbf{n}$ of non-diffusing species in the "out" solution. The effective concentration set $\mathbf{x}$ is the set of x_s defined by (compare with (8.11))

$$\zeta_s(\partial \Pi/\partial \zeta_s)_{T,\mathcal{V},\zeta - \zeta_s} \equiv kTx_s \quad (10.41)$$

From (10.38), (10.41), and (6.19) we find

$$x_s = [z_s^* - z_s]c_s^*/z_s^* \quad (10.42)$$

where $c_s^* = c_s(\mathbf{z}^* + \mathbf{y})$ is the physical concentration of species s in the "in" region. The cluster function cannot directly be identified with the excess Helmholz free energy as was done in Section 8 but all we need for the present purpose is the relation (8.16) which is still valid in the form

$$x_s = \zeta_s \exp\,(\partial \mathfrak{S}/\partial x_s) \quad (10.43)$$

which is equivalent to

$$c_s^*/c_s = [z_s^*/z_s] \exp(\partial\mathfrak{S}/\partial x_s) \tag{10.44}$$

For a non-diffusing solute z_s^* and z_s are independent variables. The condition for a diffusing solute is $z_s = z_s^*$ and with this condition (10.44) gives the concentration ratio for a diffusing solute as well as for a non-diffusing solute for which the experimenter has adjusted z_s and z_s^* to the same value.

Now we divide the solute species into two classes, n = non-diffusing and d = diffusing. Furthermore we specify that the non-diffusing species are present only in the "in" region: for each n

$$c_n = z_n = 0$$

For a species of this class equation (10.42) reduces to

$$x_n = c_n^*$$

and (10.44) reduces to

$$c_n^* = a_n^* \exp(\partial\mathfrak{S}/\partial c_n^*) \tag{10.45}$$

where a_n^* is the activity. These relations are the same as for the simple osmotic system. For the diffusing species we have $z_d = z_d^*$ so $x_d = 0$, as in the simple osmotic system, and

$$c_d^* = c_d \exp(\partial\mathfrak{S}/\partial x_d) \tag{10.46}$$

which is the desired relation. The limit, $\mathbf{x} \to \mathbf{0}$, must of course be taken after the differentiation is performed.

To make it clear how (10.46) is to be used we rewrite (10.40) in a way that explicitly represents the two solute classes:

$$\mathfrak{S} = \sum_{u+t\geq 2} \mathbf{c}^{\mathbf{u}}\mathbf{x}^{\mathbf{t}} B_{\mathbf{u+t}} \tag{10.47}$$

where $\mathbf{c}$ is the set of c_n^*, $\mathbf{x}$ is the set of x_d, $\mathbf{u}$ is a set of non-diffusing species, and $\mathbf{t}$ is a set of diffusing species. We differentiate with respect to x_d, the effective concentration of a particular diffusing species,

$$\partial\mathfrak{S}/\partial x_d = \sum_{u+t\geq 2} [t_d/x_d]\mathbf{c}^{\mathbf{u}}\mathbf{x}^{\mathbf{t}} B_{\mathbf{u+t}} \tag{10.48}$$

and then let $\mathbf{x} \to 0$ to obtain

$$\lim_{\mathbf{x}=0} \partial \mathfrak{S}/\partial x_d = \sum_{u \geq 1} \mathbf{c}^{\mathbf{u}} B_{\mathbf{u}+1_d} \tag{10.49}$$

which is the proper exponent to use in (10.46). For the problem considered above in which there is a single non-diffusing species having concentration ρ in the "in" region, the concentration ratio of any diffusing species a is given by

$$\ln (c_a^*/c_a) = \sum \rho^j B_{j+1_a} \tag{10.50}$$

where B_{j+1_a} is the irreducible cluster integral for j non-diffusing ions and one a ion. The simplest such coefficient is

$$B_{1+1_a} = \int_0^\infty [\exp(-u_{na}/kT) - 1] 4\pi r^2 \, dr \tag{10.51}$$

where u_{na} is the potential of average force for the interaction of one a ion and one non-diffusing ion in the "out" solution. It is clear that for the point-ion model B_{1+1_a} differs from the second virial coefficient of the osmotic pressure only with respect to the charge factors.

The membrane potential.‡ The condition for equilibrium with respect to the diffusion of a species d through the membrane may be written as

$$\tilde{\mu}_d^* = \tilde{\mu}_d$$

where $\tilde{\mu}_d$ is the electrochemical potential[7] of species d

$$\tilde{\mu}_d = \mu_d^0 + kT \ln c_d + kT \ln \gamma_d + \Psi z_d \epsilon \tag{10.52}$$

where μ_d^0 is the chemical potential in the reference state, γ_d is the usual activity coefficient, and Ψ is the electrical potential of the phase. We define the reference states so that $\mu_d^{0*} = \mu_d^0$ and then the membrane potential, $\Psi - \Psi^*$, is given by

$$\epsilon z_d(\Psi - \Psi^*) = kT \ln (c_d^*/c_d) + kT \ln (\gamma_d^*/\gamma_d) \tag{10.53}$$

The first term on the right may be calculated by the methods just described. The second term on the right may be obtained from the Debye-Hückel theory or, to higher approximation, from the Mayer ionic solution theory (Section 13).

Stigter and Hill have employed the McMillan-Mayer method with

‡ In this subsection z_d is the ionic charge.

direct potentials obtained from several interesting but more elaborate models for the interaction of the non-diffusing ions.[8, 9]

Notes and References

1. J. O. Hirschfelder, C. F. Curtiss, and R. B. Bird *Molecular Theory of Gases and Liquids*, (John Wiley & Sons, Inc., New York, 1950).
2. E. A. Guggenheim, *Discussions Faraday Soc.* **15**, 66 (1953).
3. T. L. Hill, *Statistical Mechanics* (McGraw-Hill Book Company, Inc., New York, 1956).
4. T. L. Hill, *Discussions Faraday Soc.* **21**, 31 (1956).
5. T. L. Hill, *J. Am. Chem. Soc.* **78**, 4281 (1956).
6. T. L. Hill, *J. Am. Chem. Soc.* **80**, 2923 (1958).
7. E. A. Guggenheim, *Thermodynamics* (Interscience Publishers, Inc., New York, 1950), second ed.
8. D. Stigter and T. L. Hill, *J. Phys. Chem.* **63**, 551 (1959).
9. D. Stigter, *J. Phys. Chem.* **64**, 838 (1960).

CHAPTER 3

Cluster Theory of Ionic Solutions

11. Introduction

In Section 2 where we apply the rudimentary cluster theory to ionic systems we find that all of the cluster integrals diverge in the limit $\mathcal{V} \to \infty$. As will be shown below the same is true for the complete and general cluster theory. In the general case, as in Section 2, the divergences can be overcome by rearranging terms in a certain way (Fig. 2.3) as shown by Mayer.

In a mathematically rigorous procedure infinite terms must not appear even in the intermediate steps. Mayer himself proceeded by first replacing the $1/r$ Coulomb potential by the Yukawa potential, $e^{-\alpha r}/r$, for which the cluster integrals converge even at $\mathcal{V} = \infty$, provided $\alpha > 0$. The rearrangement is then carried out and finally one can let $\alpha \to 0$, so this parameter need not appear in the final equations. Nevertheless the resulting theory is rigorous only for systems for which one has a physical basis for expecting $\alpha > 0$. This is the case for electrolyte solutions in a solvent which undergoes partial self ionization, e.g.,

$$H_2O \rightleftharpoons H^+ + OH^-$$

in which case α may be identified with κ_0, the Debye parameter of the pure solvent. Unfortunately there is a further difficulty with this procedure, as emphasized in Section 12.

In the development of the general ionic solution theory in Section 13 we proceed as far as we are able with the equations for a system of finite volume because this seems to be the simplest way to avoid the appearance of infinities. By this approach, together with a careful treatment of the Coulomb potential in Section 12, it appears that infinite terms are completely avoided.

The remainder of the present section is devoted to an introductory discussion of the form of the pairwise potential of average force at infinite dilution in ionic systems and to showing that the irreducible cluster integrals for general ionic systems diverge as $\mathcal{V} \to \infty$ just as they do for the hypothetical one-component ionic system discussed in Section 2.

We begin with the exact equation

$$- \mathcal{F}^{ex}/\mathcal{V}kT \equiv \mathfrak{S} = \sum_{n}{}'' \mathbf{c}^{\mathbf{n}}[B_{\mathbf{n}} + \sigma_{\mathbf{n}}] \tag{11.1}$$

As is customary we shall neglect the surface terms $\sigma_{\mathbf{n}}$ (equation 5.51) although they also contain divergent integrals. The justification for this is discussed at the end of Section 13.

Next it must be repeated that equation (11.1) may be construed as applying to either an imperfect gas or to the solute in a solution, as discussed at the end of Section 8. In the present case this is either an ionic gas or the solute in an electrolyte solution. It is sufficient here to consider only the latter case as it reduces to the former on letting the equilibrium pressure P_0 of the solvent in its reference state vanish.

We recall that $B_{\mathbf{n}}$ may be written in the form (Section 5)

$$B_{\mathbf{n}} = [\mathbf{n}!\mathcal{V}]^{-1} \int_{\mathcal{V}} S_{\mathbf{n}}(\{\mathbf{n}\})\, d\{\mathbf{n}\} \tag{11.2}$$

where $S_{\mathbf{n}}$ is the sum of all products of cluster functions $\gamma_{\mathbf{m}}$, $\mathbf{m} \leq \mathbf{n}$, that correspond to ALDC graphs on the skeleton of $\mathbf{n}$. If (11.1) applies to the solute in a solution then the sum is over all sets $\mathbf{n}$ of two or more solute molecules (or ions) and we have

$$1 + \gamma_{\mathbf{m}}(\{\mathbf{m}\}) \equiv \exp\,[-u_{\mathbf{m}}(\{\mathbf{m}\})/kT] \tag{11.3}$$

where $u_{\mathbf{m}}$ is a component of the direct potential of average force

$$W_{\mathbf{n}}(\{\mathbf{n}\},\mathbf{0}) \equiv U_{\mathbf{n}}(\{\mathbf{n}\})$$

This is the potential for the set $\mathbf{n}$ of solute molecules in the solvent in its reference state T,P_0. We have, as discussed in Section 4,

$$U_{\mathbf{n}}(\{\mathbf{n}\},\mathbf{0}) = \sum_{\{\mathbf{m}\}\subseteq\{\mathbf{n}\}}{}'' u_{\mathbf{m}}(\{\mathbf{m}\}), \tag{11.4}$$

the inverse of which is (see Section 3)

$$u_{\mathbf{m}}(\{\mathbf{m}\}) = \sum_{\{\mathbf{n}\}\subseteq\{\mathbf{m}\}}{}'' [-1]^{m-n} U_{\mathbf{m}}(\{\mathbf{n}\}) \tag{11.5}$$

These equations assure that if we are given the direct potentials $U_{\mathbf{m}}$

then we can calculate the component potentials $u_{\mathbf{m}}$ and hence, using (11.1), we can calculate $\mathfrak{S}$—except for the divergence problem in the case of ionic systems.

If particular functions $U_{\mathbf{m}}$ are assumed at the outset this set of functions corresponds to what may be called a mathematical model. But in order to study the influence of molecular interactions on macroscopic properties we must calculate the potentials $U_{\mathbf{m}}$ from a *physical* model, an idealization of the real physical system chosen for investigation. This calculation itself involves difficult mathematical and statistical problems, depending of course on the complexity of the model, but there is one feature of importance that is simply deduced and that is the same for $U_{\mathbf{m}}$ for wide classes of systems. This is the limiting form of $U_{\mathbf{m}}$ for $m = 2$.

For $m = 2$ we shall often write i,j in place of $\mathbf{m}$. Here i and j represent molecular (or ionic) species and the set i,j corresponds to the composition set $\mathbf{m} = 1_i, 1_j$ while the set i,i corresponds to the composition set $\mathbf{m} = 2_i, 0_j$. For instance we have, in view of (11.4), $u_{ij}(r_{ij}) = U_{ij}(r_{ij})$ where we also write r_{ij} instead of $\{i,j\}$.

In general the potential $U_{\mathbf{m}}$ is defined in such a way (Section 4) that it approaches zero as the mutual separations of the particles given by $\{\mathbf{m}\}$ increase without limit. As a special case, $u_{ij} \to 0$ as $r_{ij} \to \infty$. The *way* in which u_{ij} vanishes in this limit depends on the character of the molecules i and j. If i and j are uncharged non-polar molecules then at large r_{ij} we have $u_{ij} \propto r_{ij}^{-6}$ corresponding to van der Waals forces. The same behavior results if i and j are both uncharged but have permanent electric dipole moments. (This is the result after averaging over the orientational coordinates of both molecules (cf. Section 4). If i is an ion and j an uncharged dipolar molecule then $u_{ij} \propto r_{ij}^{-4}$ at large r_{ij}. Finally if i and j are both ions then $u_{ij} \propto r_{ij}^{-1}$ at large r_{ij}. So u_{ij} becomes longer in range as we go from pairs of non-polar uncharged molecules to pairs of ions in this way.

In order to go further in the treatment of general ionic systems it is necessary to write u_{ij} in a form which correctly represents the long-range part without making a commitment regarding the choice of a detailed model. This is done by writing

$$u_{ij}(r_{ij}) = z_i z_j \epsilon^2 / D r_{ij} + u_{ij}^*(r_{ij}) \tag{11.6}$$

which serves as the definition of u_{ij}^* (Fig. 11.1). Here $-\epsilon$ is the elec-

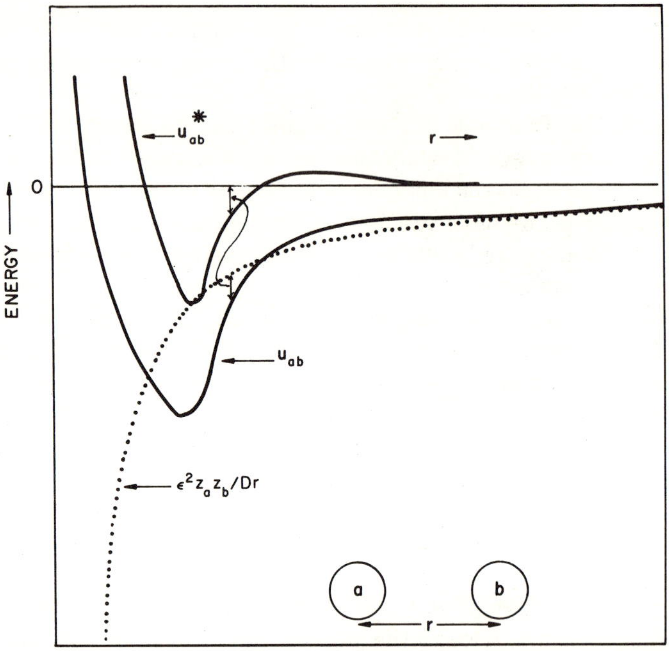

Fig. 11.1(a). Definition of u^*_{ab}.

tronic charge, $z_i\epsilon$ is the charge on an ion of species i, and D is the ordinary low-field dielectric constant of the solvent in its reference state T,P_0. This is the state in osmotic equilibrium with the solution, whose excess free energy is $\mathfrak{F}^{ex}$ in equation (11.1). (Note that z_i has previously been used to represent the fugacity of species i but from this point on will be used only to represent the ionic charge in protonic units as we have just indicated.) The first term in (11.6), the Coulomb potential, is the total potential one would observe for two charged brass balls at wide separation in the solvent. Even in this case the potential deviates from $z_i z_j \epsilon^2/Dr_{ij}$ unless r_{ij} is large because at small separations the mutual polarization of the balls is not negligible.[1]

The remainder u^*_{ij} enables us to take into account the following effects which are present in real systems and which should be allowed for in any reasonably sophisticated physical model of electrolyte solutions:

(1) Even for two ideal point charges i and j, the potential u_{ij} deviates from the Coulomb potential because of the molecular structure of the solvent.

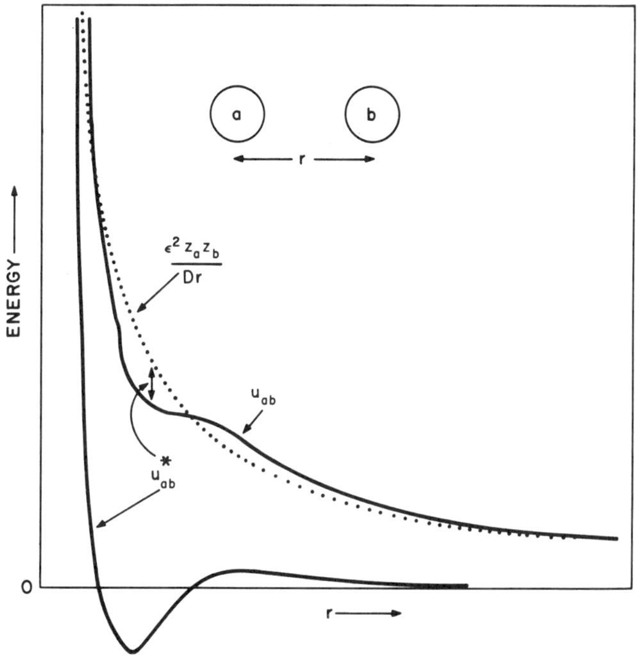

Fig. 11.1(b). Definition of u^*_{ab}.

(2) Even for two ions in a vacuum u^*_{ij} contains attractive contributions from the mutual polarization of the ions and from van der Waals forces. If the ions are dipolar there are attractive contributions to u^*_{ij} from the dipole interactions as well. Repulsive contributions to u^*_{ij} from quantum mechanical effects become important at small separations and assure that as $r_{ij} \to 0$, $u^*_{ij} \to \infty$ for all real systems.

It is of interest to discuss the effect of solvent structure in more detail here. When the separation of two ions is great enough so that there are more than two water molecules between them it is probably adequate to express the structural effects by treating the solvent as a uniform medium with a field-strength dependent dielectric constant. When the separation is this large the structure of the layer of solvent next to the ions is probably nearly independent of the separation and then the layers next to the ions act merely as regions of low dielectric constant in contributing to u_{ij}.

Let D_l be the local dielectric constant in an idealized solvent that is completely characterized by a field-strength dependent dielectric constant. According to Ritson and Hasted[2] the observations on the concentration dependence of the dielectric constant in aqueous electrolyte solutions may be understood in terms of a model in which D_l is about 5 within a sphere about three water molecules in diameter around the center of each ion and $D_l = D$, the low-field bulk dielectric constant, in regions farther removed from any ion. (This is an oversimplification of their result, but it seems adequate for the present purpose.) With the aid of Fig. 11.2 we may discuss qualitatively the the contribution to u_{ij}^* from an electrically saturated region around each ion.

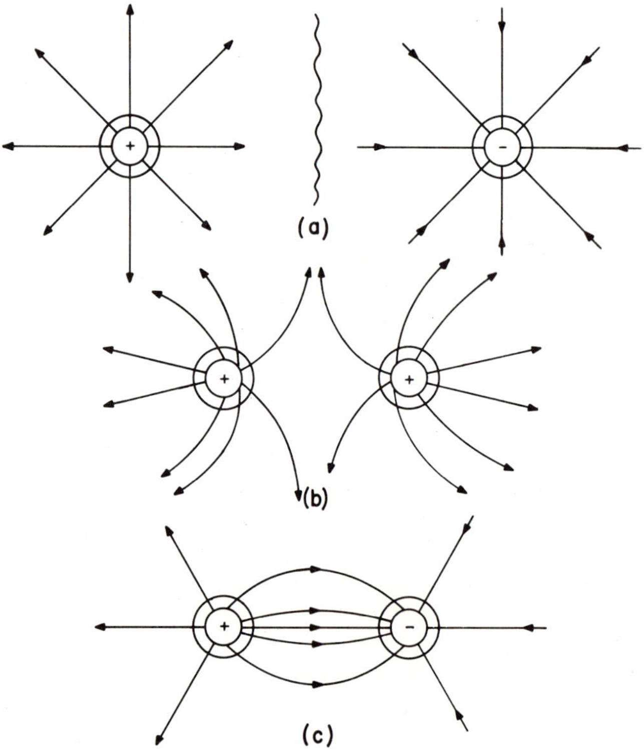

Fig. 11.2. To illustrate that the lines of force traversing a given region around an ion become more parallel to each other when a second ion is brought into the neighborhood. (a) The lines of force radiate symmetrically from isolated ions. (b) When two ions of the same charge approach, the lines of force crowd away from the region between the ions. (c) When two ions of opposite charge approach, the lines of force crowd into the region between the ions.

First it must be remembered that u_{ij} is *defined* to approach zero as $r_{ij} \to \infty$. Therefore the existence of a sphere of electrically saturated solvent around the ions when they are widely separated does not contribute to u_{ij}^* and we must consider the subtle question of how the contribution of these spheres to the free energy changes as the ions approach each other.

This qualitative discussion is based on the fact that the contribution of any geometric region of the ion-solvent system to the free energy is

$$\int [\mathbf{D} \cdot \mathbf{E}/8\pi] \, d\tau$$

where $\mathbf{D}$ is the electric displacement vector, $\mathbf{E}$ is the electric field vector, $d\tau$ is the element of volume, and the integral is over the region of interest. For this problem let the region be the sphere of electrically saturated solvent around a given ion. We also note the relation $\mathbf{E} = \mathbf{D}/D_l$ so the integral becomes

$$[8\pi D_l]^{-1} \int \mathbf{D} \cdot \mathbf{D} \, d\tau$$

since D_l is assumed to be constant within the sphere. Instead of attempting to do the integral we shall examine the behavior of the "tubes of displacement" (analogous to lines of force) as the ions approach each other. The number of tubes of displacement is constant: it is $4\pi z_i \epsilon$. However as the two ions approach each other the tubes of displacement become denser and more parallel on one side of each ion and so the integral of $\mathbf{D} \cdot \mathbf{D}$ over the spherical region increases. This means that as the ions approach each other the saturated regions around each ion make a *more positive* contribution to the free energy of the system. So, according to this model, the structure of the solvent makes a positive contribution to u_{ij}^*, whether or not ions i and j have the same sign, as long as r_{ij} is not too small.

Levine and Wrigley[3] have made calculations of u_{ij}^* for a model which is in effect similar to that just described, but in which a detailed model is used for the molecular structure within the saturated spheres. The behavior of u_{ij} according to their model is shown in Fig. 11.3.

When ions i and j are closer than twice the diameter of a solvent molecule it seems likely that the detailed molecular structure of the solvent becomes very important and may contribute oscillatory behavior to u_{ij} (as pointed out in another connection by Guggenheim[4]).

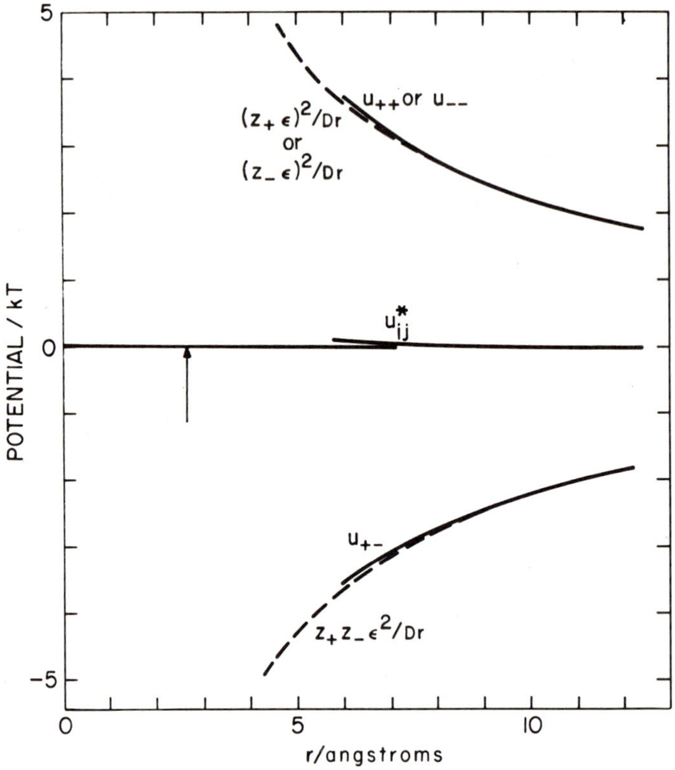

Fig. 11.3. Pairwise potentials of average force at infinite dilution for aqueous KF solutions at 25°C. The values of u^*_{ij} have been calculated by Levine and Wrigley, who find that in this range u^*_{ij} is nearly the same for K^+–K^+, K^+–F^-, and F^-–F^- interactions. They assumed that each ion was a sphere, 1.33 A in radius, with the charge at its center, and that there were four water molecules comprising the hydration layer next to each ion. These water molecules were represented as spheres, 1.38 A in radius, with dipoles 1.84 debyes at their centers. The remainder of the water was represented as a homogeneous dielectric.

The arrow indicates the ionic diameter. The calculations extend only down to 6 A, below which the hydration spheres intersect and the method of calculation becomes invalid.

Although it is obvious that further progress in understanding ionic solutions depends on being able to calculate u^*_{ij} for small r_{ij} from reasonable physical models, there has been little work done on this problem and we shall not be able to do more with it in this book.

Irreducible Cluster Integrals

The properties of u^*_{ij} that are important for a general discussion are the repulsive potential at small r_{ij} and the relatively short range of u^*_{ij} compared to the Coulomb potential. We expect, in fact, that u^*_{ij} is not longer range than r^{-4}, as found by Levine and Wrigley.[3] In this case we may represent u_{ij} at large r_{ij} as

$$u_{ij}(r_{ij}) = A/r_{ij} + B/r^4_{ij} + C/r^5_{ij} + \cdots, \qquad r_{ij} \text{ large} \tag{11.7}$$

This is an assumption which seems reasonable, but which must be justified for each physical model to which the following equations are applied. However if we write the second term as Br_{ij}^{-n} then we may state that no change is required in the theory as long as n is greater than three—a slightly less restrictive condition than that implied by (11.7). The basis of these restrictions will become clear in the following discussion.

Now by making use of (11.7) we may begin to investigate the behavior of the integrals in (11.1) for ionic systems. For B_{ij} we have

$$\begin{aligned} B_{ij} &= [2\mathcal{V}]^{-1} \int_{\mathcal{V}} \gamma_{ij}\, d\{ij\} \\ &= 2\pi \int_0^R \gamma_{ij}\, r^2_{ij}\, dr_{ij} \end{aligned} \tag{11.8}$$

where R is the radius of the vessel, assumed to be spherical. Using (11.3) and making an exponential expansion we have

$$B_{ij} = 2\pi \sum{}' \, [-kT]^{-n} \int_0^R [u_{ij}(r)]^n r^2\, dr/n! \tag{11.9}$$

Next we break the range of each integral into two parts, $0 \leq r < L$ and $L \leq r \leq R$, where L is a small, but macroscopic length, say 0.01 cm. Most experimental systems are large compare to this length so this is no real restriction on R, the radius of the system. We also substitute (11.6) for u_{ij} and then we have

$$B_{ij} = B_{ij}(L) + 2\pi \sum{}' \, [-z_i z_j \epsilon^2/DkT]^n \int_L^R r^{2-n}\, dr/n! + \cdots \tag{11.10}$$

where $B_{ij}(L)$ contains the corresponding integrals from 0 to L and where the omitted terms contain u^*_{ij}.

We recall that the left side of (11.1) must become independent of

$\mathcal{V}$ as $\mathcal{V} \to \infty$ at constant c. The same must be true of each term on the right unless there is a cancellation among various contributions that *do* depend on $\mathcal{V}$ as $\mathcal{V} \to \infty$. It is now readily seen that B_{ij} does not become independent of $\mathcal{V}$ as $\mathcal{V} \to \infty$:

Obviously $B_{ij}(L)$ is independent of R. The leading term of u_{ij}^* gives (cf. (11.7))

$$\int_L^R r^{-4} r^2 \, dr = L^{-1} - R^{-1} \to 1/L \qquad \text{as} \qquad R \to \infty$$

plus smaller terms, and in the same way any other term with $(u_{ij}^*)^n$, $n \geq 1$, approaches a constant value as $R \to \infty$. However the terms which are purely Coulombic become independent of R for large R only for $n > 3$. The three integrals

$$\int_L^R r \, dr, \qquad \int_L^R dr, \qquad \text{and} \qquad \int_L^R r^{-1} \, dr,$$

all increase without limit as $R \to \infty$,

In a similar way one may show that all of the $B_{\mathbf{n}}$ for which $\mathbf{n}$ corresponds to a set of ions diverge as $\mathcal{V} \to \infty$ rather than approach a constant limit. The argument now becomes more complicated.

We first note that the most singular contributions to $B_{\mathbf{n}}$ (i.e., the terms with the most divergent integrals) come from

(1) The terms of the integrand which correspond to simple cycles of γ_{ij} bonds, because the other terms contain either more γ_{ij} bonds, each of which tends to zero at large r_{ij}, or higher order $\gamma_{\mathbf{m}}$ bonds in place of some of the γ_{ij} bonds.

(2) The part of the configuration space for which all of the interparticle distances are greater than L, just as in the B_{ij} case.

(3) The first Coulomb term of the exponential expansion of γ_{ij}, again just as in the B_{ij} case.

Thus we assert that the largest contribution to $B_{\mathbf{n}}$ comes from the integral

$$C_n(\mathcal{V}) = \int_{\substack{\mathcal{V} \\ \text{all } r_{ij} > L}} g_{12}\, g_{23} \cdots g_{n,1} \, d\{n-1\} \tag{11.11}$$

where, as in Section 2, the distance dependence of the Coulomb potential is assumed to be

$$g_{ij} \equiv g_{ij}(r_{ij}) = 1/4\pi r_{ij}\,. \tag{11.12}$$

For a spherical container, again with radius R, we have $g_{ij} \geq 1/4\pi 2R$ and hence

$$\begin{aligned} C_n(\mathcal{V}) &\geq [8\pi R]^{-n} \int d\{n-1\} \\ &= [8\pi R]^{-n} \left[\left[\frac{4}{3}\pi R^3\right]^{n-1} - f(R,L)\right] \end{aligned} \tag{11.13}$$

where $f(R,L)$ is the correction term to account for the fact that the integral does not extend over the entire configuration space corresponding to the sphere of radius R; interparticle distances less than L are excluded. However at large R this term is certainly small compared to the R^{3n-3} term and therefore $C_n(\mathcal{V})$ certainly diverges as $\mathcal{V} \rightarrow \infty$.

The fact that every B_n diverges as $\mathcal{V} \rightarrow \infty$ in an ionic system is the basis of the possibility that in the summation process in (11.1) these singularities may cancel. In order to demonstrate that they do indeed cancel we must rearrange the series for $\mathfrak{S}$ in a way that was first shown by Mayer to remove the singularities (Sections 2 and 13). But first, in the following section some relevant mathematical relations are discussed.

Notes and References

1. See for instance J. Jeans, *The Mathematical Theory of Electricity and Magnetism* (Cambridge University Press, New York, 1925) fifth ed., Chapter VIII, Problem 27.
2. D. M. Ritson and J. B. Hasted, *J. Chem. Phys.* **16**, 11, (1948).
3. S. Levine and H. E. Wrigley, *Discussions Faraday Soc.* **24**, 43 (1957). A more fundamental approach to the same problem has recently been described by A. Bellemans and J. Stecki, *Bulletin de l'Academie Polonaise des Sciences*, **9**, 339, 343, 349, 429 (1961). However, these calculations are only carried to the point of finding the asymptotic form of $u(r)$ for large r in a non-polar solvent.
4. E. A. Guggenheim, *Discussions Faraday Soc.* **15**, 66 (1953).

12. Some Mathematical Aspects

This section begins with an exposition of some parts of the theory of Fourier transforms which are useful in the cluster theory of ionic solutions. Most of the derivations are given only in a formal way. An exposition on about the same level, but in greater detail and with exercises, has been given by Tranter[1] for one-dimensional transforms. For more sophisticated treatments reference may be made to Titchmarsh[2] and Sneddon.[3] Fourier transforms in many dimensions are discussed by Bochner[4] and Bochner and Chandrasekharan.[5]

The use of Fourier transforms in the cluster theories was introduced by Montroll and Mayer[6] while the discovery that they lead to a solution of the problem of applying the cluster theory to ionic systems is due to Mayer.[7] Such methods have since been used in many papers on the cluster theories but we only mention here that they form the basis for additional developments in the theory of ionic systems due to Haga,[8] Meeron,[9] Salpeter,[10] and Friedman.[11]

After the introductory paragraphs on Fourier transforms we consider in detail two mathematical problems connected with the application of the cluster theory to ionic solutions. They are both connected with the long range of the Coulomb potential. This part of the presentation is rather discursive because it is not yet perfectly clear how to solve these problems in such a way that the fewest restrictions are put on the range of validity of the resulting theory.

In Section 12 we use bold-face letters to represent only ordinary vectors, rather than composition sets as in most of the rest of this book. We also use r to represent $|\mathbf{r}|$, the magnitude of the vector $\mathbf{r}$ as usually defined. We use the usual notation, $i = [-1]^{1/2}$. As in other sections every integral is over the infinite range of the variables of integration unless explicitly restricted.

(a) *Some General Properties of Fourier Transforms*

The Fourier transform of $f(\mathbf{r})$ is defined by the equation

$$\tilde{f}(\mathbf{t}) \equiv \int f(\mathbf{r}) \exp(i\mathbf{r}\cdot\mathbf{t})\, d\mathbf{r} \qquad (12.1)$$

where $\mathbf{r}\cdot\mathbf{t}$ is the scalar product. If $\mathbf{r}$ is a vector in k dimensions, so is $\mathbf{t}$. The equation uniquely defines $\tilde{f}(\mathbf{t})$ provided that the integral converges. In some applications this transform is associated with a transformation from coordinate space to momentum space or from a real lattice to a reciprocal lattice but for our purpose there does not seem to be any corresponding significance to the transformation. It is introduced only for mathematical convenience.

Fourier transforms are useful for evaluating integrals of the form

$$p_n(\mathbf{r}_{1,n+1}) \equiv \int f_1(\mathbf{r}_{1,2}) f_2(\mathbf{r}_{2,3}) \cdots f_n(\mathbf{r}_{n,n+1})\, d\mathbf{r}_{1,2}\, d\mathbf{r}_{1,3} \cdots d\mathbf{r}_{1,n} \qquad (12.2)$$

where $\mathbf{r}_{a,b}$ is the vector from a point a to another point b in the k-dimensional space. Such an integral is called a *convolution* (or faltung or folding). If the vectors are three-dimensional then the integrand corresponds to a graph of the kind introduced in Section 5. In this case the integrand corresponds to a chain of bonds: an f_1 bond from vertex 1 to vertex 2, an f_2 bond from vertex 2 to vertex 3, $\cdots$, and an f_n bond from vertex n to vertex $n + 1$. The integration is over all coordinates of the intermediate vertices (vertices 2 to n) with vertices 1 and $n + 1$ fixed. Note that in a convolution the factors $f_1, f_2, \cdots$ may be different functions of $\mathbf{r}$, but each is pairwise; that is, each depends on a single k-vector or on the relative positions of two points in the k space. Functions that correspond to the higher order bonds, $\gamma_3, \gamma_4, \cdots$, are not considered in this section.

If in (12.2) $r_{1,n+1} > 0$ then the integrand of the convolution corresponds to a chain graph as we have just shown. But if $r_{1,n+1} = 0$ then the integrand of the convolution corresponds to a *cycle* (ring) of f bonds. Both cases are of interest in the cluster theories but only the former case, which is the more general, need be discussed here.

Our applications of Fourier transforms mostly depend on the following equation, known as the *convolution theorem.*

$$\begin{aligned} \tilde{p}_n(\mathbf{t}) &\equiv \int p_n(\mathbf{r}) \exp(i\mathbf{r}\cdot\mathbf{t})\, d\mathbf{r} \\ &= \prod_{j=1}^{n} \tilde{f}_j(\mathbf{t}) \end{aligned} \tag{12.3}$$

To derive this theorem we begin by taking the Fourier transform of both sides of (12.2) to get

$$\begin{aligned} \tilde{p}_n(\mathbf{t}) = \int &\exp(i\mathbf{t}\cdot\mathbf{r}_{1,n+1}) \\ &\cdot\left[\int f_1(\mathbf{r}_{1,2}) \cdots f_n(\mathbf{r}_{n,n+1})\, d\mathbf{r}_{1,2} \cdots d\mathbf{r}_{1,n}\right] d\mathbf{r}_{1,n+1} \end{aligned} \tag{12.4}$$

Next we factor the exponential into n factors corresponding to the n bonds of the convolution,

$$\exp(i\mathbf{t}\cdot\mathbf{r}_{1,n+1}) = \exp(i\mathbf{t}\cdot\mathbf{r}_{1,2}) \exp(i\mathbf{t}\cdot\mathbf{r}_{2,3}) \cdots \exp(i\mathbf{t}\cdot\mathbf{r}_{n,n+1})$$

and rearrange the order of integration on the basis of the following considerations. First the integrand as a whole is separable into n factors, each of the form $f(\mathbf{r}) \exp(i\mathbf{t}\cdot\mathbf{r})$. Second, the integration over

$\mathbf{r}_{1,2}$, $\mathbf{r}_{1,3}$, $\cdots$, $\mathbf{r}_{1,n}$ *and* $\mathbf{r}_{1,n+1}$ is equivalent to integrating over $\mathbf{r}_{1,2}$, $\mathbf{r}_{2,3}$, $\cdots$, and $\mathbf{r}_{n,n+1}$ *independently*. This is different from the integral in (12.2) in which $\mathbf{r}_{1,n+1}$ is fixed because only if $\mathbf{r}_{1,n+1}$ sweeps over the entire configuration space is every combination of $\mathbf{r}_{1,2}$, $\mathbf{r}_{2,3}$, and $\mathbf{r}_{n,n+1}$ possible. Therefore we may write (12.4) in the form

$$\tilde{p}_n(\mathbf{t}) = \int \exp(i\mathbf{t}\cdot\mathbf{r}_{1,2}) f_1(\mathbf{r}_{1,2})\, d\mathbf{r}_{1,2} \cdot \int \exp(i\mathbf{t}\cdot\mathbf{r}_{2,3}) f_2(\mathbf{r}_{2,3})\, d\mathbf{r}_{2,3} \cdots \tag{12.5}$$

which, in view of (12.1), is the convolution theorem.

Next we derive the *inversion* formula for Fourier transforms. If $\tilde{f}(\mathbf{t})$ is the transform of $f(\mathbf{r})$ then

$$[2\pi]^k f(\mathbf{r}) = \int \tilde{f}(\mathbf{t}) \exp(-i\mathbf{r}\cdot\mathbf{t})\, d\mathbf{t} \tag{12.6}$$

is the formula for the inversion. The first step of the proof is the substitution of (12.1) in (12.6) to obtain (after a slight change in notation)

$$\begin{aligned}[2\pi]^k f(\mathbf{a}) &= \int \exp(-i\mathbf{a}\cdot\mathbf{t})\, d\mathbf{t} \int f(\mathbf{r}) \exp(i\mathbf{t}\cdot\mathbf{r})\, d\mathbf{r} \\ &= \int d\mathbf{t} \int f(\mathbf{r}) \exp(i\mathbf{t}\cdot[\mathbf{r} - \mathbf{a}])\, d\mathbf{r} \\ &= \int d\mathbf{t} \int f(\mathbf{r} + \mathbf{a}) \exp(i\mathbf{t}\cdot\mathbf{r})\, d\mathbf{r}\end{aligned} \tag{12.7}$$

The last member follows from the one preceding because the integration is over the entire $\mathbf{r}$ space and therefore the integral depends only on the *difference* of the independent variables of the functions $f(\mathbf{r})$ and $\exp(i\mathbf{t}\cdot\mathbf{r})$.

Now we interchange the order of integration in (12.7) to obtain

$$f(\mathbf{a}) = \int f(\mathbf{r} + \mathbf{a})\, \delta(\mathbf{r})\, d\mathbf{r} \tag{12.8}$$

where

$$\delta(\mathbf{r}) \equiv [2\pi]^{-k} \int \exp(i\mathbf{r}\cdot\mathbf{t})\, d\mathbf{t} \tag{12.9}$$

is the Dirac delta function in k dimensions.[12, 13] Making use of the property of the Dirac delta function

$$\int f(\mathbf{r})\, \delta(\mathbf{r})\, d\mathbf{r} = f(\mathbf{0})$$

we have at once

$$\int f(\mathbf{r} + \mathbf{a}) \delta(\mathbf{r})\, d\mathbf{r} = f(\mathbf{a}) \tag{12.10}$$

which completes the proof of the inversion formula.

For the reader who is not familiar with the Dirac delta function we offer the following more detailed (although still only formal!) derivation of (12.10). Beginning with (12.9) we have

$$\begin{aligned} [2\pi]^k \delta(\mathbf{r}) &= \prod_{j=1}^{k} \int_{-\infty}^{\infty} e^{ir_j t_j}\, dt_j \\ &= \prod \int_{-\infty}^{\infty} [\cos (r_j t_j) + i \sin (r_j t_j)]\, dt_j \\ &= 2^k \prod \int_0^{\infty} \cos (r_j t_j)\, dt_j \\ &= 2^k \lim_{m\to\infty} \prod_{j=1}^{k} [\sin (m r_j)/r_j] \end{aligned} \tag{12.11}$$

This is substituted in the left side of (12.10) to obtain

$$\begin{aligned} &\int f(\mathbf{r} + \mathbf{a}) \delta(\mathbf{r})\, d\mathbf{r} \\ &= \pi^{-k} \lim_{m\to\infty} \int f(\mathbf{r} + \mathbf{a}) \prod_{j=1}^{k} [\sin (m r_j)/r_j]\, dr_j \\ &= \pi^{-k} \lim_{m\to\infty} \int f\left(\frac{x_1}{m} + a_1, \frac{x_2}{m} + a_2, \cdots\right) \prod [\sin x_j/x_j]\, dx_j \\ &= \pi^{-k} f(\mathbf{a}) \left[\int_{-\infty}^{\infty} [\sin (x)/x]\, dx\right]^k \\ &= f(\mathbf{a}) \end{aligned} \tag{12.12}$$

which completes the proof.

By applying the inversion formula to the convolution theorem, (12.3), we may obtain the latter in another form:

$$[2\pi]^k p_n(\mathbf{r}) = \int \exp(i\mathbf{r}\cdot\mathbf{t}) \prod_{j=1}^{n} \tilde{f}_j(\mathbf{t})\, d\mathbf{t} \tag{12.13}$$

In this form of the convolution theorem its value is clear, namely, if we can obtain the transforms $\tilde{f}_j(\mathbf{t})$ then we may replace the convolution integral having $[n - 1]k$ variables of integration by an integral having only k variables.

(b) *Specialization to Radial Functions in Three Dimensions*

The above theorems are valid for very general functions $f(\mathbf{r})$ (subject to certain restrictions to assure the convergence of the integrals,[5, 13]) but now it is convenient to specialize for the case of interest, namely, $k = 3$ and $f(\mathbf{r}) = f(r)$: In our applications we shall only be concerned with three-dimensional space and with functions that depend only on the magnitude of the vector from one particle to another.

If $f(\mathbf{r}) = f(r)$ then also $\tilde{f}(\mathbf{t}) = \tilde{f}(t)$ and we have, for $k = 3$,

$$\tilde{f}(t) = 4\pi \int_0^\infty f(r)[r/t] \sin(rt)\, dr \tag{12.14}$$

$$f(r) = [1/2\pi^2] \int_0^\infty \tilde{f}(t)[t/r] \sin(rt)\, dt \tag{12.15}$$

and (12.13) becomes

$$p_n(r) = [1/2\pi^2] \int_0^\infty \prod_{j=1}^{n} \tilde{f}_j(t)[t/r] \sin(rt)\, dt \tag{12.16}$$

Equation (12.14) may be derived from (12.1) as follows.[4, 6] We begin with

$$\begin{aligned} \tilde{f}(t) &= \int f(r) \exp(i\mathbf{r}\cdot\mathbf{t})\, d\mathbf{r} \\ &= \int_0^\infty f(r) r^2\, dr \int_0^{2\pi} d\varphi \int_0^\pi \exp(irt\cos\theta) \sin\theta\, d\theta \end{aligned} \tag{12.17}$$

where we have chosen θ, the polar angle, to be the angle between $\mathbf{r}$ and the fixed vector, $\mathbf{t}$. More elementary manipulations follow.

$$\tilde{f}(t) = 2\pi \int_0^\infty f(r) r^2 \, dr \int_0^\pi [-irt]^{-1} \, d[\exp(irt \cos \theta)]$$

$$= 4\pi \int_0^\infty f(r)[r/t] \, dr [\exp(irt) - \exp(-irt)]/2i \quad (12.18)$$

$$= 4\pi \int_0^\infty f(r)[r/t] \sin(rt) \, dr.$$

The same procedure is applicable to the derivation of (12.15). Equation (12.16) follows directly from (12.13) and (12.15).

The usual definition of an infinite one-dimensional sine transform is

$$\int_0^\infty f(r) \sin(rt) \, dr$$

so tables of such transforms may be of assistance in evaluating integrals like those in (12.14) to (12.16). Extensive tables of sine transforms have been published by Erdélyi and his collaborators.[14] A short table of transforms defined by (12.14) is given in Table 12.1. This includes all such transforms that have so far been employed in the cluster theory of ionic solutions.

In the remainder of this book the term Fourier transform is used in the sense of the transforms defined by (12.14) and (12.15). The reader is cautioned that in the literature the same term is used for

TABLE 12.1. Three-dimensional Fourier transforms of some radial functions.

$f(r)$	$4\pi \int_0^\infty f(r)[r/t] \sin(rt) \, dr$
$1/4\pi r$	t^{-2} [a]
$e^{-kr}/4\pi r$	$[k^2 + t^2]^{-1}$ [b]
$[e^{-kr}/4\pi r]^2$	$[4\pi t]^{-1} \tan^{-1}(t/2k)$ [b]
$[e^{-kr}/4\pi r][1 - kr/2]$	$t^2/[k^2 + t^2]^2$ [b]
$\begin{matrix} 0 & r < L \\ e^{-kr}/4\pi r & L > r \end{matrix} \Big\}$	$[k^2 + t^2]^{-1} - t^{-2}[1 - \cos(Lt)]$ $- kt^{-3}[\sin(Lt) - Lt \cos(Lt)] + \cdots$ [c]
$\begin{matrix} 0 & r < L \\ [e^{-kr}/4\pi r]^2/2 & L > r \end{matrix} \Big\}$	$[8\pi t]^{-1}[\tan^{-1}(t/2k) - \text{Si}(Lt)] + \cdots$ [c]

[a] See text.

[b] Erdélyi, Magnus, Oberhettinger and Tricomi, *Tables of Integral Transforms* (McGraw-Hill Book Company, Inc., New York, 1953), Vol. 1.

[c] Section 15.

other types of transforms as well, the one in (12.1) being only the most general of these.

(c) *The Transform of the Coulomb Potential*

In the remainder of Section 12 we shall consider several difficulties which arise in the application of the theory of Fourier transforms to the ionic solution problem in statistical mechanics. The first such difficulty arises when we attempt to calculate the transform of $g(r) \equiv 1/4\pi r$, the function that is obtained as the leading term in the expansion of a γ_2 bond (cf. Section 2 or 13). In this case the transform is

$$\tilde{g}(t) = t^{-1} \int_0^{\infty} \sin(rt)\, dr \qquad (12.19)$$

but the integral does not converge. To see this in detail we define

$$\tilde{g}(t,\alpha) \equiv t^{-1} \int_0^{1/\alpha} \sin(rt)\, dr = t^{-2}[1 - \cos(t/\alpha)] \qquad (12.20)$$

Then from the definition of the infinite integral

$$\tilde{g}(t) \equiv \tilde{g}(t,0) = \lim_{\alpha \to 0} \tilde{g}(t,\alpha) \qquad (12.21)$$

but the limit does not exist.

In such cases it is sometimes possible to determine a limit by indirect methods. We give two examples of such methods.

Example 1. Replace $g(r) = 1/4\pi r$ by $g_1(r,\alpha) = e^{-\alpha r}/4\pi r$. It is certainly true that as $\alpha \to 0$, $g_1(r,\alpha) \to g(r)$. The transform is

$$\tilde{g}_1(t,\alpha) = t^{-1} \int_0^{\infty} e^{-\alpha r} \sin(rt)\, dr = [\alpha^2 + t^2]^{-1} \qquad (12.22)$$

and the limit of this as $\alpha \to 0$ is

$$\tilde{g}_1(t,0) = t^{-2}$$

Example 2. We define $g_2(r,\alpha) = 1/4\pi r^{1+\alpha}$. This too approaches $g(r)$ as $\alpha \to 0$. We also have

$$\tilde{g}_2(t,\alpha) = t^{-1} \int_0^{\infty} r^{-\alpha} \sin(rt)\, dr = t^{\alpha-2}\Gamma(1 - \alpha) \cos(\pi\alpha/2) \qquad (12.23)$$

and taking the limit as $\alpha \to 0$ gives

$$\tilde{g}_2(t,0) = t^{-2}$$

(In (12.23) $\Gamma(x)$ is the gamma function ($\Gamma\ (1 + x) = x\Gamma(x)$ and $\Gamma(1) = 1$).)

Example 1 is closely related to the method used by Mayer[7] in connection with this problem. Example 2 is included to show that Example 1 is not unique. This seems to be an important point because the "convergence factor" in Example 1 is confusingly similar to the factor $e^{-\kappa r}$ which appears in the Debye potential.

It is nice that $\tilde{g}_1(t,0)$ and $\tilde{g}_2(t,0)$ are equal, and perhaps we can derive other $\tilde{g}_n(t,0)$ that are also equal to t^{-2} without finding any that give other functions of t, but we cannot by such methods *prove* that any $\tilde{g}_n(t,0) = \tilde{g}(t)$, and it is $\tilde{g}(t)$ that we need. Proofs of this sort are called Tauberian theorems in mathematics and many have been carefully investigated (see for example reference 5) but it does not appear that any has been established for the present case.

Before considering a clearer method of deriving $\tilde{g}(t) = t^{-2}$ we note one test of consistency that is passed by this $\tilde{g}(t)$. Namely its inversion is

$$[2\pi^2]^{-1} \int_0^\infty t^{-2}[t/r] \sin(rt)\, dt = 1/4\pi r = g(r) \qquad (12.24)$$

as expected.

Examples 1 and 2 suggest that if the Coulomb potential decreased even infinitesimally more rapidly than $1/r$ as $r \to \infty$ then there would be no difficulty in obtaining its Fourier transform by a direct method. On the other hand the following derivation shows unambiguously that the most decisive experimental evidence concerning the distance dependence of the Coulomb potential leads to $\tilde{g}(t) = 1/t^2$.[15]

The fundamental experiment of electrostatics is the attempted detection of an electric field within a cavity in a conductor (the Cavendish experiment).[16] The most delicate measurements have failed to detect such a field.[17] This result is most often interpreted as showing that the exponent of r in the Coulomb potential is $-1[1 + \alpha]$ where $|\alpha|$ is not larger than 10^{-9}. We prefer first to make the more direct deduction that the experiment establishes the validity of Gauss's law of induction:

$$\int_S \nabla\Psi \cdot d\mathbf{S} = -[4\pi/D] \int_\mathcal{V} \rho\, d\mathbf{r}$$

where Ψ is the local potential, ρ is the local charge density, D is the dielectric constant of the medium, and the integral on the left is over the surface of the region of volume $\mathcal{V}$. If the charge is localized at point sources then the integral on the right becomes $\epsilon \sum_i z_i$. The effects of the various sources on the surface integral are clearly additive so if we have a single charge in the volume Gauss's law reduces to

$$\int_S \nabla\Psi \cdot d\mathbf{S} = -[4\pi/D]z_i\epsilon$$

while if there is no charge in the volume we have

$$\int_S \nabla\Psi \cdot d\mathbf{S} = 0$$

Furthermore we may apply Gauss's divergence theorem to express Gauss's law as a volume integral

$$\int_{\mathcal{V}} \nabla^2\Psi \, d\mathbf{r} = -[4\pi/D]z_i\epsilon \qquad \text{one point source}$$

$$= 0 \qquad \text{no source}$$

Both of these possibilities are expressed by the differential equation

$$\nabla^2\Psi(\mathbf{r}) = -(4\pi/D)z_i\epsilon\delta(\mathbf{r} - \mathbf{r}_i) \tag{12.25}$$

where $\delta(\mathbf{r})$ is the Dirac delta function (see equation (12.9)) and where there is a point source with charge ϵz_i at $\mathbf{r}_i$. This is immediately verified by integrating (12.25) over the volume. Thus (12.25), which is the form of Poisson's equation appropriate to systems with point charges, expresses the information about $\Psi(\mathbf{r})$ that is contained in the Cavendish experiment.

We continue by taking the Fourier transform (equation (12.1), $k = 3$) of both sides of equation (12.25).

$$\int \nabla^2\Psi(r) \exp(i\mathbf{r}\cdot\mathbf{t}) \, dr = -[4\pi/D] \int \delta(\mathbf{r} - \mathbf{r}_i) \exp(i\mathbf{r}\cdot\mathbf{t}) \, d\mathbf{r} \tag{12.26}$$

The left side is given by $-t^2\tilde{\Psi}(\mathbf{t})$ as may be seen in the following way. Writing $\mathbf{r} = x,y,z$ (Cartesian components) we have

$$\int [d\Psi/dx] \exp(i\mathbf{r}\cdot\mathbf{t}) \, d\mathbf{r}$$

$$= \int \lim_{\xi\to 0} [\Psi(x + \xi,y,z) - \Psi(x,y,z)]\xi^{-1} \exp(i\mathbf{r}\cdot\mathbf{t}) \, d\mathbf{r}$$

Then we interchange the order of integration and limit process and note that

$$\int \Psi(x + \xi,y,z) \exp(i\mathbf{r}\cdot\mathbf{t})\, d\mathbf{r} = \int \Psi(\mathbf{r}) \exp(i\mathbf{t}\cdot[x - \xi,y,z])\, d\mathbf{r}$$

$$= \exp(-i\xi a)\tilde{\Psi}(t)$$

where a is the Cartesian component of $\mathbf{t}$ that corresponds to x. That is

$$\mathbf{t}\cdot\mathbf{r} = ax + by + cz$$

Then we have

$$\int [d\Psi/dx] \exp(i\mathbf{r}\cdot\mathbf{t})\, d\mathbf{r} = \lim_{\xi\to 0} [\tilde{\Psi}(\mathbf{t})[\exp(-i\xi a) - 1]/\xi] = -ia\tilde{\Psi}(\mathbf{t})$$

The differentiation may be repeated to obtain

$$\int [d^2\Psi/dx^2] \exp(i\mathbf{r}\cdot\mathbf{t})\, d\mathbf{r} = -a^2\tilde{\Psi}(t)$$

and repeated again for y and z as variables to finally obtain

$$\int \nabla^2\Psi \exp(i\mathbf{r}\cdot\mathbf{t})\, d\mathbf{r} = -[a^2 + b^2 + c^2]\tilde{\Psi}(\mathbf{t}) = -t^2\tilde{\Psi}(\mathbf{t}) \qquad (12.27)$$

From (12.10) we see that the transform of $\delta(r)$ is unity and so if the charge is at the origin then the transform of (12.25) is

$$t^2\,\tilde{\Psi}(t) = 4\pi z_i\epsilon/D$$

or

$$\tilde{\Psi}(t) = 4\pi z_i\epsilon/Dt^2 \qquad (12.28)$$

Our procedures in the cluster theory are equivalent to the relation $\tilde{g}(t) = \tilde{\Psi}(t)/[4\pi z\epsilon/D]$ so (12.25) and (12.28) are equivalent to the equations,

$$\nabla^2 g(r) = -\delta(r) \qquad (12.29)$$

$$\tilde{g}(t) = t^{-2} \qquad (12.30)$$

respectively. The latter was obtained above by another method. The Poisson equation in the form (12.29) is also useful in the sequel.

Of course this procedure leads, via the Fourier inversion of (12.30), to the expected $1/r$ dependence of the Coulomb potential. A consistent point of view for this book is to regard $g(r)$ as *defined* to be the Fourier transform of $\tilde{g}(t)$, which is in turn determined by the foregoing procedure. Then it is clear that in equations in which $\tilde{g}(t)$ appears (12.30) is the correct function to use. It would perhaps be even clearer if the whole development of the cluster theory were carried through in t space rather than in r space as has indeed been done for the quantum mechanical case,[18] but the next mathematical problem to be discussed would still remain.

(d) *Sum over g-Bond Chains*

By a process which is described in Section 2 and more fully in Section 13, the cluster theory of ionic solutions leads to the problem of determining sums of the form

$$q(r,\mathcal{V}) = \sum{}' \, [-\kappa^2]^{n-1} p_n(r,\mathcal{V}) \tag{12.31}$$

where

$$p_n(r_{1,n+1}, \mathcal{V}) \equiv \int_{\mathcal{V}} g(r_{12}) g(r_{23}) \cdots g(r_{n,n+1}) \, d\mathbf{r}_{12} \cdots d\mathbf{r}_{1n} \tag{12.32}$$

is the convolution integral for a finite volume, and where

$$\kappa^2 \equiv \lambda \sum c_s z_s^2$$

where λ is a positive number (equations (2.30) and (13.2)). To express $q(r,\mathcal{V})$ as a function that approaches zero faster than r^{-3} as $r \to \infty$, even in the limit of infinite volume, is to overcome the principle difficulty in the cluster theory of ionic solutions, namely, the divergence of the irreducible cluster integrals B_n at infinite volume. The remainder of Section 12 is devoted to a presentation of various approaches to this problem.

On careful examination of (12.31) and (12.32) one finds that the notation is incomplete. That is, at finite volume p_n depends not only on $r_{1,n+1} = |\mathbf{r}_1 - \mathbf{r}_{n+1}|$ but rather on both $\mathbf{r}_1$ and $\mathbf{r}_{n+1}$. We discuss the implications of this at the end of this section, but until then we use the simpler notation and neglect the dependence of p_n on the positions of the ends of the chain.

It is natural to attempt to evaluate q with the help of the convolution theorem. We write

$$p_n(r) \equiv p_n(r, \infty)$$

$$q(r) \equiv q(r, \infty)$$

and use $\tilde{g}(t) = t^{-2}$ as determined in Section 12(c). Then the convolution theorem can be applied to $p_n(r)$ to get the transform of $q(r)$:

$$\tilde{q}(t) = -\kappa^{-2}\sum{}' [-\kappa^2/t^2]^n \tag{12.33}$$

If $|\kappa^2/t^2| < 1$ the infinite sum converges and we have

$$\tilde{q}(t) = -\kappa^{-2}\left[\frac{1}{1+\kappa^2/t^2} - 1\right] = [\kappa^2 + t^2]^{-1} \tag{12.34}$$

but in order to invert this expression to obtain $q(r)$ we would need to know that it represents $\tilde{q}(t)$ *throughout* the range of the integration in the inversion formula, that is, the range $0 \leq t < \infty$. It can be shown that (12.34) is the analytic continuation[19, 20] of the sum in (12.33), which implies that in a mathematical sense $[\kappa^2 + t^2]^{-1}$ represents this sum even for $\kappa > t$. However, the physical meaning of analytic continuation is obscure and there seems to be a possibility that the ensuing equations can be valid at most in the limit of $\kappa \to 0$. Before continuing we note that if we neglect the difficulty and transform (12.34) without further analysis we obtain

$$q(r) = e^{-\kappa r}/4\pi r,$$

the Debye potential.

It is remarkable that mathematical difficulties of the sort encountered here do not appear in the Debye-Hückel theory which leads to the same potential function (although its significance is not precisely the same in the two theories). This aspect is discussed in more detail in Section 12(f).

(e) *Review of the Sum Problem*

The necessity of using a continuation procedure to evaluate $q(r)$ raises the possibility that the two limiting processes involved in the definitions of this sum may be incorrectly handled. These limiting processes are for $n \to \infty$ in the infinite sum (12.31) and for $\mathcal{V} \to \infty$ in the convolution integrals. If the volume is infinite then the sums certainly are not convergent in the ordinary sense because each p_n is infinite. Moreover the convergence of the sum $\tilde{q}(t,\mathcal{V})$ is questionable even for finite macroscopic volumes although in this case every term of the sum is finite.

We note first that one of Salpeter's‡ procedures appears to lead to the evaluation of $q(r)$ by a Fourier transform method without re-

‡ Reference 10, pp. 208–209.

sorting to analytical continuation, but that in fact he is only using a *particular* technique of continuation. It is well known that if analytical continuation is possible in a given case then every continuation technique leads to the same result.[19]

The following method of avoiding the continuation procedure is due to Mayer[7] (although he did not interpret the method in this way). We begin with the convolution integral for a finite volume (12.32) and the corresponding sum (12.31). The convolution theorem for $p_n(r,\mathcal{V})$ is much more complicated than for $p_n(r)$ in which the volume is infinite. In order to be able to use the simpler convolution theorem we replace $g(r)$ by

$$g(r,\alpha) \equiv e^{-\alpha r}/4\pi r \tag{12.36}$$

in (12.32) and let $\mathcal{V} \rightarrow \infty$ to get $p_n(r_{1,n+1}, \alpha)$. This corresponds roughly to replacing the sharp boundaries of the physical container, which puts an upper limit on the intermolecular distances in the physical system, by a factor $e^{-\alpha r}$ in the potential. This factor reduces the contribution of widely separated molecules to the partition function. To the extent that this correspondence is valid we have $\alpha \sim \mathcal{V}^{-1/3}$.

We also have

$$q(r,\alpha) \equiv \sum{}' [-\kappa^2]^{n-1} p_n(r,\alpha) \tag{12.39}$$

which approaches $q(r)$ as $\alpha \rightarrow 0$ and $\mathcal{V} \rightarrow \infty$. We proceed just as before, taking transforms of both sides of (12.39), noting (12.22), to get

$$\begin{aligned} \tilde{q}(t,\alpha) &= -\kappa^{-2} \sum{}' [-\kappa^2/[\alpha^2 + t^2]]^n \\ &= [\alpha^2 + \kappa^2 + t^2]^{-1}, \qquad \kappa^2/[\alpha^2 + t^2] < 1 \end{aligned} \tag{12.40}$$

The condition for convergence of the sum is now met even for $t = 0$ if $\alpha > \kappa$. Furthermore, inversion yields

$$q(r,\alpha) = e^{-r[\kappa+\alpha]}/4\pi r$$

Letting $\alpha \rightarrow 0$, which corresponds to $\mathcal{V} \rightarrow \infty$, we obtain the Debye potential as before.

Were this process satisfactory from a physical point of view the magnitude of α necessary to insure convergence in (12.40) even for

$t = 0$ would correspond to a reasonable volume of the physical system. However the convergence condition is $\mathcal{V}^{-1/3} \sim \alpha > \kappa$ or $\mathcal{V} < \kappa^{-3}$. This is so small a volume that there is only one ion in the system, so Mayer's procedure cannot be interpreted physically in the way we have attempted. We also remark that Mayer's[7] physical interpretation of this procedure is not satisfactory because it implies that α is so small that any $\kappa' < \alpha$ is in the Debye-Hückel limiting law region, whereas within the circle of convergence of (12.40) the summation procedure requires that $\kappa < \alpha$, where κ is the parameter for the solution under investigation.

A similar approximate procedure for evaluating $q(r)$ suffers from the same difficulty of physical interpretation. It is based on the assumption that the Coulomb potentials have a definite range, α^{-1}, beyond which they vanish. The transform of this potential is (12.20) but the sum $\tilde{q}(t)$ converges for $t = 0$ only if $\kappa < \alpha\sqrt{2}$.

Another procedure is based on the fact that $q(r,\mathcal{V})$ may be identified[6, 10, 21] as the Neumann-Liouville expansion of a Fredholm integral equation of the second kind. However we find from the theory of integral equations[22] that this identification is *assured* only if $\mathcal{V} < \kappa^{-3}$.

We note that all of these methods of evaluating $q(r)$ do lead to the same result. Still another method that also leads to this result, but perhaps in a more satisfactory way than any of the others, is discussed in Section 12(g).

(f) *Comparison with the Debye-Hückle Theory*

It is interesting to compare the derivation of the Debye-Hückel limiting law from Debye's hypothesis with the derivation from the cluster theory to see what differences are responsible for the mathematical difficulty in the cluster theory that we have been discussing.

Differential equation method. This is the now familiar method used by Debye and Hückel. An outline will suffice here.

The average potential at a point r distant from an ion of species α in the solution is given by the Poisson equation,

$$\nabla^2\Psi^{\alpha}(r) = -[4\pi\epsilon/D]\sum z_s c_s(r)$$

where the sum is over species. The *average* potential at a point would of course vanish unless at least one ion is held at a fixed distance from the given point. Hence, the superscript α is used to indicate the species of the one ion so distinguished in the present calculation.

An approximate expression for the average concentration of species s at r, $c_s(r)$, is given by the Boltzmann equation

$$c_s(r) = c_s \exp\left[-\epsilon\Psi^\alpha(r)z_s/kT\right]$$

where c_s is the bulk concentration of species s. The approximation lies in the fact that the exact expression has the pairwise potential of average force in the solution $W_{\alpha s}(r)$ in place of $\Psi^\alpha(r)z_s\epsilon$. The serious effect of this approximation is well known[23, 24] and has been more recently discussed by Kirkwood and Poirier[25] and Frank and Thompson.[26]

Combining the two equations and expanding the exponential we have

$$\nabla^2\Psi^\alpha(r) = -[4\pi\epsilon/D][\sum c_s z_s - \Psi^\alpha(r)\epsilon^2 \sum c_s z_s^2/kT + \cdots]$$

The first term vanishes in electrically neutral systems and we neglect terms higher than the second. We define

$$\kappa^2 = [4\pi\epsilon^2/DkT]\sum c_s z_s^2$$

and then we have the linearized Poisson-Boltzmann equation

$$\nabla^2\Psi^\alpha(r) = \kappa^2\Psi^\alpha(r) \tag{12.41}$$

The solution of this equation is

$$\Psi^\alpha(r) = A[e^{\kappa r}/r] + B[e^{-\kappa r}/r]$$

Applying the boundary conditions, $\Psi^\alpha(r) \to 0$ as $r \to \infty$ and $\Psi^\alpha(r) \to \epsilon z_\alpha/rD$ as $r \to 0$ we find $A = 0$, $B = \epsilon z_\alpha/D$ and hence obtain the Debye potential:

$$\Psi^\alpha(r) = \epsilon z_\alpha e^{-\kappa r}/Dr$$

Integral equation method.[27] Fixing ion α at the origin of coordinates we have as the potential at a point r distant from the origin

$$\Psi^\alpha(r) = \epsilon z_\alpha/Dr + \epsilon \int [\sum c_s(r_1)z_s/D\,|\,\mathbf{r} - \mathbf{r}_1\,|]\, d\mathbf{r}_1$$

This integral equation is equivalent to the Poisson equation. Using the Debye approximation for $c_s(r)$ and linearizing as before, we obtain

$$\Psi^\alpha(r) = [\epsilon z_\alpha/Dr] - [\kappa^2/4\pi] \int [\Psi^\alpha(r_1)/|\,\mathbf{r} - \mathbf{r}_1\,|]\, d\mathbf{r}_1$$

$$= [4\pi\epsilon z_\alpha/D]g(r) - \kappa^2 \int \Psi^\alpha\,(|\mathbf{r}_1\,|)g(|\,\mathbf{r} - \mathbf{r}_1\,|)\, d\mathbf{r}_1$$

where $g(r) = 1/4\pi r$. Now we take the Fourier transform of each term of this equation, noting that the integral is a convolution. The result is

$$\tilde{\Psi}^\alpha(t) = [4\pi\epsilon z_\alpha/D]\tilde{g}(t) - \kappa^2\tilde{\Psi}^\alpha(t)\tilde{g}(t)$$
$$= [4\pi\epsilon z_\alpha/D]\tilde{g}(t)/[1 + \kappa^2\tilde{g}(t)]$$

Inversion now gives the formal solution

$$\Psi^\alpha(r) = [2\epsilon z_\alpha/\pi D] \int_0^\infty \tilde{g}(t)[1 + \kappa^2 g(\bar{t})]^{-1}[t/r] \sin(rt)\, dt$$

and introducing $\tilde{g}(t) = t^{-2}$ gives the Debye potential as a result:

$$\Psi^\alpha(r) = [2\epsilon z_\alpha/\pi D] \int_0^\infty t \sin(rt)\, [t^2 + \kappa^2]^{-1}\, dt = \epsilon z_\alpha e^{-\kappa r}/Dr$$

Discussion. A summary of the differences in the three ways of arriving at the Debye potential follows.

Cluster theory. There are no statistical mechanical approximations. Mathematical difficulties occur in obtaining the transform of $q(r)$ and with the summation of the power series in κ for $q(r)$ or $\tilde{q}(t)$.

Debye-Hückel theory, differential equation method. A statistical approximation is made, namely, $\epsilon z_s \Psi^\alpha(r)$ = potential of average force, and a mathematical approximation, the neglect of the higher terms in the expansion of the exponential. Neither transform nor summation problem appears.

Debye-Hückel theory, integral equation method. The same statistical and mathematical approximations are made. The problem of evaluating the Fourier transform of the Coulomb potential appears, but not the summation problem.

Now we attempt an interpretation of these observations. In the differential equation method of solving the Debye-Hückel problem the Coulomb potential appears only implicitly, as the potential that satisfies Poisson's equation. The operation of solving the Poisson-Boltzmann equation is equivalent to *assuming* that the Debye potential is the solution, and then verifying this by differentiation. These operations do not explicitly involve the Coulomb potential and no divergence difficulties at all are encountered.

The other extreme is represented in the cluster theory method in which we attempt to *construct* the Debye potential by adding together, bit by bit, contributions from Coulomb potentials. The divergence difficulties all arise from the long range of the latter. The integral equation method of solving the Debye-Hückel problem is illustrative of an intermediate case: *both* the Debye potential and the Coulomb potential appear in the integral equation and here too we find a divergence problem (in obtaining $\tilde{g}(t)$) associated with the $1/r$ potential, but it is only part of the problem encountered in the cluster theory.

Another aspect of this comparison is illustrated in Fig. 12.1 where we represent by graphs the operations in the three theories. In the cluster theory we deal only with g-bond chains, as indicated. On the other hand if we make the corresponding interpretation of the integral in the integral equation method of solving the Debye-Hückel problem, we must consider the integrand as a chain of one g bond and one Ψ^α bond (Debye potential). This correspondence is of course not exact because the integrals have rather different significance in the two cases but the comparison does show that the Debye-Hückel formulation, when expressed in integral equation form, corresponds to a *prior* partial summation of the graphs of the cluster theory, and it is this prior summation which reduces the divergence difficulty. Finally in the Debye-Hückel theory, differential equation method, the only interactions which are counted correspond to Ψ^α bonds ($\sim q$ bonds). The $1/r$ dependence that is characteristic of the Coulomb potential now enters only implicitly, through the use of the Poisson equation. Of course in this last case the graphical interpretation is looser and

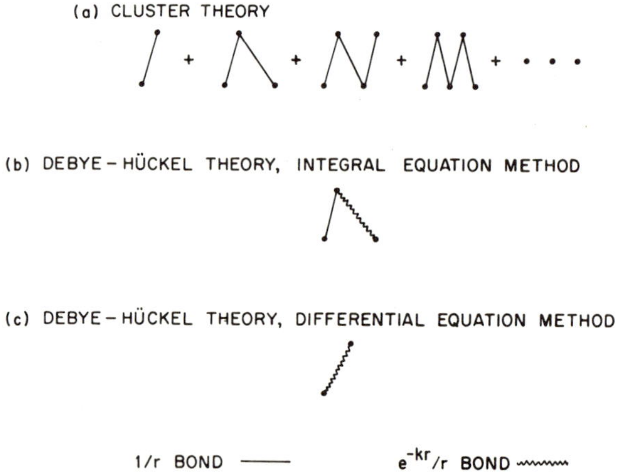

Fig. 12.1. Graphic analysis of different approaches to an ionic solution theory.

one could just as well say that the Poisson-Boltzmann equation corresponds to a graph like that in Fig. 12.1(b) and that the divergence difficulty does not appear at all in the differential equation method because the solution is obtained by verifying that an assumed function satisfies the differential equation, rather than by integration. However the following calculation shows the merit of the first point of view, namely, that *mathematical difficulties may be avoided by using Poisson's equation rather than the* $1/r$ *dependence as the fundamental property of the Coulomb potential.*

(g) *δ Function Method to Determine $q(r)$*

We recall again that all of the methods discussed above lead to the same function $q(r)$ but that all are mathematically questionable because the sum form (12.31) of $q(r)$ with which we must begin is a sum of infinite terms if $\mathcal{V} = \infty$ and is not easily shown to be convergent even if $\mathcal{V}$ is finite (though macroscopic). These deductions about $q(r)$ are based on the definition $g(r) = 1/4\pi r$. It is proposed to resolve the difficulty in the same way that was used to establish the Fourier transform of the Coulomb potential in Section 12(c). Thus we accept at the outset that integrals over the $1/r$ potential are poorly behaved and therefore that we had best regard the basic property of the Coulomb potential to be Poisson's equation. This corresponds to defining $g(r)$ by the equation (12.29)

$$\nabla^2 g(r) = -\delta(r)$$

(The constants of integration are fixed by the physical requirement that $g(r)$ and its derivatives with respect to r vanish as $r \to \infty$.)

Now we examine $q(r, \mathcal{V})$ with this definition of $g(r)$. The following procedure for doing this, which was suggested to the author by D. C. Mattis,[28] is in some ways similar to one of the ways of evaluating $q(r)$ that was proposed by Salpeter.‡

The first step is to take the Laplacian derivative of each term of (12.31) to get

$$\begin{aligned}\nabla^2 q(r,\mathcal{V}) &= \sum{}' [-\kappa^2]^{n-1} \nabla^2 p_n(r,\mathcal{V}) \\ &= -\delta(r) + \sum{}'' [-\kappa^2]^{n-1} \nabla^2 p_n(r,\mathcal{V})\end{aligned} \tag{12.42}$$

For the last member we have used the definition $p_1(r,\mathcal{V}) = g(r)$.

‡ Reference 10, equations (5.9) to (5.11).

For $n > 1$ we may write the convolution integral in the form

$$p_n(\mathbf{r}_{n+1}, \mathbf{r}_1, \mathcal{V}) = \int_{\mathcal{V}} p_{n-1}(\mathbf{r}_n, \mathbf{r}_1, \mathcal{V}) g(|\mathbf{r}_{n+1} - \mathbf{r}_n|)\, d\mathbf{r}_n \qquad (12.43)$$

where $\mathbf{r}_n$ is the vector from the origin to the vertex n. We assume that $\mathbf{r}_1$ is fixed and then we have

$$\begin{aligned} \nabla^2 p_n(\mathbf{r}_{n+1}, \mathbf{r}_1, \mathcal{V}) &= -\int_{\mathcal{V}} p_{n-1}(\mathbf{r}_n, \mathbf{r}_1, \mathcal{V}) \delta(\mathbf{r}_{n+1} - \mathbf{r}_n)\, d\mathbf{r}_n \\ &= -p_{n-1}(\mathbf{r}_{n+1}, \mathbf{r}_1, \mathcal{V}) \end{aligned} \qquad (12.44)$$

The Dirac δ function in the integrand "picks" the value of

$$p_{n-1}(\mathbf{r}_n, \mathbf{r}_1, \mathcal{V})$$

for which $\mathbf{r}_n = \mathbf{r}_{n+1}$ just as it would for the infinite integral (cf. equation (12.8) and those following it). The reason for this is that $\mathbf{r}_{n+1}$ is independently constrained to lie within the volume $\mathcal{V}$ and therefore, even if the range of the integral were extended, the range of $\mathbf{r}_n$ outside of the original $\mathcal{V}$ would make no contribution to the integral.

We combine (12.43) with (12.42) to get

$$\begin{aligned} \nabla^2 q(r,\mathcal{V}) &= -\delta(r) - \sum\nolimits'' [-\kappa^2]^{n-1} p_{n-1}(r,\mathcal{V}) \\ &= -\delta(r) + \kappa^2 q(r,\mathcal{V}) \end{aligned} \qquad (12.45)$$

If we now let $\mathcal{V} \to \infty$ we have

$$\nabla^2 q(r) = -\delta(r) + \kappa^2 q(r) \qquad (12.46)$$

which is the differential equation that is equivalent to (12.31). This equation is a form of the linearized Poisson-Boltzmann equation (12.41) that is valid for all r, even $r = 0$. Therefore (12.46), unlike (12.41), may be solved by a Fourier transform method. The transform of (12.46) is

$$-t^2 \tilde{q}(t) = -1 + \kappa^2 \tilde{q}(t) \qquad (12.47)$$

This leads to

$$\tilde{q}(t) = 1/[\kappa^2 + t^2] \qquad (12.48)$$

and finally

$$q(r) = e^{-\kappa r}/4\pi r \qquad (12.49)$$

This method of evaluating $q(r)$ suggests the following interpretation of the procedures of Mayer's ionic solution theory. (These procedures are developed in detail in Section 13 but have already been illustrated in Section 2.) First we recall that it has been demonstrated[25] that the linearized Poisson-Boltzmann equation (12.41) is rigorous in a system in which it is permissible to linearize the exponential of the pairwise potential, $\exp(-W_{ij}/kt) \to 1 - W_{ij}/kT$. On the other hand in the cluster theory one proceeds by expanding $\exp(-W_{ij}/kT)$ without rejecting any terms, but this is followed by systematically collecting terms in such a way that $q(r)$ is a sum of contributions of *linear* terms of the expansion of the exponential. We have shown that $q(r)$ is equivalent to the linearized Poisson-Boltzmann equation, and this appears to be a reasonable result. Moreover Mayer's procedure (with some simplifications) expresses $\mathfrak{S}$ as a sum of products of terms $q(r)$ and therefore gives an expansion of the properties of a real ionic system in terms of sums of products of hypothetical ionic systems having linearized interactions. A similar discussion has been given by Meeron.[9] As he pointed out it must be possible to get different functions that play the role of $q(r)$ if one modifies the way in which terms are collected in Mayer's rearrangement, Fig. 2.3.

We now return to a discussion of $q(r,\mathcal{V})$ at finite $\mathcal{V}$. It is convenient to write out the more complete notation, $q(\mathbf{r}_a, \mathbf{r}_b, \mathcal{V})$, where $\mathbf{r}_a$ and $\mathbf{r}_b$ are the coordinates of the vertices joined by all of the g-bond chains, p_n (cf. equation (12.43)). If we write

$$q(\mathbf{r}_a, \mathbf{r}_b, \mathcal{V}) = g(|\mathbf{r}_a - \mathbf{r}_b|) + \sum{}'' [-\kappa^2]^{n-1} p_n(\mathbf{r}_a, \mathbf{r}_b, \mathcal{V})$$

then the first term on the right represents the direct contribution of the ions at $\mathbf{r}_a$ and $\mathbf{r}_b$ to the potential while the sum represents the shielding effect of all the other ions.

The direct potential certainly depends only on $|\mathbf{r}_a - \mathbf{r}_b|$ but the shielding will be different if $\mathbf{r}_a$ and $\mathbf{r}_b$ are near the surface of the system than if they are deep in the interior. The reason for the distinction is, in terms of Faraday's "visual integration," that the lines of force that contribute to the direct potential are not restricted to lie within the confines of the system but the ions which contribute to the shielding are.

Thus (12.45) is a partial differential equation whose complete solution must be of the form

$$q(\mathbf{r}_a, \mathbf{r}_b, \mathcal{V}) = e^{-\kappa r}/4\pi r + \varphi(\mathbf{r}_a, \mathbf{r}_b, \kappa, \mathcal{V}) \qquad (12.50)$$

where $r = |\mathbf{r}_a - \mathbf{r}_b|$ and where φ must vanish as $\mathcal{V} \to \infty$. In the cluster theory of the bulk properties of systems the comparison with experiment may always be made after taking the limit of the cluster integrals as $\mathcal{V} \to \infty$ and in this limit the contribution of φ vanishes. On the other hand it is sometimes more convenient to manipulate the cluster integrals for finite volume. To simplify the notation in such integrals we shall neglect the φ term in (12.50), but it must then be understood that the equations are only exact in the limit $\mathcal{V} \to \infty$.

Notes and References

1. C. J. Tranter, *Integral Transforms in Mathematical Physics* (John Wiley & Sons, Inc., New York, 1956), second ed.
2. E. C. Titchmarsh, *Introduction to the Theory of Fourier Integrals* (Oxford University Press, New York, 1937).
3. Sneddon, "Functional Analysis," *Handbuch der Physik* (Springer-Verlag, Berlin, 1955) Band 11.
4. Bochner, *Vorlesungen uber Fourier Transformen* (Akademische Verlaksgesellschaft MBH, Leipzig, 1932).
5. S. Bochner and K. Chandrasekharan, *Fourier Transforms* (Princeton University Press, Princeton, N. J., 1949).
6. E. W. Montroll and J. E. Mayer, *J. Chem. Phys.* **9,** 626 (1941).
7. J. E. Mayer, *J. Chem. Phys.* **18,** 1426 (1950).
8. E. Haga, *J. Phys. Soc. Japan* **8,** 714 (1953).
9. E. Meeron, *J. Chem. Phys.* **28,** 630 (1958).
10. E. E. Salpeter, Annals of Physics **5,** 183 (1958).
11. H. L. Friedman, *Molecular Phys.* **2,** 23 (1959); H. L. Friedman, *Molecular Phys.* **2,** 190, 436 (1959).

12a. W. Heitler, *The Quantum Theory of Radiation,* (Oxford University Press, New York, 1954).

12b. I. N. Sneddon, *Special Functions of Mathematical Physics and Chemistry* (Oliver & Boyd, Ltd., London, 1956).

13. M. J. Lighthill, *Fourier Analysis and Generalized Functions* (Cambridge University Press, New York, 1958).
14. Erdelyi, Magnus, Oberhettinger and Tricomi, *Tables of Integral Transforms* (McGraw-Hill Book Company, Inc., New York, 1953), Vol. 1.
15. The author is obliged to E. W. Montroll for suggesting this procedure.
16. J. C. Maxwell, *A Treatise on Electricity and Magnetism* (Dover Publications, Inc., New York, 1954), Vol. 1.
17. S. J. Plimpton and W. E. Lawton, *Phys. Rev.* **50,** 1066 (1936).
18. E. W. Montroll and J. C. Ward, *Phys. Fluids* **1,** 55 (1958).
19. E. T. Whittaker and G. N. Watson, *A Course of Modern Analysis* (Cambridge University Press, New York, 1935), p. 96.
20. E. T. Whittaker and G. N. Watson, *A Course of Modern Analysis* (Cambridge University Press, New York, 1935), p. 140.
21. F. P. Buff and F. H. Stillinger, Jr., *J. Chem. Phys.* **25,** 312 (1956), Equations (42)-(43).

22. F. G. Tricomi, *Integral Equations* (Interscience Publishers, Inc., New York, 1957).
23. L. Onsager, *Chem. Revs.* **13**, 73 (1933).
24. R. Fowler and E. A. Guggenheim, *Statistical Thermodynamics* (Cambridge University Press, New York, 1952).
25. J. G. Kirkwood and J. Poirier, *J. Phys. Chem.* **58**, 591 (1954).
26. H. S. Frank and P. T. Thompson, *J. Chem. Phys.* **31**, 1086 (1959).
27. This method is due to I. Prigogine and A. Bellemans (unpublished work). It could be used to obtain some of the higher terms of the Debye-Hückel theory by proceeding in a way similar to that followed by S. Kaneko, *Researches of the Electrotechnical Laboratory (Tokyo)*, No. 403 (1937).
28. D. C. Mattis, private communication.

13. The Excess Free Energy of Ionic Systems[1]

(a) *Introduction.* Beginning with equation (8.20)

$$\mathfrak{S} = \sum_{m}{}'' \mathbf{c}^{\mathrm{m}} B_{\mathrm{m}} \tag{13.1}$$

we now proceed to rearrange the terms according to the procedures depicted in Fig. 2.3 in order to get modified irreducible cluster integrals which do not diverge as $\mathcal{V} \to \infty$, even for ionic systems.

The first step is to expand the cluster functions for two particles. We have, according to Section 11,

$$\gamma_{ij}(r_{ij}) = \exp\left[-u_{ij}^{*}/kT - z_i z_j \lambda g(r_{ij})\right] - 1 \tag{13.2}$$

where

$$\lambda = 4\pi\epsilon^2/DkT$$

is the Coulomb length, a characteristic length of the system and where $4\pi g(r_{ij})$ is the distance dependence of the Coulomb potential. It is convenient also to define the function

$$k_{ij} \equiv \exp\left[-u_{ij}^{*}(r_{ij})/kT\right] - 1 \tag{13.3}$$

Then we may write (13.2) in the form ($g_{ij} \equiv g(r_{ij})$)

$$\gamma_{ij} = k_{ij} + [1 + k_{ij}] \sum_{p}{}' \left[-z_i z_j \lambda g_{ij}\right]^p / p! \tag{13.4}$$

This expansion of $\gamma_{i,j}$ in terms of k_{ij} and g_{ij} corresponds to replacing each graph of a term of B_{m} in which there is a γ_2 bond by an infinity of graphs with zero or one k bond and any number of g bonds in place of the γ_2 bond. This expansion of the graphs corresponding to

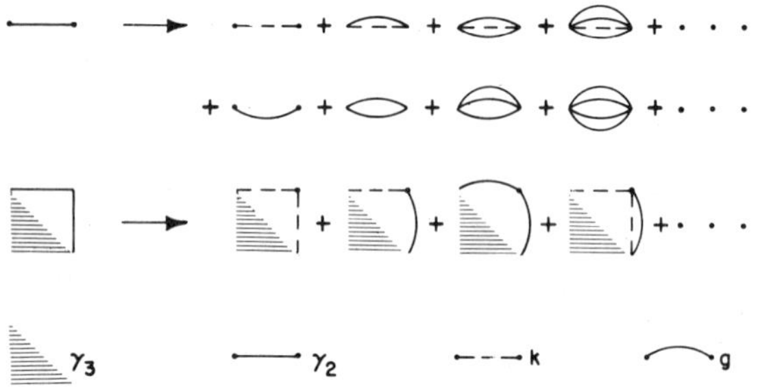

Fig. 13.1. Illustrating the expansion of graphs corresponding to the expansion of γ_2 bonds. (From H. L. Friedman, *Molecular Phys.* **2,** 23 (1959).)

terms of the B_{m} is illustrated for two examples in Fig. 13.1. The coefficients (numerical and powers of λ and $z_i z_j$) which result from the expansion are not shown in the figure. Each of the graphs resulting from the expansion contributes to the same irreducible cluster integral B_{m} as the parent graph.

In the expanded graphs those vertices which lie at the junction of exactly two g bonds are called *g-bond nodes.* A sequence of g bonds connected by g-bond nodes is said to be a g-bond *chain.* Just two of the graphs in Fig. 13.1 have any g-bond chains.

For a given graph on the skeleton of $\mathbf{m}$ the vertices of the skeleton which are g-bond nodes correspond to the subset $\mathbf{n}$ of $\mathbf{m}$. The remainder of $\mathbf{m}$ is designated $\mathbf{u}$. Thus $\mathbf{m} = \mathbf{n} + \mathbf{u}$. Now the cluster integral sum is rewritten as

$$\mathfrak{S} = \mathfrak{S}_l + \mathfrak{S}_c + \sum{}'' \mathbf{c}^{\mathbf{u}} B_{\mathbf{u}}(\kappa) \tag{13.5}$$

where

$\mathfrak{S}_l$ is the sum over all terms corresponding to two vertices connected only by a single g bond.

$\mathfrak{S}_c$ is the sum over all terms corresponding to simple cycles of g bonds.

$\mathbf{c}^{\mathbf{u}} B_{\mathbf{u}}(\kappa)$ is the sum over all terms corresponding to graphs in which there are u vertices assigned composition set $\mathbf{u}$ that are not g-bond nodes (but not counting those graphs for $u = 2$ that are already included in $\mathfrak{S}_l$).

The modified irreducible cluster integral $B_{\mathbf{u}}(\kappa)$ is similar to the original irreducible cluster integral $B_{\mathbf{u}}$, but it turns out to be a function of κ, the Debye parameter. Therefore, this notation is suitable to distinguish them.

Evaluation of $\mathfrak{S}_l$

The contributions to $\mathfrak{S}_l$ can come only from the $m = 2$ terms of (13.1). We have

$$\begin{aligned}\mathfrak{S}_l &= \sum_{i=1}^{\sigma}\sum_{j=1}^{\sigma}[c_ic_j/[i,j]!]\mathcal{V}^{-1}\int_{\mathcal{V}} z_iz_j\lambda^2 g_{ij}\, d\{i,j\} \\ &= \frac{1}{2}\left[\lambda\sum_{s=1}^{\sigma}c_sz_s\right]^2\int_0^R g(r)4\pi r^2\, dr \\ &= 0\end{aligned} \tag{13.6}$$

because of the condition for electrical neutrality, and because the integral is independent of the composition set ij. The factorial $[i,j]!$ is $\mathbf{u}!$ for $\mathbf{u}$ equal to either 1_i, 1_j; 2_i, 0_j; or 0_i, 2_j. Hence $[i,j]! = 2$ if $i = j$, but $= 1$ if $i \neq j$.

Evaluation of $\mathfrak{S}_c$

Every term of (13.1) contributes to $\mathfrak{S}_c$. For $m = 2$ the g-bond cycle is derived from the $p = 2$ term of equation (13.4) and the contributions to $\mathfrak{S}_c$ are of the form

$$[c_ic_j/[i,j]!][[-z_iz_j\lambda]^2/2!]\mathcal{V}^{-1}\int_{\mathcal{V}} g_{ij}g_{ji}\, d\{i,j\}$$

(cf. Fig. 2.3). But for $m > 2$ every g-bond cycle is derived from a γ_2-bond cycle *via* the $p = 1$ term of (13.4) for each γ_2 bond.

For a given m the integrand of every $B_{\mathbf{m}}$ has the same number of terms that correspond to γ_2-bond cycles, namely, the number of distinguishable permutations of m numbered vertices in a cycle. This number is $[m - 1]!/2$. Before expansion of the γ_2 bonds it is not generally true that all of these $[m - 1]!/2$ cycle terms of the integrand will contribute the same amount to the integral. As an example, in general we have

$$\int \gamma_{ab}\gamma_{bc}\gamma_{cd}\gamma_{ad}\, d\{4\} \neq \int \gamma_{ac}\gamma_{bc}\gamma_{bd}\gamma_{ad}\, d\{4\}$$

if a, b, c, and d represent different molecular species. However if we expand the γ_2 bonds, then for the leading terms, i.e., for the g-bond cycles, this inequality becomes an equality. So we see that the contribution to $\mathfrak{S}_c$ from a given $\mathbf{m}$, $m \geq 3$, is

$$[\mathbf{c}^{\mathbf{m}}/\mathbf{m}!]\mathbf{z}^{2\mathbf{m}}[-\lambda]^m[[m - 1]!/2\mathcal{V}]\int_{\mathcal{V}} g_{1,2}g_{2,3}\cdots g_{m-1,m}g_{m,1}\, d\{m\}$$

where

$$\mathbf{z}^{\mathbf{m}} \equiv z_1^{m_1} z_2^{m_2} \cdots z_\sigma^{m_\sigma} \tag{13.7}$$

and z_i is the ionic charge, not the fugacity. The factor $\mathbf{z}^{\mathbf{m}}$ appears twice because each vertex in the cycle is intersected by two γ_2 bonds in the unexpanded graph and according to (13.4) this introduces the factor z_i^2 for a vertex of composition i; z_i for each of the two bonds. We also note that the integral no longer depends on the composition set $\mathbf{m}$ but only on m.

These contributions to $\mathfrak{S}_c$ may be combined to give

$$\mathfrak{S}_c = \sum{}'' [\mathbf{c}^{\mathbf{m}}/\mathbf{m}!]\mathbf{z}^{2\mathbf{m}}[-\lambda]^m[m-1]!C_m/2$$

where we have included the contribution from $m = 2$ and where C_m is the convolution integral for a cycle of m g-bonds:

$$C_m(\mathcal{V}) \equiv \mathcal{V}^{-1} \int_{\mathcal{V}} g_{12}g_{23} \cdots g_{m-1,m}\,,\ g_{m,1}\ d\{m\} \tag{13.8}$$

The equation for $\mathfrak{S}_c$ can be further simplified by defining

$$\kappa^2 \equiv \lambda \sum_s c_s z_s^2 \tag{13.9}$$

and then since the multinominal expansion gives

$$\kappa^{2m} = \lambda^m \sum_{m_1+m_2+\cdots+m_\sigma=m} [m!/\mathbf{m}!]\mathbf{c}^{\mathbf{m}}\mathbf{z}^{2\mathbf{m}}$$

we have

$$\mathfrak{S}_c = \tfrac{1}{2} \sum{}'' [-\kappa^2]^m C_m(\mathcal{V})/m \tag{13.10}$$

This sum for $\mathfrak{S}_c$ is similar to that for $q(r,\mathcal{V})$ discussed in Section 12. It can be expressed in terms of q by the following procedure. We first write $\mathfrak{S}_c$ as an integral

$$\mathfrak{S}_c = -\int_0^\kappa \kappa \sum{}'' [-\kappa^2]^{n-1} C_n(\mathcal{V})\ d\kappa \tag{13.11}$$

and note that

$$C_n(\mathcal{V}) = \lim_{r\to 0} p_n(r,\mathcal{V}) \tag{13.12}$$

where $p_n(r,\mathcal{V})$ is the convolution integral defined in (12.32). So we have, after changing the order of limit and sum over n, (cf. equation (12.31))

$$\mathfrak{S}_c = \int_0^{\kappa} \kappa \, d\kappa \lim_{r\to 0} [g(r) - q(r,\mathfrak{V})] \tag{13.13}$$

If we use the result of Section 12,

$$q(r,\mathfrak{V}) = e^{-\kappa r}/4\pi r$$

we find

$$\lim_{r\to 0} [g(r) - q(r,\mathfrak{V})] = \kappa/4\pi \tag{13.14}$$

and

$$\mathfrak{S}_c = \kappa^3/12\pi \tag{13.15}$$

This is identical with the Debye-Hückel limiting law expression for the thermodynamic function $\mathfrak{S}$.

Classification of the Terms of $B_{\mathbf{u}}(\kappa)$

The terms of the integrand of $B_{\mathbf{u}}(\kappa)$ correspond to an infinite number of expanded graphs of g bonds and k bonds, γ_3 bonds, $\cdots$, γ_u bonds on a skeleton of $u + n$ vertices. In order to evaluate $B_{\mathbf{u}}(\kappa)$ we must classify these graphs in a much greater detail than has been done.

The basic category in the classification is the *protograph*, a specification of the number and interconnections of k bonds, γ_3 bonds, $\cdots$, and γ_u bonds, but not g bonds, on the skeleton of u vertices. In other words if one is given an expanded graph that corresponds to a term of $B_{\mathbf{u}}(\kappa)$ and deletes all of the g bonds, all of the vertices of n, and all of the labels of the vertices of u indicating chemical species, then what remains is a protograph.

It is clear that many distinguishable expanded graphs will reduce to the same protograph. Some protographs are illustrated in Fig. 13.2. It is noteworthy that a protograph may contain no bonds at all—only the u vertices. Any expanded graph in which each vertex of the skeleton of $\mathbf{u}$ intersects three or more g bonds, but no k bonds, γ_3 bonds, etc., reduces to such a protograph. The other protographs may be obtained by adding $k, \gamma_3, \gamma_4, \cdots, \gamma_u$ bonds to this simplest protograph. Proceeding in this way one finally obtains the most highly connected protograph in which every pair of vertices is connected by

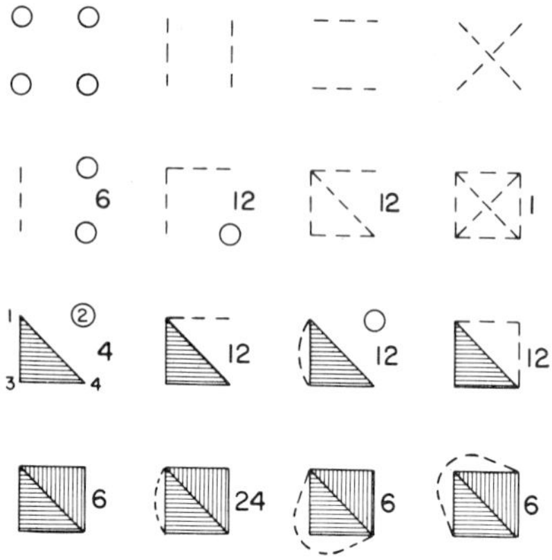

Fig. 13.2. Some protographs on a skeleton of four vertices. In the top row there are two protographs, one having one element and the other having three. In the remaining rows the number of elements in each protograph is given by the number to the right of the graph. The bonds are represented in the same way as in Fig. 13.1. Vertices which are not connected by any bonds are represented as circles. The protograph with numbered vertices is referred to in the text. (From H. L. Friedman, *Molecular Phys.* **2,** 23 (1959).)

a k bond, every three vertices by a γ_3 bond, $\cdots$, and the u vertices by a γ_u bond.

We next define the group of elements of a protograph. This comprises all of the distinguishable graphs that can be generated from a protograph by numbering its vertices from 1 to u. The elements are therefore distinguishable from each other only by the labeling of the vertices. By way of explanation we note that we might make a similar classification of the graphs that correspond to the terms of the integrand $S_{\mathbf{k}}$ of $B_{\mathbf{k}}$. We would say that $S_{\mathbf{k}}$ is obtained by first summing over all topologically distinguishable ALDC graphs on a skeleton of k unlabeled vertices, and next summing over the group of elements of each of these graphs. Now we have $\mathbf{k} = k_1, k_2, \cdots, k_\sigma$. We have already numbered the vertices of each element from 1 to k in order to perform the sum over elements. So for each element of each ALDC graph we assign species 1 to vertices 1 to k_1, species 2 to vertices $k_1 + 1$ to $k_1 + k_2$, $\cdots$, and species σ to vertices $k - k_\sigma$ to k. If we write down the graphs as we carry out the summation and species assignment then we make a complete paper and pencil representation of the terms of $S_{\mathbf{k}}$.

It is more complicated to represent the terms of $B_{\mathbf{u}}(\kappa)$, and indeed we cannot do so by finite processes alone because there is an infinite number of terms even in $B_{i,j}(\kappa)$. But if we were to begin to do so a logical first step would be to write down the finite number of protographs on the skeleton of u. The second step would be to write down the group of elements of each protograph, and the third step would be to assign species to the vertices of these graphs, just as we have done above for the S_κ graphs. A logical fourth step would be to begin to add to these graphs all possible numbers of g-bond chains connecting vertex 1 of an element of a protograph to vertex 2, or vertex 1 to vertex 3, or $\cdots$. A notation is simply developed if we first number the *pairs* of vertices of the element of the protograph $1, 2, \cdots, j, \cdots u[u-1]/2$. (This need not involve an arbitrary numbering process in addition to the arbitrary numbering of the vertices of the elements of the protograph. For instance let $j = 1$ correspond to vertices $1, 2$; $j = 2$ to vertices $1, 3$; $\cdots$; $j = u - 1$ to vertices $1, u$; $j = u$ to vertices $2, 3$; etc.) Let ν_j be the number of g-bond chains that connect one vertex of the jth pair to the other. Then the set

$$\boldsymbol{\nu} = \nu_1, \nu_2, \cdots, \nu_j, \cdots \nu_{u[u-1]/2} \tag{13.16}$$

completely specifies the number and connection of g-bond chains on the skeleton of u.

Now if we designate the protograph as τ, and the element of the protograph as τ_i then a graph that is specified by $\mathbf{u}$, τ_i, $\boldsymbol{\nu}$ corresponds to what Mayer[2] has called a *prototype*. In order to specify a particular expanded graph we must still specify the number of nodes in each g-bond chain and the composition at each of these nodes. We note, however, that these additional specifications do not change the *connectivity* of the graph on the skeleton of u if in defining the connectivity we regard each g-bond chain merely as a connection between two vertices of the skeleton. Therefore the prototype graph must be an ALDC graph on the skeleton of $\mathbf{u}$. Also for a given protograph on u the smallest $\boldsymbol{\nu}$ to give a prototype is not the set $\boldsymbol{\nu} = \mathbf{0}$ unless the protograph itself is ALDC on u.

For example, referring to the graph with numbered vertices in Fig. 13.2, and assigning j values to the pairs of vertices as suggested above (e.g., ν_1 is the number of chains connecting vertex 1 to vertex 2) we can see that the prototypes derived from this protograph conform to the following conditions

$$\nu_6 \geq 0, \qquad \nu_3 \geq 0, \qquad \nu_2 \geq 0$$

together with either

$$\nu_1\nu_5\nu_4 \geq 1,$$
$$\nu_1 = 0,\ \nu_4\nu_5 \geq 2,$$
$$\nu_5 = 0,\ \nu_1\nu_4 \geq 2,$$

or

$$\nu_4 = 0,\ \nu_1\nu_5 \geq 2$$

The fact that the smallest $\boldsymbol{\nu}$ associated with a given protograph is not in general the set $\boldsymbol{\nu} = \mathbf{0}$ is important because it introduces some troublesome complications in the sequel.

In order to specify the length and composition of each of the g-bond chains we must now number them. We use the index α for this, $1 \leq \alpha \leq \nu$. Let the number n_s^α be the number of vertices of species s in the αth g-bond chain. Then the length and composition of all of the g-bond chains is specified by the two-dimensional array

$$\langle n_s^{(\alpha)} \rangle \equiv \begin{matrix} n_1^{(1)} n_2^{(1)} & \cdots & n_\sigma^{(1)} \\ n_1^{(2)} n_2^{(2)} & \cdots & n_\sigma^{(2)} \\ \vdots \quad \vdots & & \vdots \\ n_1^{(\nu)} n_2^{(\nu)} & \cdots & n_\sigma^{(\nu)} \end{matrix} \tag{13.18}$$

It is convenient to define the *set* of elements in a row or column of this array as follows:

$$\text{row:} \qquad \mathbf{n}^{(\alpha)} \equiv n_1^{(\alpha)}, n_2^{(\alpha)}, \cdots n_\sigma^{(\alpha)} \tag{13.19}$$

$$\text{column:} \quad \mathbf{n}_s \equiv n_s^{(1)}, n_s^{(2)}, \cdots n_s^{(\nu)} \tag{13.20}$$

The composition set of the αth chain is $\mathbf{n}^{(\alpha)}$. The other set $\mathbf{n}_s$ also has the main numerical property of a composition set: It is a set of non-negative integers. The total number of nodes in the αth chain is $n^{(\alpha)}$ and the total number of vertices of species s in all ν chains is

$$n_s = \sum_\alpha n_s^{(\alpha)} \tag{13.21}$$

Finally

$$\mathbf{n} = n_1, n_2, \cdots, n_\sigma \tag{13.22}$$

is the composition set of all ν g-bond chains collectively.

It is convenient here to describe the criterion by which we shall determine whether two different arrays $\langle n_s^{(\alpha)}\rangle$ and $\langle n_s^{(\alpha)}\rangle^*$ are distinguishable. We shall use the simple criterion that the arrays are distinguishable if their written representations (13.18) are. For example

$$\begin{matrix} 0 & 0 \\ 2 & 1 \end{matrix} \quad \text{and} \quad \begin{matrix} 2 & 1 \\ 0 & 0 \end{matrix}$$

are distinguishable, one being obtained from the other by interchanging rows.

Now if we specify $\mathbf{u}$, τ_i, $\mathbf{v}$, $\langle n_s^{(\alpha)}\rangle$, and the order of species within each g-bond chain, we completely specify a particular expanded graph, which is a particular term of B_{m}, $\mathbf{m} = \mathbf{u} + \mathbf{n}$. However the integral $\int \{dm\}$ is the same for all expanded graphs which differ only in the order of species in the g-bond chains. We call all such graphs a particular *class* of graphs.

We recapitulate here the hierarchy of the present classification.

Protograph :τ

Element of a protograph: τ_i

Prototype: $\mathbf{u}$, τ_i, $\mathbf{v}$

Class of graphs: $\mathbf{u}$, τ_i, $\mathbf{v}$, $\langle n_s^{(\alpha)}\rangle$

Expanded graph: $\mathbf{u}$, τ_i, $\mathbf{v}$, $\langle n_s^{(\alpha)}\rangle$, specified order of species in each g-bond chain.

Now we may express $B_{\mathrm{u}}(\kappa)$ in terms of this classification. We write

$$\mathcal{V}B_{\mathrm{u}}(\kappa) = \sum_{\tau_i}\sum_{\nu}\sum_{\langle n_s^{(\alpha)}\rangle} \mathbf{c}^{\mathrm{n}}K(\mathbf{u}, \tau_i, \mathbf{v}, \langle n_s^{(\alpha)}\rangle)I(\mathbf{u}, \tau_i, \mathbf{v}, \langle n_s^{(\alpha)}\rangle) \tag{13.23}$$

where K is a combinatorial factor and I is the integral $\int d\{n + u\}$ on a graph of specified class. The concentration factors are now explicitly accounted for. Thus if $\mathbf{m} = \mathbf{n} + \mathbf{u}$, then $\mathbf{c}^{\mathrm{m}} = \mathbf{c}^{\mathrm{n}}\mathbf{c}^{\mathrm{u}}$ and $\mathbf{c}^{\mathrm{u}}$ appears as a coefficient of $B_{\mathrm{u}}(\kappa)$ in equation 13.5.

In (13.32) the sum over τ_i is really a double sum $\sum_\tau \sum_i$, first over all distinguishable protographs on a skeleton of u unlabeled vertices, and second over all elements of each protograph. The sum over $\mathbf{v}$ is an infinite sum but, depending on the protograph, a number of the smallest sets may be excluded as discussed above. The last sum over all distinguishable arrays $\langle n_s^{(\alpha)}\rangle$ is also an infinite sum, the first term of which corresponds to the $\sigma \times \nu$ array, $\langle 0\rangle$.

The Combinatorial Factor, $K(\mathbf{u},\tau,\mathbf{v}, \langle n_s^{(\alpha)}\rangle)$

There are three contributions to this factor. The first is $1/\mathbf{m}!$ from the definition of the cluster integral $B_{\mathbf{m}}$. The second comes from the factorials generated by the exponential expansion in equation (13.4). If we define $\bar{\nu}_j$, as the number of *direct* g bonds connecting the pair j of vertices in a given class of graphs, then this second contribution is $1/\bar{\mathbf{v}}!$. The third contribution to K is N_e, the number of distinguishable expanded graphs in a given class of graphs. This requires more elaborate consideration.

All of the expanded graphs corresponding to a given class of graphs have the same composition set

$$\mathbf{m} = \mathbf{u} + \mathbf{n}^{(1)} + \cdots + \mathbf{n}^{(\alpha)}$$

If the elements of the set $\mathbf{m}$ are all labeled (e.g., they are numbered from 1 to m) then the number of distinguishable ways of decomposing $\mathbf{m}$ into subsets of the specified composition is

$$\mathbf{m}!/\mathbf{u}!\prod_{\alpha} \mathbf{n}^{(\alpha)}! \tag{13.24}$$

Therefore this is a factor in N_e. There is also a factor $\prod_\alpha n^{(\alpha)}!$, the number of expanded graphs that differ only in the order of the labeled vertices within each g-bond chain.

The expression (13.24) corresponds to dividing the set $\mathbf{m}$ into various subsets, with a subset of composition $\mathbf{u}$ in a first box, a subset of composition $\mathbf{n}^{(1)}$ in a second box, etc. When the order of the boxes makes no difference, appropriate additional combinatorial factors must be introduced. In calculating N_e the order in which we choose, say $\mathbf{n}^{(\alpha_1)}$ and $\mathbf{n}^{(\alpha_2)}$, makes no difference if α_1 and α_2 are chains of the same j, that is, with the same endpoints on the skeleton of $\mathbf{u}$. The reason is that interchanging two such chains in an expanded graph does not make a new expanded graph.

If there are ν_j chains connecting the pair j of vertices of $\mathbf{u}$, and if the sets $\mathbf{n}^{(\alpha_1)}, \mathbf{n}^{(\alpha_2)}, \cdots$, corresponding to these chains are all nonzero, then there must be a factor $1/\nu_j!$ in N_e to correct (13.24). On the other hand if the sets corresponding to these chains are all zero no correction is needed. In general the correction factor is $\bar{\nu}_j!/\nu_j!$ where $\bar{\nu}_j$ is the number of direct bonds (zero chains) connecting j and where ν_j is the total number of chains connecting j. So we find

$$N_e = \mathbf{m}!\bar{\mathbf{v}}![\prod_{\alpha} n^{(\alpha)}!/\mathbf{n}^{(\alpha)}!]/\mathbf{u}!\mathbf{v}! \tag{13.25}$$

With the notation

$$\prod \mathbf{n}^{(\alpha)}! = \prod_{\alpha} \prod_{s} n_s^{(\alpha)}! \equiv \langle n_s^{(\alpha)} \rangle!$$

we have the final expression for the combinatorial factor,

$$K(\mathbf{u},\tau_i,\mathbf{v},\langle n_s^{(\alpha)}\rangle) = [\prod_{\alpha} n^{(\alpha)}!]/\mathbf{u}!\mathbf{v}!\langle n_s^{(\alpha)}\rangle! \tag{13.26}$$

Integration over g-Bond Chains

The integral in (13.23) is of the form

$$I(\mathbf{u}, \tau_i, \mathbf{v}, \langle n_s^{(\alpha)}\rangle) = \int_{\mathcal{V}} F_{\tau_i}(\{\mathbf{u}\}) \prod_{\alpha=1}^{\nu} \Psi(\{\mathbf{n}^{(\alpha)}\},\{\mathbf{u}_\alpha\})\, d\{\mathbf{u}+\mathbf{n}\} \tag{13.27}$$

The first factor in the integrand is a product of k, γ_3, γ_4, $\cdots$ bonds on the skeleton of $\mathbf{u}$, as specified by τ_i, the element of the protograph. The next product is over chains of g bonds, as specified by the class of graphs. The functions Ψ are themselves products, each corresponding to a chain of g bonds. Each depends on the composition and coordinates $\{\mathbf{n}^{(\alpha)}\}$ of the intermediate nodes of the chain as well as on the composition and coordinates $\{\mathbf{u}_\alpha\}$ of the two vertices of the skeleton of $\mathbf{u}$ which the chain connects.

Let r_α be the separation corresponding to $\{\mathbf{u}_\alpha\}$ and let z_α and $z_{\alpha'}$ be the charges of the species $\mathbf{u}_\alpha$. Then, recalling the abbreviation (13.7), we have

$$\int \Psi(\{\mathbf{n}^{(\alpha)}\}, \{\mathbf{u}_\alpha\})\, d\{\mathbf{n}^{(\alpha)}\} = -\lambda z_\alpha z_{\alpha'}[-\lambda]^{n(\alpha)} \mathbf{z}^{2n(\alpha)} p_\alpha(r_\alpha, \mathcal{V}) \tag{13.28}$$

where $p_\alpha(r_\alpha, \mathcal{V})$ is the convolution integral for a chain of g bonds having $n^{(\alpha)} + 1$ bonds in the chain. As discussed in Section 12, p_α depends not only on r_α but on the complete set of spatial coordinates $\{\alpha,\alpha'\}$ at finite $\mathcal{V}$, but for the theory of bulk properties no error is made by neglecting the distinction at this point. With this (13.27) becomes

$$I(\mathbf{u}, \tau_i, \mathbf{v}, \langle n_s^{(\alpha)}\rangle) = \int_{\mathcal{V}} F_{\tau_i}(\{\mathbf{u}\}) \left[\prod_{\alpha=1}^{\nu} z_\alpha z_{\alpha'}[-\lambda]^{n(\alpha)+1} \times \mathbf{z}^{2n(\alpha)} p_\alpha(r_\alpha, \mathcal{V})\right] d\{\mathbf{u}\} \tag{13.29}$$

$$= [-\lambda]^{\nu} \prod_{j} [z_j z_{j'}]^{\nu_j} \int_{\mathcal{V}} F_{\tau_i}(\{\mathbf{u}\}) \prod_{\alpha=1}^{\nu} [-\lambda]^{n(\alpha)} \mathbf{z}^{2n(\alpha)} p_\alpha\, d\{\mathbf{u}\}$$

where z_j and $z_{j'}$ are the charges of the two species associated with pair j of the vertices of the skeleton of $\mathbf{u}$.

With equations (13.26) for K and (13.29) for I we can now write the equation (13.23) for $B_{\mathrm{u}}(\kappa)$ in a more manageable form. The required manipulation is based on the identity

$$\sum_{\langle n_s^{(\alpha)}\rangle} \prod_{\alpha=1}^{\nu} f(\mathbf{n}^{(\alpha)}) = \prod_{\alpha=1}^{\nu} \sum_{\mathbf{n}^{(\alpha)}} f(\mathbf{n}^{(\alpha)}) \tag{13.30}$$

This is proved by noting that every distinguishable *ordered* product of factors

$$[f(\mathbf{n})]_1[f(\mathbf{n})]_2 \cdots [f(\mathbf{n})]_\nu$$

appears exactly once on each side of the equation. Here $\mathbf{n}$ is a composition set $n_1, n_2, \cdots, n_\sigma$. By distinguishable *ordered* products we mean that if in a given product the factors $f(\mathbf{n})$ and $f(\mathbf{n}')$ are interchanged in position then a new distinguishable ordered product results if and only if $\mathbf{n} \neq \mathbf{n}'$. It is because we count ordered products that combinatorial coefficients do not appear in this proof.

The identity (13.30) can be applied to (13.23) because both K and I have been expressed as products of functions of $\mathbf{n}^{(\alpha)}$ times another product that depends on the prototype graph. Thus with (13.26) and (13.29) we have

$$\mathcal{V}B_{\mathrm{u}}(\kappa) = [\mathbf{u}!]^{-1} \sum_{\tau_i} \sum_{\boldsymbol{\nu}} [-\lambda]^{\nu} [\prod_j [z_j z_{j'}]^{\nu_j}/\nu_j!] J(\mathbf{u}, \tau_i, \boldsymbol{\nu}) \tag{13.31}$$

$$\begin{aligned} J(\mathbf{u}, \tau_i, \boldsymbol{\nu}) \equiv \int_{\mathcal{V}} F_{\tau_i}(\mathbf{u}) [\sum_{\langle n_s^{(\alpha)}\rangle} \prod_{\alpha} \\ \cdot [n^{(\alpha)}! \mathbf{c}^{\mathbf{n}^{(\alpha)}} \mathbf{z}^{2\mathbf{n}^{(\alpha)}} [-\lambda]^{n^{(\alpha)}} / \mathbf{n}^{(\alpha)}!] p_\alpha] \, d\{\mathbf{u}_\alpha\} \end{aligned} \tag{13.32}$$

This expression for J may be rearranged with the help of (13.30). We also recall that $n^{(\alpha)}!/\mathbf{n}^{(\alpha)}!$ is the multinominal coefficient (see equation (3.20)) and the definition of κ(13.9). Then (13.32) becomes

$$\begin{aligned} J(\mathbf{u}, \tau_i, \boldsymbol{\nu}) &= \int_{\mathcal{V}} F_{\tau_i}(\mathbf{u}) \prod_{\alpha} \sum [-\kappa^2]^{n^{(\alpha)}} p_\alpha(r_\alpha, \mathcal{V}) \, d\{\mathbf{u}\} \\ &= \int_{\mathcal{V}} F_{\tau_i}(\mathbf{u}) \prod_{\alpha} q(r_\alpha, \mathcal{V}) \, d\{\mathbf{u}\} \end{aligned} \tag{13.33}$$

where, as in Section 12,

$$q(r, \mathcal{V}) \equiv \sum{}' [-\kappa^2]^{n-1} p_n(r, \mathcal{V}) \tag{13.34}$$

As shown there we have, with negligible error,

$$q(r,\mathcal{V}) = e^{-\kappa r}/4\pi r \tag{13.35}$$

the distance dependence of the Debye potential.

Now it is convenient to define a "Debye potential" for a pair of ions, a and b, as

$$q_{ab}(r_{ab}, \mathcal{V}) \equiv -\lambda z_a z_b q(r,\mathcal{V}) \tag{13.36}$$

If ions a and b are the pair j of the skeleton of $\mathbf{u}$ then we shall write q_j instead of q_{ab} and (13.31) becomes

$$\mathbf{u}!\mathcal{V}B_{\mathbf{u}}(\kappa) = \sum_{\tau_i}\sum_{\boldsymbol{\nu}}\int_{\mathcal{V}} F_{\tau_i}(\mathbf{u})[\textstyle\prod_j [q_j]^{\nu_j}/\nu_j!]\, d\{\mathbf{u}\} \tag{13.37a}$$

The κ dependence of $B_{\mathbf{u}}(\kappa)$ enters through the q bonds (equation (13.35)) and has its origin in the summation over all lengths and compositions of g-bond chains since the terms in such sums are composition-dependent.

It is interesting to note that the form of $B_{\mathbf{u}}(\kappa)$ is very similar to that of the ordinary irreducible cluster integral, $B_{\mathbf{u}}$, except that in the former multiple q bonds appear where in the latter only single γ_2 bonds are found. In fact the following graphical interpretation of the terms of $B_{\mathbf{u}}(\kappa)$ is readily deduced.

If we write $B_{\mathbf{u}}(\kappa)$ in the form

$$\mathbf{u}!\mathcal{V}B_{\mathbf{u}}(\kappa) = \int_{\mathcal{V}} S_{\mathbf{u}}(\kappa)\, d\{\mathbf{u}\} \tag{13.37b}$$

then the terms of $S_{\mathbf{u}}(\kappa)$ correspond, one-to-one, to all of the distinguishable graphs of q, k, γ_3, γ_4, $\cdots$ bonds on the skeleton of $\mathbf{u}$ that meet the following conditions:

(1) Each graph is ALDC on the skeleton of $\mathbf{u}$.

(2) Any pair of vertices is intersected by at most one k bond, any m vertices by at most one γ_m bond, but a pair of vertices may be intersected by any number of q bonds.

(3) No vertex of $\mathbf{u}$ is a q-bond node.

The last condition is established by the observation that a graph with a q-bond node in the skeleton of $\mathbf{u}$ would be derived by summing expanded graphs having a g-bond node in the skeleton of $\mathbf{u}$, which is contrary to the definition of $\mathbf{u}$.

Next we sum over terms which differ only in number of q bonds and also combine the k and q components of the pairwise bonds. This gives a form of $B_{\mathrm{u}}(\kappa)$ which is even more similar to B_{u}.

Summation over Prototypes

We note at once that q_j appears in (13.37) in a form that suggests the power series expansion of the exponential function. Thus we have

$$e^{q_j} = \sum_{\nu_j \geq 0} q_j^{\nu_j}/\nu_j!$$

and

$$\prod_j e^{q_j} = \sum_{\mathbf{v}} \prod_j q_j^{\nu_j}/\nu_j! \tag{13.38}$$

In *this* sum over $\mathbf{v}$ there is one term for every distinguishable set $\mathbf{v}$ in which the elements are non-negative integers. As has already been pointed out this is not in general true of the sum over $\mathbf{v}$ that appears in an equation for $B_{\mathrm{u}}(\kappa)$, such as (13.37), where only values of $\mathbf{v}$ are allowed that characterize graphs that are ALDC and have no q-bond nodes. It is clear that in a certain sense the excluded values of $\mathbf{v}$ are small values. In fact if we write

$$\prod_j [e^{q_j} - 1 - q_j] = \sum_{\mathbf{v}} \prod_j q_j^{\nu_j}/\nu_j!, \tag{13.39}$$

in which the sum over $\mathbf{v}$ includes all sets for which every $\nu_j \geq 2$, then the range of $\mathbf{v}$ is in general smaller than the range in (13.37). That is, if in (13.37) we sum only over all $\mathbf{v}$ for which every $\nu_j \geq 2$ then we include only ALDC graphs with no q-bond nodes, but we do not include *all* such graphs. This is true only if $u \geq 3$. The case $u = 2$ must be considered separately.

If $u = 2$ then there are only two protographs, one with a k bond and one without, and each has only a single element. Therefore (13.37) becomes

$$B_{ab}(\kappa) = [\mathcal{V}[a,b]!]^{-1} \int_{\mathcal{V}} [\sum_{\nu} q_{ab}^{\nu}/\nu! + k_{ab} \sum_{\nu} q_{ab}^{\nu}/\nu!]\, d\{a,b\} \tag{13.40}$$

The range of ν in these sums is not the same. In the first sum we only meet the conditions for ALDC graphs with no q-bond nodes if $\nu \geq 3$. In the second sum this condition is met if $\nu \geq 0$. (Recall that ALDC has a singular meaning for a skeleton of two vertices; Section 5.) Therefore we have

$$
\begin{aligned}
B_{ab}(\kappa) &= [\mathcal{V}[a,b]!]^{-1} \int_{\mathcal{V}} [[1 + k_{ab}]e^{q_{ab}} - 1 - q_{ab} - q_{ab}^2/2]\, d\{a,b\} \\
&\equiv [\mathcal{V}[a,b]!]^{-1} \int_{\mathcal{V}} \Phi_{ab}'''\, d\{a,b\}
\end{aligned}
\tag{13.41}
$$

where the last equation serves as the definition of Φ'''.

Now we return to the general case $u \geq 3$. We notice that if in an element of a protograph there is a k bond on a given pair j' then in the sum over $\mathbf{v}$ we have all terms for which $\nu_{j'} \geq 0$. All such terms together result in a bond ke^q on pair j'. If on the other hand there is no k bond on j' then in the sum over $\mathbf{v}$ the *lowest* value of $\nu_{j'}$ is dependent on the number of q bonds on the other pairs of vertices, but is certainly not larger than two. This is illustrated in Fig. 13.3. If we count only such terms as have $\nu_{j'} \geq 2$ and combine them we get a bond $e^q - 1 - q$ on pair j'. Furthermore, for every element of a protograph in which there is no k bond on j' there is just one element of some other protograph that is the same as the first element, except that it has a k bond on j'. Therefore we may add their contributions to get a graph with the bond

$$\Phi'' \equiv [1 + k]e^q - 1 - q$$

on pair j.

The same process can be carried out independently on every pair of vertices because in the specified range of ν_j every set $\mathbf{v}$ occurs. This gives a Φ'' bond on every pair of vertices of the skeleton of $\mathbf{u}$.

We next must include the terms which correspond to some $\nu_j < 2$ even when there is no k bond on j. (See Fig. 13.3). This is not easily done if we continue to use the form of equation (13.37) because for these omitted terms the value of ν_j is dependent on $\nu_{j'}$, etc. It is, however, easily managed by using the less detailed form that was originally used to define the integrand of $B_{\mathbf{u}}$. That is, we must allow for ν_j to be 0 or 1 whenever this does not imply a q-bond node or a less than doubly connected graph, and this is most easily done by simply stating this as the condition. For example, we may proceed in the following way.

The integrand $S_{\mathbf{u}}(\kappa,\{\mathbf{u}\})$ is a sum of terms corresponding, one-to-one, to all of the distinguishable graphs on the skeleton of $\mathbf{u}$ that may be formed from q bonds, Φ'' bonds, γ_3 bonds, $\cdots$, and a γ_u bond subject to the following conditions:

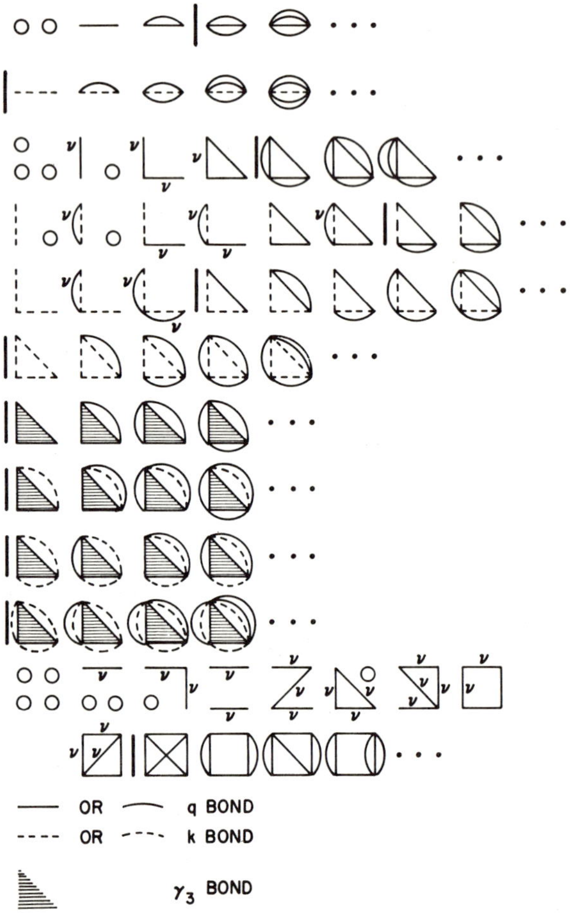

Fig. 13.3. Allowed values of $\mathbf{v}$ correspond to graphs to the right of the vertical bar in each sequence. A $\mathbf{v}$ next to a q bond indicates that the graph retains its position relative to the vertical bar for any multiplicity of this bond from 1 to ∞. (From H. L. Friedman, *Molecular Phys.* **2,** 23 (1959).)

(1) Every such graph is ALDC on $\mathbf{u}$.

(2) There are no q-bond nodes.

(3) On a given pair of vertices there may be at most one q bond or one Φ'' bond, but not both. On any m vertices there may be at most one γ_m bond.

Note that any pair of vertices that is intersected by the edge of a

γ_m bond occurs once with a Φ'' bond, once with a q bond, and once with neither, hence effectively with a Φ bond:

$$\Phi \equiv [1 + k]e^q$$

The definitions of the new pairwise bonds of the ionic solution theory which are derived from γ_2 bonds of the non-ionic theory are summarized here:

$$\begin{aligned} \Phi_{ab} &\equiv [1 + k_{ab}]e^{q_{ab}} = \exp\,[-u^*_{ab}/kT - \lambda z_a z_b e^{-\kappa r}/4\pi r] \\ \Phi'_{ab} &\equiv \Phi_{ab} - 1 \\ \Phi''_{ab} &\equiv \Phi_{ab} - 1 - q_{ab} \\ \Phi'''_{ab} &\equiv \Phi_{ab} - 1 - q_{ab} - q^2_{ab}/2 \end{aligned} \tag{13.42}$$

A more systematic way to specify the graphs that enter into the definition of the terms of $S_{\mathbf{u}}(\kappa,\{\mathbf{u}\})$ is the following.

(1) Form all distinguishable configurations of γ_3, γ_4, $\cdots$, γ_u bonds on a skeleton of u unlabeled vertices, beginning with the skeleton itself as the first such configuration.

(2) In each graph made in (1) every pair of vertices that is at least doubly connected by edges of γ_m bonds must now be connected by a Φ bond.

(3) Every pair of vertices which remains without a direct connection may be either left this way, connected by a q bond, or connected by a Φ'' bond.

Every ALDC graph on u unlabeled vertices that is formed by this sequence is called a kappagraph. The group of elements of a kappagraph is the collection of distinguishable graphs that may be formed by numbering its vertices from 1 to u. The terms of $S_{\mathbf{u}}(\kappa,\{\mathbf{u}\})$ are obtained by assigning species of the composition set $\mathbf{u}$ to the numbered vertices of the elements of the kappagraphs on u.

The final result for the expansion of $\mathfrak{S}$ for ionic solutions of finite volume is obtained by combining this definition of $S_{\mathbf{u}}(\kappa)$ with the equations

$$B_{\mathbf{u}}(\kappa) = [\mathbf{u}!\mathcal{V}]^{-1} \int_{\mathcal{V}} S_{\mathbf{u}}(\kappa,\{\mathbf{u}\})\, d\{\mathbf{u}\} \tag{13.43a}$$

and

$$\mathfrak{S} = \kappa^3/12\pi + \sum{}'' \mathbf{c}^{\mathbf{u}} B_{\mathbf{u}}(\kappa) + \sum{}'' \mathbf{c}^{\mathbf{u}} \sigma_{\mathbf{u}} \tag{13.43b}$$

where we have formally included the surface terms which have been neglected up to this point in this section.

Systems with Negligible Surface

The customary way to get statistical mechanical expressions for bulk thermodynamic properties is to consider the limit as $\mathcal{V} \to \infty$ at constant $\mathbf{c}$. In this limit, for instance, $\mathfrak{S}$ becomes independent of the surface of the system. However the behavior of the surface terms in this limit has not yet been adequately investigated in the case that the pairwise direct potentials have the form (11.7)

$$u_{ij} = Ar_{ij}^{-1} + Br_{ij}^{-4} + Cr_{ij}^{-5} + \cdots$$

By considering the simplest contribution to the surface terms for this case it is easy to see (equation (5.65)) that $\sigma_3 \to \infty$ as $\mathcal{V} \to \infty$, and it is possible that every other $\sigma_{\mathbf{n}}$ also diverges as $\mathcal{V} \to \infty$. This is in contrast to the behavior of the equations for non-ionic systems where every $\sigma_{\mathbf{n}} \to 0$ as $\mathcal{V} \to \infty$.

It is also possible that these infinities among the $\sigma_{\mathbf{n}}$ terms cancel each other and even that $\sum \mathbf{c}^{\mathbf{n}}\sigma_{\mathbf{n}} \to 0$ as $\mathcal{V} \to \infty$ for ionic systems just as for non-ionic systems. In fact if one handled the surface terms as we have handled the $B_{\mathbf{u}}$ terms in this section the chains of g bonds in the surface terms would reduce to q bonds. The evaluation of these in Section 12 is applicable and, as shown in Section 5, σ_3 and presumably the other surface terms vanish for a potential of this form.

A less general but easier procedure for handling the surface terms is to restrict the application of the theory to ionic systems in a solvent which has a non-vanishing value of the Debye parameter, κ_0. This is the case for an ordinary solvent which undergoes self-dissociation, e.g.,

$$H_2O \rightleftharpoons H^+ + OH^-$$

If the systems of interest is an ionic gas, we modify it slightly by incorporating a small concentration of an ionizing gas, $AB \rightleftharpoons A^+ + B^-$, that is not present in the original system. The pure gas AB at the same temperature and fugacity as the AB in the mixture then plays the role of the solvent.

To express the direct potentials of average force in such a solvent we invoke the Debye-Hückel limiting law result[3] that for low enough κ_0 we have

$$u_{ij} = z_i z_j \lambda e^{-\kappa_0 r}/4\pi r + u_{ij}^*$$

With *this* pairwise potential the σ_n certainly vanish as $\mathcal{V} \rightarrow \infty$. The expression for $q(r,\mathcal{V})$ in this case is readily found by any of the procedures in Section 12 to be

$$q(r,\mathcal{V}) = \exp\,(-[\kappa_0 + \kappa]r)/4\pi r$$

In the experimentally accessible concentration range this differs to a negligible extent from $e^{-\kappa r}/4\pi r$. Therefore letting $\mathcal{V} \rightarrow \infty$ we now have, with negligible error,

$$\mathfrak{S} = \kappa^3/12\pi + \sum_{n}'' \mathbf{c}^{\mathbf{n}} B_{\mathbf{n}}(\kappa) \tag{13.44}$$

where $\mathfrak{S}$ is now the true bulk thermodynamic variable and where each $B_{\mathbf{n}}(\kappa)$ is defined as the limit of (13.43a) as $\mathcal{V} \rightarrow \infty$.

All of the thermodynamic functions for ionic solutions can be deduced from (13.44). The comparison of these functions with experiment is discussed in Chapter 4.

It remains to discuss briefly some of the salient features of (13.44). In the first place at any $\kappa > 0$ the modified irreducible cluster integrals, $B_{\mathbf{n}}(\kappa)$ are all convergent, even in the limit of infinite volume. This is assured because of the effect of the $e^{-\kappa r}$ factors in the pairwise bonds. Of course Φ bonds are not short range, they approach unity at infinite separation, but in the terms of the integrand of $B_{\mathbf{n}}(\kappa)$ one finds Φ bonds only on the edges of γ_m bonds ($m \geq 3$), and a product such as $\gamma_{abc}\Phi_{ab}\Phi_{bc}\Phi_{ac}$ *is* a short-range bond because of the short range of γ_{abc}.

Equation (13.1) is entirely equivalent to the virial expansion of the pressure. On the other hand (13.44) is not equivalent to a virial expansion because in (13.44) the coefficients $B_{\mathbf{n}}(\kappa)$ are not independent of composition. In fact (13.44) is not a power series expansion at all. This observation raises the possibility that (13.44), even though correct, may not be a reasonable way to organize the contributions to $\mathfrak{S}$. One criterion of reasonableness, having to do with the way successive terms of (13.44) tend to zero as $\kappa \rightarrow 0$, is defined and investigated in detail in Section 15. Perhaps a more interesting question is whether the terms of (13.44) are at least roughly in order of decreasing numerical magnitude in the experimental concentration range. This is also much more difficult and apparently can be handled only on the basis of numerical calculations based on particular models for the ionic system. Another question is whether all of the terms together for which n is greater than some small integer, say 3 or 4, can be neglected in the experimental concentration range. Again a general answer seems to be extraordinarily difficult to obtain and probably the best one can do is to compare calculated values of $\mathfrak{S}$ based only on the first few terms with experimental values. Even in this way it is difficult to decide whether the higher terms collectively are indeed negligible, because calculated values of $\mathfrak{S}$ are found to depend strongly on the choice of the model on which the direct potentials are based. However, as discussed in Section 18, it seems that by comparison of calculated and experimental quantities for mixed electrolyte solutions we can get a little closer to an answer to this question.

One other general aspect of (13.44) is worth mentioning here. At moderate to high concentrations of ionic species the limiting law term, $\kappa^3/12\pi$, does not even have the same sign as $\mathfrak{S}$. This suggests that for applications in this concentration range it might be convenient to rearrange (13.44) by redistributing the contributions to $\kappa^3/12\pi$ among the terms of the sum over $\mathbf{u}$. In fact Meeron[4] has done just this for the terms of $u = 2$ in order to get a $B_{ik}(\kappa)$ having more nearly the form of the second virial coefficient. However the choice of whether or not to rearrange the equation in this way is just one of the details that one must attend to in planning numerical evaluations.

If one supposes that the solute includes some non-ionic species and then lets $\kappa \to 0$, equation (13.44) reduces to (13.1) for the non-ionic species, just as expected. On the other hand the essential part of Mayer's rearrangement procedure can also be applied to non-ionic systems although then there is no particular advantage in "expanding" the graphs before summing over graphs of the same topological classification. In the case of ionic solutions the Mayer rearrangement gives an enormous improvement in the convergence of the cluster expansion of $\mathfrak{S}$: The improvement is from divergence to convergence. The corresponding rearrangement may be expected to give an increase in convergence even when applied to non-ionic systems. This procedure has been investigated by Salpeter,[5] Meeron,[6, 7, 8] van Leeuwen, Groeneveld, and de Boer,[9] and Morita.[10, 11] As Meeron has also shown the rearrangement procedure can be applied repeatedly, each time collecting contributions from the least connected graphs that remain from the preceding rearrangement, until finally all convolution integrals—like our $p_n(r,\mathfrak{V})$—have been summed. A sum of terms corresponding to highly connected graphs remains unaffected by this procedure and Meeron proposes that the *convolution approxi-*

mation, in which this sum is neglected, is likely to be useful for treating systems of high particle number density, c, perhaps even to phase transitions. These procedures are applicable to multicomponent and ionic systems as well. However the derivations to date are restricted to the case in which the direct potential is a sum of only pairwise components. The higher components of the direct potential will appear only in the terms which are neglected in the convolution approximation.

Notes and References

1. This section closely follows H. L. Friedman, *Molecular Phys.* **2,** 23 (1959). It is a generalization of Mayer's[2] procedure to systems that are not restricted to pairwise-additive direct potentials.
2. J. E. Mayer, *J. Chem. Phys.* **18,** 1426 (1950).
3. J. G. Kirkwood and J. Poirier, *J. Phys. Chem.* **58,** 591 (1954).
4. E. Meeron, *J. Chem. Phys.* **28,** 630 (1958).
5. E. E. Salpeter, *Annals of Physics* **5,** 183 (1958).
6. E. Meeron, *Phys. Fluids* **1,** 139 (1958).
7. E. Meeron, *Phys. Fluids* **1,** 246 (1958).
8. E. Meeron, *J. Math. Phys.* **1,** 192 (1960).
9. J. van Leeuwen, J. Groeneveld and J. de Boer, *Physica* **25,** 792 (1959).
10. T. Morita, *Prog. Theoret. Phys.* (*Kyoto*) **20,** 920 (1958); **23,** 829 (1960).
11. T. Morita and K. Hiroike, *Progr. Theoret. Phys.* (*Kyoto*) **23,** 1003 (1960).

14. Spatial Correlation Functions for Ionic Solutions

Introduction

The procedure of Section 13 for rearranging the expansion of $\mathfrak{S}$ to make the coefficients converge for ionic systems may also be applied to the expansions of the spatial correlation functions that we derived in Section 9. In this way we would get new expansions of these functions in powers of $\mathbf{c}$, but in which the coefficients are functions of κ. It is easy to see that the main effect of the Mayer rearrangement must again be to substitute the term $e^{-\kappa r}/r$ in place of the $1/r$ term of the direct potential of each pair of ions. This direct procedure has been followed by Meeron[1] for the important case of the simplest correlation function g_{ab} in a system in which the direction potential $U_{\mathbf{n}}$ is a sum of pairwise components.

In order to get the expansion of the general correlation function $g_{\mathbf{n}}(\{\mathbf{n}\},\mathbf{c})$ that corresponds to the expansion of $\mathfrak{S}$:

$$\mathfrak{S} = \kappa^3/12\pi + \sum{}'' \mathbf{c}^{\mathbf{n}} B_{\mathbf{n}}(\kappa) \tag{14.1}$$

It appears to be simplest to start with the latter and proceed just as we did in Section 9 for non-ionic systems.

Derivation

From Section 9 and equation (14.1) we have

$$g_{\mathbf{k}} = -\frac{\mathcal{V}\mathbf{k}!}{\beta \mathbf{c}^{\mathbf{k}}} \sum{}'' \mathbf{c}^{\mathbf{n}} \frac{\partial}{\partial u_{\mathbf{k}}} B_{\mathbf{n}}(\kappa) \tag{14.2}$$

because κ is independent of $u_{\mathbf{k}}$. We consider first the case $\mathbf{k} = a, b$. The cluster integral for $n = 2$ is of a different form from the others and will be considered first. Its integrand is

$$\Phi'''_{ab} \equiv \Phi_{ab} - 1 - q_{ab} - \tfrac{1}{2}q^2_{ab}$$

We shall assume that u_{ab} is being varied by changing $\overset{*}{u}_{ab}$ rather than the Coulomb part, $z_a z_b \lambda/4\pi r$. Then the derivative of the integrand with respect to u_{ab} is effectively, as in the procedure on page 95,

$$-\beta\Phi_{ab}$$

and

$$\frac{\partial B_{ab}(\kappa)}{\partial u_{ab}} = \frac{-\beta}{[a,b]!\mathcal{V}} \Phi_{ab} \tag{14.3}$$

For $n \geq 3$ we write the integrand of $B_{\mathbf{n}}(\kappa)$ as

$$S_{\mathbf{n}}(\kappa) = [\prod_{\{a,b\}\subset\{\mathbf{n}\}} [1 + q_{ab} + \Phi''_{ab}] \prod_{\{\mathbf{m}\}\subseteq\{\mathbf{n}\}}''' [1 + \gamma_{\mathbf{m}}]]:\mathfrak{S} \tag{14.4}$$

where the notation $:\mathfrak{S}$ instructs us to retain only those terms of the product that correspond to terms of $\mathfrak{S}$ in the ionic solution theory. The derivative of the integrand with respect to u_{ab} is effectively

$$-\beta\Phi_{ab}[\prod_{\substack{\{i,j\}\subset\{\mathbf{n}\}\\ \{i,j\}\neq\{a,b\}}} [1 + q_{ij} + \Phi''_{ij}] \prod_{\{\mathbf{m}\}\subseteq\{\mathbf{n}\}}''' [1 + \gamma_{\mathbf{m}}]]:\mathfrak{S} - \Phi''_{ab} \tag{14.5}$$

where now the notation $:\mathfrak{S} - \Phi''_{ab}$ instructs us to retain those terms that correspond to terms of $\mathfrak{S}$ from which a Φ'' bond has been deleted. This sum of terms we designate $S_\kappa(a,b:\mathbf{n} - a,b)$. We have

$$\frac{\partial B_{\mathbf{n}}}{\partial u_{a,b}} = -\frac{\beta}{\mathbf{n}!\mathcal{V}} \Phi_{ab} \binom{\mathbf{n}}{a,b} \int S_\kappa(a,b:\mathbf{n} - a,b)\, d\{\mathbf{n} - a,b\} \tag{14.6}$$

and, with (14.2)

$$g_{a,b} = \Phi_{ab} \sum \mathbf{c}^{\mathbf{m}} P_\kappa(a,b\mathbf{:m}) \tag{14.7}$$

where

$$\begin{aligned} P_\kappa(a,b\mathbf{:m}) &\equiv [\mathbf{m}!]^{-1} \int S_\kappa(a,b\mathbf{:m})\, d\{\mathbf{m}\} \qquad &\text{for } m > 0 \\ &\equiv 1 &\text{for } m = 0 \end{aligned} \tag{14.8}$$

The general correlation function for an ionic system is readily obtained by the same method. We define

$S_\kappa(\mathbf{k:m})$ = The sum of all graphs on the skeleton of $\mathbf{k} + \mathbf{m}$ with no bonds intersecting only the skeleton of $\mathbf{k}$ and such that the addition of a [$\Phi_{\mathbf{k}}''$ bond if $k = 2$, $\gamma_{\mathbf{k}}$ bond if $k > 2$] makes a term of $\mathfrak{S}$ of the ionic solution theory.

This is completely analogous to $S(\mathbf{k:m})$ except that the former has q, Φ'', and Φ bonds where the latter has γ_2 bonds. No term of $S_\kappa(\mathbf{k:m})$ has any q-bond chains.

We also define the modified pairwise potential

$$-kT \ln \Phi_{a,b} = \bar{u}_{a,b} \equiv \overset{*}{u}_{a,b} + kTz_az_b\lambda e^{-\kappa r}/4\pi r \tag{14.9}$$

and corresponding to this the general modified direct potential,

$$\bar{U}_{\mathbf{n}} = \sum_{\{a,b\}\subseteq\{\mathbf{n}\}} \bar{u}_{ab}(\{a,b\}) + \sum_{\{\mathbf{m}\}\subseteq\{\mathbf{n}\}}{}''' u_{\mathbf{m}}(\{\mathbf{m}\}) \tag{14.10}$$

and then we obtain

$$g_{\mathbf{k}}(\{\mathbf{k}\},\mathbf{c}) = \exp(-\beta\bar{U}_{\mathbf{k}}) \sum_m \mathbf{c}^{\mathbf{m}} P_\kappa(\mathbf{k:m}) \tag{14.11}$$

where

$$\begin{aligned} P_\kappa(\mathbf{k:m}) &\equiv [\mathbf{m}!]^{-1} \int S_\kappa(\mathbf{k:m})\, d\{\mathbf{m}\} \qquad &\text{for } m > 0 \\ &\equiv 1 &\text{for } m = 0 \end{aligned} \tag{14.12}$$

Equation (14.11) correctly reduces to (14.7) for the case $k = 2$.

By taking the logarithm of (14.11) we obtain the potential of average force

$$\begin{aligned} W_{\mathbf{k}}(\{\mathbf{k}\},\mathbf{c}) &= -kT \ln g_{\mathbf{k}}(\{\mathbf{k}\},\mathbf{c}) \\ &= \bar{U}_{\mathbf{k}} - kT \sum_m{}' \mathbf{c}^{\mathbf{m}} Q_\kappa(\mathbf{k:m}) \end{aligned} \tag{14.13}$$

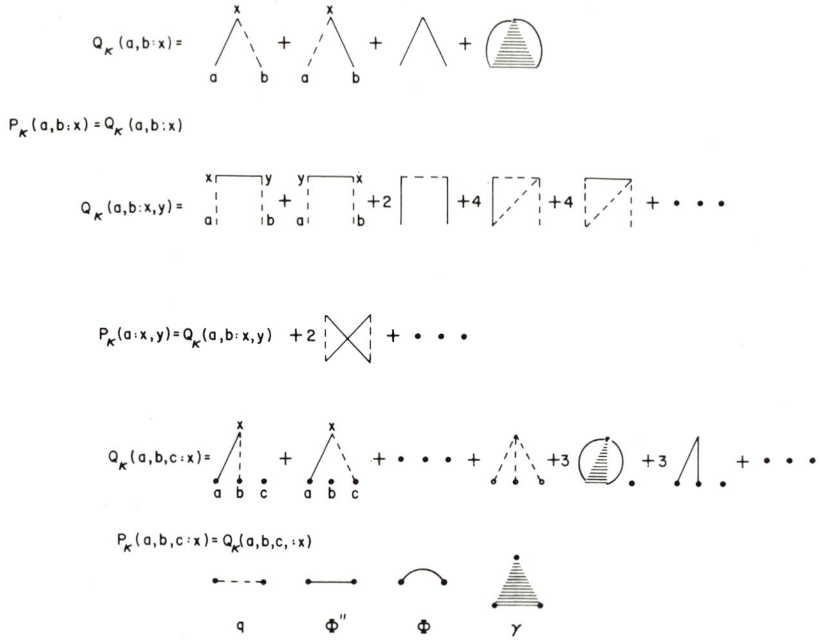

Fig. 14.1. Graphical representation of some contributions to correlation functions in ionic systems.

where the integrand of Q_κ is the same as that of P_κ except for the following additional restriction of the Q graphs:

Every pair of the vertices of the skeleton of **m** is connected by at least one path that does not pass through the skeleton of **k**.

The graphs corresponding to the integrands of the simplest P_κ and Q_κ functions are shown in Fig. 14.1.

Correction to the Debye-Hückel Potential

For point-charge ions in an ideal dielectric solvent the term u_{ab}^* vanishes and the modified potential $\bar{u}_{ab}$ of equation (14.9) is the Debye-Hückel potential

$$\epsilon^2 z_a z_b e^{-\kappa r}/Dr$$

For this model $W_{ab}(r)$ approaches the Debye-Hückel potential at low κ. It is of great interest to use (14.13) to find the next term in W_{ab} at low κ. We shall find that the next term is independent of the model but depends on whether or not μ_3 vanishes, where μ_n is the nth moment of the concentration of charge types,

$$\mu_n \equiv \sum_s c_s z_s^n$$

For the present purpose equation (14.13) may be written, for any model,

$$W_{ab} = \bar{u}_{ab} - kT \sum{}' \mathbf{c}^{\mathbf{n}} Q_\kappa^{(0)}(a,b:\mathbf{n}) - kT \sum{}' \mathbf{c}^{\mathbf{n}} Q_\kappa^{(1)}(a,b:\mathbf{n}) - \cdots \tag{14.14}$$

The coefficient $Q_\kappa^{(0)}(a,b:\mathbf{n})$ has all of the terms of $Q_\kappa(a,b:\mathbf{n})$ with the smallest number of q bonds. For n small these terms all correspond to chains of alternating q and q^2 bonds. The coefficient $Q_\kappa^{(1)}(a,b:\mathbf{n})$ has all of the terms of $Q_\kappa(a,b:\mathbf{n})$ with one more than the minimum number of q bonds.

These definitions are fairly transparent if in the integrand of $Q_\kappa(a,b:\mathbf{n})$ we write out every Φ' bond as $\Phi''' + q + q^2/2$ and every Φ'' bond as $\Phi''' + q^2/2$. Then some of the products will contain only q and q^2 bonds as factors. The terms of $Q_\kappa^{(0)}$ and $Q_\kappa^{(1)}$ are among these products. For instance we have

$$\begin{aligned}\Phi''_{ax}\Phi'_{xb} &= [\Phi'''_{ax} + q_{ax}^2/2][\Phi'''_{xb} + q_{xb} + q_{xb}^2/2] \\ &= \tfrac{1}{2}q_{ax}^2 q_{xb} + \tfrac{1}{4}q_{ax}^2 q_{xb}^2 + \cdots\end{aligned}$$

Thus $Q_\kappa^{(0)}$ and $Q_\kappa^{(1)}$ are independent of the model. Furthermore the integrals converge if $r_{ab} > 0$ and $\kappa > 0$.

From these definitions and (13.36) we readily deduce the following equations.

$$2 \sum_{n=1} \mathbf{c}^{\mathbf{n}} Q_\kappa^{(0)}(a,b:\mathbf{n}) = -[z_a + z_b] z_a z_b \mu_3 \lambda^3 \int q(r_{ax}) q^2(r_{xb})\, d\{x\} \tag{14.15}$$

$$2 \sum_{n=2} \mathbf{c}^{\mathbf{n}} Q_\kappa^{(0)}(a,b:\mathbf{n}) = z_a z_b \mu_3^2 \lambda^4 \int q(r_{ax}) q^2(r_{xy}) q(r_{yb})\, d\{x,y\} \tag{14.16}$$

$$4 \sum_{n=1} \mathbf{c}^{\mathbf{n}} Q_\kappa^{(1)}(a,b:\mathbf{n}) = z_a^2 z_b^2 \mu_4 \lambda^4 \int q^2(r_{ax}) q^2(r_{xb})\, d\{x\} \tag{14.17}$$

Thus for a solution in which $\mu_3 \neq 0$ the limiting form of the pair potential of average force is

$$W_{ab}(r)/kT = u^*_{ab}(r)/kT + z_a z_b \lambda q(r) - \frac{1}{2} z_a z_b \mu_3 \lambda^3 [\mu_3 \lambda \int q(r_{ax}) q^2(r_{xy}) q(r_{yb})\, d\{xy\} - [z_a + z_b] \int q(r_{ax}) q^2(r_{xb}) d\{x\}] + \cdots \tag{14.18}$$

while if $\mu_3 = 0$ the limiting form is

$$W_{ab}(r)/kT = u^*_{ab}(r)/kT + z_a z_b \lambda q(r) - \frac{1}{4} z_a^2 z_b^2 \mu_4 \lambda^4 \int q^2(r_{ax}) q^2(r_{xb})\, d\{x\} + \cdots \tag{14.19}$$

By the methods of Section 15 it is readily shown that in each case if κ is small enough, then the omitted terms are negligible compared to those that are written out. Under these circumstances we also have $W_{ab} \sim \bar{u}_{ab}$ and the correction to the Debye-Hückel potential is small if u^*_{ab} is also negligible.

Discussion

Equation (14.7) was first derived by Meeron for the special case in which the direct potential consists of only pairwise components.[1] He also obtained the corresponding expansion for the potential of average force and discussed some aspects of the convergence of these series. It seems that more could be done along the lines of Section 15.

The cluster expansions of $g_{\mathbf{k}}$ and $W_{\mathbf{k}}$ for ionic systems are not necessary for the calculation of thermodynamic properties‡ but nevertheless are expected to be useful in several types of problems. One is the direct comparison of the cluster theory with others, such as the extended Debye-Hückel theory, for which comparison of free energy expressions is more difficult. Another is the quantitative treatment of the problem of ion association in electrolyte solutions.

The correlation functions appear in current statistical theories of transport phenomena in ionic solutions. Although there may already be some error in using equilibrium correlation functions, there may still be some advantage in using the results of the cluster theory in

‡ If equations (9.19) to (9.26) are employed to calculate thermodynamic properties for ionic solutions then the sum over composition set $\mathbf{n}$ at $n = 2$ must be performed before the integration in each of these equations in order to avoid divergence of the integrals.

place of less accurate equilibrium correlation functions. We also note that theories of the Gouy double layer and other surface phenomena in ionic solutions can be based on cluster expansions of the correlation functions.[2]

As has been remarked before the summation procedure that is characteristic of Mayer's ionic solution theory can be used repeatedly to get series which in some sense are more rapidly convergent than the original cluster expansions, and one may finally arrive at an integral equation whose solution represents a partial sum of large classes of graphs.[3,4,5] Such procedures have yet to be applied in the ionic solution theory although it seems most likely that only by repeated summation will the cluster theory yield potentials of average force that exhibit damped spatial oscillation at high κ. This property of the $W_{\mathbf{n}}$ corresponds to a short-range order of the ions in the solution as expected at high concentrations on intuitive grounds and also on the basis of some other statistical approaches.[6, 7, 8, 9]

Notes and References

1. E. Meeron, *J. Chem. Phys.* **28,** 630 (1958).
2. S. Ono, *J. Phys. Soc. Japan* **6,** 10 (1951).
3. E. Meeron, *J. Math. Phys.* **1,** 192 (1960).
4. J. van Leeuwen, J. Groeneveld, and J. de Boer, *Physica* **25,** 792 (1959).
5. T. Morita and K. Hiroike, *Progr. Theoret. Phys.* (*Kyoto*) **23,** 1003 (1960).
6. F. H. Stillinger, Jr., J. G. Kirkwood, and P. J. Wojtowicz, *J. Chem. Phys.* **32,** 1837 (1960).
7. F. H. Stillinger, Jr., and J. G. Kirkwood, *J. Chem. Phys.* **33,** 1282 (1960).
8. J. G. Kirkwood, *Chem. Revs.* **19,** 275 (1936).
9. J. G. Kirkwood and J. C. Poirier, *J. Phys. Chem.* **58,** 591 (1954).

15. Questions of Convergence

(a) *Introduction*

The equation derived in Section 13

$$\mathfrak{S} = \kappa^3/12\pi + \sum{}'' \mathbf{c}^{\mathbf{n}} B_{\mathbf{n}}(\kappa) \tag{15.1}$$

is not a power series expansion because $B_{\mathbf{n}}(\kappa)$ depends on the composition through κ. Furthermore if the composition set $\mathbf{n}$ is a set of ions then $B_{\mathbf{n}}(\kappa)$ diverges in the limit as $\kappa \to 0$. We express this divergence as

$$B_{\mathbf{n}}(\kappa) < O(\kappa^0) \tag{15.2}$$

This notation is used here in the sense that if $g(x)$ is of order $f(x)$,

$$g(x) = O(f(x))$$

then as $x \to 0$ both $|g(x)/f(x)|$ and its reciprocal are of bounded variation. But if

$$g(x) \gtrless O(f(x))$$

then as $x \to 0$, $|g(x)/f(x)|$ approaches zero for the upper sign and infinity for the lower. For example $x^3 > O(x^2)$ and $\ln x < O(x)$. Thus the comparison of orders has the same sense as the comparison of exponents, but the opposite sense to the comparison of magnitudes.

Now a reasonable way to group the terms of (15.1) is

$$\mathfrak{S} = \kappa^3/12\pi + \sum_u \mathfrak{S}_u \tag{15.3}$$

$$\mathfrak{S}_u \equiv \sum_{n=u} \mathbf{c}^{\mathbf{n}} B_{\mathbf{n}}(\kappa) \tag{15.4}$$

As a result of (15.2) we have

$$\mathfrak{S}_u < O(c^u) \tag{15.5}$$

In this section we seek to find whether

$$\mathfrak{S}_u \leq O(\mathfrak{S}_{u'}), \qquad u < u' \tag{15.6}$$

and

$$\mathfrak{S}_u > O(\kappa^3) = O(c^{3/2}), \qquad u \geq 2 \tag{15.7}$$

We expect equation (15.6) to be satisfied for any physically reasonable grouping of the terms of the cluster expansion of the excess free energy. Unless equation (15.7) is satisfied, the theory is not consistent with experiment, for the $\kappa^3/12\pi$ term of equation (15.1) is the Debye-Hückel limiting law for $\mathfrak{S}$. These problems were investigated by Mayer[1] to find all of the terms of (15.1) whose contribution to $\mathfrak{S}$ was of order not greater than $c^{5/2}$. Haga[2] used a different mathematical approach to the same problem and obtained slightly different results, while Friedman[3] determined the order of all of the terms that correspond to the more loosely connected graphs and searched for certain cancellations among the most divergent terms. These cancellations are based on stoichiometry conditions, of which electroneutrality is a special case. This section closely follows the last of these papers.

The determination of the singularities that result from those terms of the integrals which correspond to highly connected graphs of q and Φ'' bonds proves to be difficult, but while we do not yet have a "rigorous and complete proof"[4] of the validity of equations (15.4) and (15.5), the following analysis does appear to make it "transparent and obvious"[5] that the theory passes these tests. Although the convergence questions dealt with here are rather academic in significance compared to the question of the numerical convergence of the series at any finite concentration, some of the results and methods of this section can be used to answer certain questions that arise in connection with numerical computations with the theory.

The part of the analysis which is common to all of the integrals is given in Section 15(b). The exact analysis of several terms of one class of singular integrals is given in Section 15(c) and the generality of these results is discussed in Section 15(d). The application of the same methods to the other class of singular integrals is briefly outlined in Section 15(e). The results are collected and discussed in Section 15(f).

(b) *General Analysis*

For given $n = u$ the various $B_{\mathbf{n}}(\kappa)$ differ in the factorial $\mathbf{n}!$, included in the definition of $B_{\mathbf{n}}(\kappa)$, and in the assignment of species to the vertices of the graphs corresponding to the integrand of $B_{\mathbf{n}}(\kappa)$. However in other respects, such as number and connectivity of terms, the integrands of these cluster integrals for given n are the same. To use this fact to facilitate our efforts to find the order of $\mathfrak{S}_u$ we write

$$B_{\mathbf{n}}(\kappa) = [\mathbf{n}!\mathfrak{V}]^{-1}A_{\mathbf{n}}I^{(n)} \tag{15.8}$$

where $I^{(n)}$ is the irreducible cluster integral for a set of n molecules, but without a specific assignment of species. The *assignment operator* $A_{\mathbf{n}}$ assigns species set $\mathbf{n}$ to the n vertices of the various graphs that correspond to the terms of the integrand. As discussed in Section 13 (in the paragraphs following equation (13.15)) the vertices can each be assigned a species in an arbitrary way, consistent only with the over-all composition set $\mathbf{n}$, because the vertices are all equivalent in a group of elements of a protograph. Now (15.4) may be written

$$\mathfrak{S}_u = \Big[\sum_{n=u} \mathbf{c}^{\mathbf{n}}A_{\mathbf{n}}/\mathbf{n}!\mathfrak{V}\Big]\, I^{(u)} \tag{15.9}$$

and-we may focus our attention on those mathematical properties of $I^{(u)}$ that do not depend on the assignment of a composition set.

We choose a length, L, that is great enough so that to a satisfactory approximation we have

$$\gamma_{abc\cdots}(r_{ab}, r_{bc}, \cdots, L, \cdots) = 0 \tag{15.10}$$

$$\Phi_{ab}(L) = 1 \tag{15.11}$$

$$\Phi''_{ab}(L) = q^2_{ab}(L)/2! + q^3_{ab}(L)/3! + \cdots, \tag{15.12}$$

$$\Phi'''_{ab}(L) = q^3_{ab}(L)/3! + q^4_{ab}(L)/4! + \cdots \tag{15.13}$$

where $q_{ab}(r) \equiv - z_a z_b \lambda q(r) = -z_a z_b \lambda e^{-\kappa r}/4\pi r$. Then we divide the configuration space of the integration of $I^{(u)}$ into three parts as follows:

(A) All intermolecular distances less than L.
(B) Some intermolecular distances less than L and some greater.
(C) All intermolecular distances greater than L.

Dividing the integral into terms corresponding to integration over the three parts of configuration space

$$I^{(u)} = I_A^{(u)} + I_B^{(u)} + I_C^{(u)} \tag{15.14}$$

we have at once

$$I_A^{(u)} = O(\kappa^0) \tag{15.15}$$

because this is the integral of a bounded function over a finite part of the configuration space, even when $\kappa = 0$.

On the other hand, as will be shown below, $I_C^{(u)}$ is singular at $\kappa = 0$ because of the long range of the q bonds. It is expected that

$$I_C^{(u)} \leq O(I_B^{(u)}) \tag{15.16}$$

because the part of the configuration space that leads to the divergence of $I_B^{(u)}$ is equivalent to a part of the range of integration of $I_C^{(u)}$, namely, that part in which some of the intermolecular distances are held fixed near L. On the basis of equation (15.16) we need only analyse $I_C^{(u)}$ to find the strongest singularity of $I^{(u)}$. However, we will see that in some circumstances the contribution to $\mathfrak{S}$ of the most singular terms in $I_C^{(u)}$ vanishes because of a sort of generalized electroneutrality condition and then we must turn to $I_B^{(u)}$ to find the most singular term remaining.

We will first consider the $I_C^{(u)}$ integrals in detail and will return to the $I_B^{(u)}$ integrals in Section 15(e).

The only kappagraphs with non-vanishing contributions to $I_C^{(u)}$ are those with no γ_n bonds. We now work with expansions of such kappagraphs, based upon equations (15.12) and (15.13). The lowest term of such an expansion will be that in which each Φ'' bond is replaced by a $q^2/2!$ bond, i.e., a double q bond. In the following terms all of the Φ'' bonds are replaced by either double q bonds, triple q bonds, etc., or by any combination of these. Because of the fact that $q(L) \ll 1$, the terms of the expansion of a protograph with the fewest q bonds make the largest contribution to the integral of that protograph over C. However, as will be shown, the contributions of such terms to $\mathfrak{S}$ may vanish for solutions of certain compositions, and then we may wish to consider terms with more q bonds which make a non-vanishing contribution to $\mathfrak{S}$.

The terms from the expansion of the kappagraphs have the essential properties of the kappagraphs themselves and they will be designated as τ', τ'', etc. Thus, the expansion of a kappagraph may be represented by the equation

$$\tau = \tau' + \tau'' + \cdots \tag{15.17}$$

The contribution to $\mathfrak{S}_u$ of all of the terms in $I_C^{(u)}$ which correspond to the *elements* of a single kappagraph, τ', is

$$\mathfrak{S}_u(\tau', C) = [\sum_{n=u} \mathbf{c}^{\mathbf{n}} A_{\mathbf{n}}/\mathbf{n}!] \int \sum_{\tau'_i} \prod_j [q_j^{\nu_j}/\nu_j!]\, d\{u\}' \tag{15.18}$$

where the summation over τ'_i is over the elements of the term of the kappagraph and ν_j is the number of q bonds connecting the *pair* j of vertices of the skeleton of u, and the product is over the $u(u-1)/2$ pairs of vertices of the skeleton. The differential $d\{u\}'$ represents variation of all but one of the elements of $\{u\}$; that is, we integrate over the coordinates of all but one of the molecules. The bonds q_j are not completely defined until the $A_{\mathbf{n}}$ operator acts. However, the q_j bonds depend upon the composition of the vertices only by multiplicative factors and, therefore, equation (15.18) can be written in another way in which the $A_{\mathbf{n}}$ operation is already performed.

First we define $J_u(\tau'_i)$ as the integral of equation (15.18) less the coordinate-independent coefficients:

$$J_u(\tau_i') = \int_C \prod_j [\exp(-\kappa r_j)/4\pi r_j]^{\nu_j}\, d\{u\}' \tag{15.19}$$

This is the same for all of the elements of a given term of a kappa-graph and therefore the designation, i, of the element may be eliminated and (15.18) becomes

$$\mathfrak{S}_u(\tau',C) = J_u(\tau')[-\lambda]^t \Big[\sum_{n=u} \mathbf{c}^{\mathbf{n}} A_{\mathbf{n}}/\mathbf{n}!\Big] \sum_{\tau_i'} \prod [z_j z_{j'}]^{\nu_j}/\nu_j! \tag{15.20}$$

where z_j and $z_{j'}$ represent the charges that are assigned to the two vertices of the pair j by the $A_{\mathbf{n}}$ operation and t is the number of bonds in τ'.

We note at once that for every element of the term of the kappa-graph the product $\prod \nu_j! = \boldsymbol{\nu}!$ is the same. Considerable further simplification of the product of sums appearing here can be achieved in the following way. The sum over $\mathbf{n}$ can evidently be written as

$$\frac{1}{u!} \sum_{\substack{\text{ordered}\\ \text{products}}} c_1 A_1 c_2 A_2 \cdots c_n A_n$$

where we sum over all distinguishable *ordered* products that are consistent with $n = u$ and with the composition of the system, i.e., with $s = 1, 2, \cdots, \sigma$. For example if $u = 2$ and $\sigma = 2$ (species a and b),

$$2 \sum_{n=2} \mathbf{c}^{\mathbf{n}} A_{\mathbf{n}}/\mathbf{n}! = c_a A_a c_a A_a + c_a A_a c_b A_b + c_b A_b c_a A_a + c_b A_b c_b A_b \tag{15.21}$$

To proceed we must also write the sum over τ_i' as a sum over ordered products of n factors. For the transformation we note that a term of a protograph, such as τ', is characterized by a set, $\mathbf{h}$

$$\mathbf{h} = h^{(3)}, h^{(4)}, \cdots, h^{(w)}, \cdots \tag{15.22}$$

where $h^{(w)}$ is the number of vertices of τ' that have degree w. The degree of a vertex is the number of q bonds that intersect it. Then the sum over τ_i' can be written as

$$[T(\tau')/\boldsymbol{\nu}!] \sum_{\substack{\text{ordered}\\ \text{products}}} z_1^{w_1} z_2^{w_2} \cdots z_n^{w_n}$$

where $T(\tau')$ is an integer and where the sum is over all distinguishable ordered products of n factors that are consistent with the set $\mathbf{h}$. The subscripts here are merely the order numbers of the factors and are ignored in determining distinguishability. For example for the term of a kappagraph,

the sum over τ_i' is

$$\tfrac{1}{4}[z_1^3 z_2^4 z_3^3 + z_1^4 z_2^3 z_3^3 + z_1^3 z_2^3 z_3^4]$$

with $T = 1$, while for the term of a kappagraph

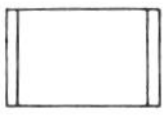

the sum over τ_i' is

$$\tfrac{6}{4}[z_1^3 z_2^3 z_3^3 z_4^3]$$

with $T = 6$ since in this case the six elements of the term of the kappagraph all have the same product of powers of z_i.

When the sum over $\mathbf{n}$ and the sum over τ_i', each expressed as a sum over ordered products, are multiplied we get a new sum over ordered products

$$[T(\tau')/u!\,\boldsymbol{\nu}!] \sum_{\substack{\text{ordered}\\ \text{products}}} c_1 z_1^{w_1} c_2 z_2^{w_2} \cdots c_n z_n^{w_n}$$

where all distinguishable ordered products appear that are consistent with the σ species in the system and with the set $\mathbf{h}$. The subscripts here are species designations and are to be taken into account in determining distinguishability. This sum of ordered products can as well be generated by

$$\boldsymbol{\mu}^{\mathbf{h}} = \prod_w \mu_w^{h_w} \tag{15.23}$$

where μ_w is the wth moment of the concentration of charge types,

$$\mu_w \equiv \sum_{s=1}^{\sigma} c_s z_s^w \tag{15.24}$$

So we obtain the compact form of (15.20),

$$\mathfrak{S}_u(\tau', C) = J_u(\tau')[-\lambda]^t T(\tau')\boldsymbol{\mu}^{\mathbf{h}}/u!\boldsymbol{\nu}! \tag{15.25}$$

In this section we are not concerned with factors that are independent of composition and concentration of the solute and it is often convenient to omit these factors. The proportionality sign will be used to indicate that this has been done. Thus equation (15.25) becomes

$$\mathfrak{S}_u(\tau',C) \propto \boldsymbol{\mu}^{\mathbf{h}} J_u(\tau') \tag{15.26}$$

As has already been pointed out, one may expect the most singular $J_u(\tau')$ to be derived from the τ' with the fewest possible bonds. This implies that in such a graph all, or all but one, of the vertices will be of the lowest possible degree, namely $w = 3$. But it is clear from equation (15.26) that the contribution to $\mathfrak{S}_u$ from such a term vanishes if $\mu_3 = 0$, as it does in solutions of electrolytes of symmetrical charge type. In this case the most singular integral which makes a non-vanishing contribution to $\mathfrak{S}_u$ must correspond to a term of the protograph in which $w \geq 4$ for every vertex.

We now proceed with the calculation of the singularities of various $J_u(\tau')$, seeking for each u both the most singular integral and the most singular integral whose contribution to $\mathfrak{S}_u$ does not vanish when $\mu_3 = 0$.

(c) *Calculation of Singularities*

The procedures used in this section for $u = 2$ have already been used by Mayer[1] and for $u = 3$ and 4 by Haga.[2] However, it seems worth while to repeat them here to show the application of equation (15.26) and as an introduction to the treatment of the J integrals in the general problem.

The notation of the Bateman manuscripts[6, 7] is employed for the higher transcendental functions. This source may also be referred to for most of the integrals in this section.

$u = 2$: The only kappagraph is a Φ''' bond connecting the two vertices. The lowest term in this kappagraph, with a triple bond, is designated α. Then

$$J_2(\alpha) = [[4\pi]^2]^{-1}\int_L^{\infty} [\exp(-3r\kappa)/r^3]r^2\,dr \propto E_1(3\kappa L) = O(\ln \kappa) \tag{15.27}$$

and

$$\mathfrak{S}_2(\alpha,C) \propto \mu_3^2 E_1(3\kappa L) = O(c^2 \ln c). \tag{15.28}$$

where $E_1(x)$ is the exponential integral,

$$E_1(x) = \int_1^\infty e^{-tx} t^{-1}\, dt$$

If $\mu_3 = 0$, the contribution from α vanishes and then we consider the next term of the kappagraph, β, in which there is a quadruple bond

$$J_2(\beta) = [4\pi]^{-3} \int_L^\infty [\exp(-4\kappa r)/r^2]\, dr = O(\kappa^0). \quad (15.29)$$

That is, this integral is convergent even if $\kappa = 0$. Then we have

$$\mathfrak{S}_2(\beta, C) \propto \mu_4^2 = O(c^2). \quad (15.30)$$

This illustrates the general observation that the contributions of the most singular integrals vanish if $\mu_3 = 0$.

$u = 3$: The kappagraphs which contribute to $I_C^{(3)}$ are shown in Fig. 15.1 (a) and (b). The corresponding integrals are most readily analyzed by a convolution method, but for this purpose it is best to redefine the bonds so that the integrals extend over the entire configuration space.

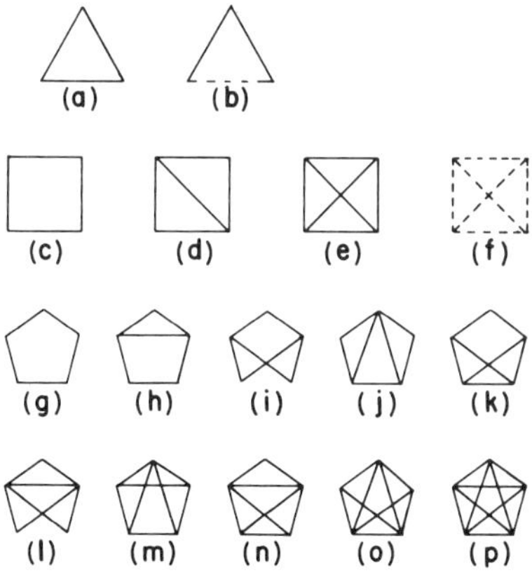

Fig. 15.1. Representations of some kappagraphs which contribute to $I_C^{(u)}$ for $u = 3, 4$, and 5. Except for (b) and (f) only kappagraphs with only Φ'' bonds (solid lines) are shown. The remaining kappagraphs for $u = 4$ and 5 can be constructed from these by substituting q bonds for Φ'' bonds (broken lines in this figure), subject to the restriction that there be no q-bond nodes. This is illustrated by the relation of (b) to (a) and (f) to (e). (From H. L. Friedman, *Molecular Phys.* **2**, 190, 436 (1959).)

We define $f(L,R)$ to be 0 if $L > R$ and to be $\exp(-\kappa R)/4\pi R$ if $R > L$. The integral corresponding to the lowest term of the protograph of Fig. 15.1(b) is then

$$J_3(\alpha) = \int f_{1,2}f_{2,3}^2 f_{1,3}^2 \, d\{1,2,3\} \tag{15.31}$$

where the integration is over the finite range of coordinates of two of the vertices, the third being chosen as the origin.

The Fourier transform of a single bond is (cf. Section 12)

$$\begin{aligned} \tilde{f}_1(t) &= 4\pi \int_0^\infty f(L,r)[r/t] \sin(rt)\, dr \\ &= [\kappa^2 + t^2]^{-1} - [1 - \cos(Lt)]/t^2 + \kappa t^{-3}[\sin(Lt) \\ &\quad - Lt\cos(Lt)] + \cdots \\ &\cong [\kappa^2 + t^2]^{-1} \end{aligned} \tag{15.32}$$

where the omitted terms are found to be without effect in determining the singularities of integrals. The Fourier transform of a double bond is

$$\begin{aligned} \tilde{f}_2(t) &= 4\pi \int_0^\infty f^2[r/t] \sin(rt)\, dr \\ &\cong [4\pi t]^{-1}[\tan^{-1}(t/2\kappa) - \mathrm{Si}(Lt)], \end{aligned} \tag{15.33}$$

where again we have omitted terms which are without effect in the present calculations. Indeed, Si (Lt) can also be omitted in most cases.

The transcendental function Si (x) is defined by[6]

$$\mathrm{Si}(x) = \int_0^x x^{-1} \sin x \, dx$$

and $\tan^{-1} x$ is the arctangent of x.

The convolution theorem is now applied to equation (15.31) to obtain

$$\begin{aligned} J_3(\alpha) &= [2\pi^2]^{-1} \int_0^\infty \tilde{f}_1 \tilde{f}_2^2 t^2 \, dt \\ &\propto \int_0^\infty [\tan^{-1}(t/2\kappa) - \mathrm{Si}(Lt)]^2 [t^2 + \kappa^2]^{-1} \, dt. \end{aligned} \tag{15.34}$$

This integral converges even at $L = 0$ if $\kappa > 0$, and therefore the

Si (Lt) term is without effect on the singularity at $\kappa = 0$. Omitting this term and making the substitution, $x = t/\kappa$, we have

$$J_3(\alpha) \propto \kappa^{-1} \int_0^\infty [\tan^{-1}(x/2)]^2[1 + x^2]^{-1}\, dx \propto \kappa^{-1} \quad (15.35)$$

because the remaining integral is convergent and independent of κ. The contribution to $\mathfrak{S}$ is

$$\mathfrak{S}_3(\alpha,C) \propto \mu_3^2\mu_4\kappa^{-1} = O(c^{5/2}), \quad (15.36)$$

which is the result obtained by Haga.[2] If $\mu_3 = 0$ this contribution vanishes and we turn to the lowest term of the kappagraph of Figure 15.1(a). This gives

$$\begin{aligned} J_3(\beta) &= \int f_{1,2}^2 f_{1,3}^2 f_{2,3}^2\, d\,\{1,2,3\} = [2\pi^2]^{-1} \int_0^\infty \tilde{f}_2^3 t^2\, dt \\ &\propto \int_0^\infty [\tan^{-1}(t/2\kappa) - \mathrm{Si}\,(Lt)]^3 t^{-1}\, dt \end{aligned} \quad (15.37)$$

This integral diverges at $L = 0$ even for $\kappa > 0$ and therefore Si (Lt) must be retained. An elementary analysis[3] shows that $J_3(\beta) = O(\ln \kappa)$. The contribution of this to $\mathfrak{S}_3$ is

$$\mathfrak{S}_3(\beta,C) \propto \mu_4^3 O(\ln \kappa) = O(c^3 \ln c) \quad (15.38)$$

$u > 3$, cycles: We first consider the terms of J_u which correspond to cycle graphs, in which each vertex is connected by bonds to exactly two other vertices. Examples of such graphs are the terms of Fig. 15.1 (c) and (g) and the terms of the kappagraphs that are related to these by replacing some of the Φ'' bonds by q bonds. The cycle-graph term of J_u with the fewest possible bonds, $J_u(\alpha)$, must have $\frac{1}{2}(u - \delta)$ single bonds and $\frac{1}{2}(u + \delta)$ double bonds where $\delta = 0$ if u is even, $\delta = 1$ if u is odd. Applying the convolution method to this integral we have

$$J_u(\alpha) = (2\pi^2)^{-1} \int_0^\infty \tilde{f}_1^{(1/2)(u-\delta)} \tilde{f}_2^{(1/2)(u+\delta)} t^2\, dt \quad (15.39)$$

and after noting that the Si (Lt) term may be omitted and then substituting $x = t/\kappa$, we obtain

$$\begin{aligned} J_u(\alpha) \propto \kappa^{3+(1/2)(\delta-3u)} &\int_0^\infty [\tan^{-1}(x/2)]^{(1/2)(u+\delta)} \\ &\cdot [1 + x^2]^{(1/2)(\delta-u)} x^{(1/2)(4-u-\delta)}\, dx \end{aligned} \quad (15.40)$$

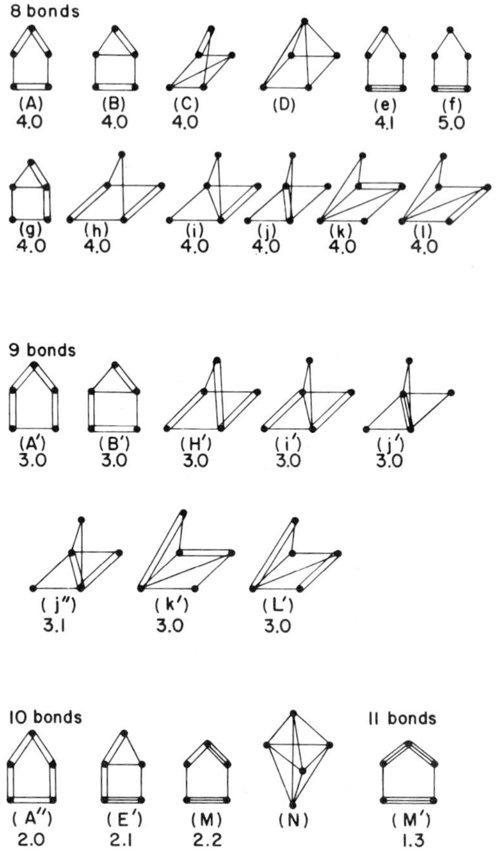

Fig. 15.2. Graphs related to J_5. The graphs designated by capital letters are among the terms of J_5. The others are included for comparison. The numbers indicate the singularity of the integral on the graph according to the following code: $O(\kappa^{-p}[\log \kappa]^q) = p,q$. The graphs designated by primed letters may be formed by adding one or two bonds to the graphs designated by the corresponding unprimed letters. Note that the singularity is never found to increase when bonds are added. The calculation of the singularity of the integral on (C) is based on Haga's calculation for the graph of figure 1(f) (E. Haga, *J. Phys. Soc., Japan* **8**, 714 (1953)). From H. L. Friedman, *Molecular Phys.* **2**, 190, 436 (1959).)

The integral is convergent and is not a function of κ so it may be omitted. Then we have

$$\mathfrak{S}_u(\alpha,C) \propto \mu_3^{u-\delta}\mu_4^{\delta}\kappa^{(1/2)(6+\delta-3u)} = O(c^{(1/4)(6+u+\delta)}) \qquad (15.41)$$

which vanishes if $\mu_3 = 0$. In this case we consider $J_u(\beta)$, the integral corresponding to the cycle with all double bonds, and $J_u(\gamma)$, the integral corresponding to the cycle with alternating single and triple bonds if u is even, and with one extra triple bond if u is odd. The usual method yields for $J_u(\beta)$

$$J_u(\beta) = [2\pi^2]^{-1} \int_0^\infty \tilde{f}_2^u t^2 \, dt \propto \kappa^{(3-u)} \tag{15.42}$$

For the other integral we need the transform of a triple bond:

$$\tilde{f}_3(t) = 4\pi \int_0^\infty f^3[r/t] \sin(rt) \, dr \tag{15.43}$$

The evaluation of $\tilde{f}_3$ and the calculation of the singularities of integrals in which $\tilde{f}_3$ appears are discussed in reference 3. The result of interest here is

$$J_u(\gamma) = [2\pi^2]^{-1} \int_0^\infty \tilde{f}_3^{(u+\delta)/2} \tilde{f}_1^{(u-\delta)/2} t^2 \, dt = O(\kappa^{3-u+\delta}[\ln \kappa]^{(u+\delta)/2}) \tag{15.44}$$

If u is even, then β and γ have the same number of bonds and $J_u(\gamma)$ is more singular than $J_u(\beta)$. But if u is odd, then β has one less bond than γ and we find that $J_u(\beta)$ is the more singular. As a result we have for the most singular contribution to $\mathfrak{S}_u$ from cycle integrals when $\mu_3 = 0$:

$$\mathfrak{S}_u(\gamma) \propto \mu_4^u O(\kappa^{3-u}[\ln \kappa]^{u/2}) = O(c^{(u-3)/2}[\log c]^{u/2}), \quad u \text{ even} \tag{15.45}$$

$$\mathfrak{S}_u(\beta) \propto \mu_4^u O(\kappa^{3-u}) = O(c^{(u-3)/2}), \quad u \text{ odd} \tag{15.46}$$

$u > 4$, bicycles: There are many more complex graphs than cycles among the terms of J_u which have a low enough connectivity to be treated readily by convolution methods. Some other examples are illustrated in Fig. 15.2. We discuss here the general case, $u > 4$, for one class of such graphs, namely, those that may be obtained by rearranging the bonds in a cycle of type α to form a bicycle in which the two cycles have in common two vertices connected by a single bond and in which $h^{(3)}$ and $h^{(4)}$ are the same as in the cycle α. An example is the rearrangement Fig. 15.2(A) to Fig. 15.2(B). The integral on a general bicycle of this kind is

$$J_u(\Delta) = \int_L^\infty f(L,r)\tilde{f}_a(L,r)\tilde{f}_b(L,r) 4\pi r^2 \, dr, \tag{15.47}$$

where $\bar{f}_a$ and $\bar{f}_b$ are effective bond functions for the two chains of the bicycle, obtained by integrating over the coordinates of the intermediate vertices in each chain. The numbers of interior vertices in the chains are a and b, respectively, and we have

$$u = a + b + 2 \tag{15.48}$$

The simplest possible such chain has one vertex of the third degree, and its effective bond function is,

$$\begin{aligned} \bar{f}_1(L,R) &= [2\pi^2]^{-1} \int_0^\infty \bar{f}_1 \bar{f}_2[t/r] \sin(rt)\, dt \\ &= [\exp(-\kappa r)/64\pi^2\kappa r]\,[\ln(3) - E_1(\kappa r) \\ &\qquad + \exp(2\kappa r) E_1(3\kappa r) + \cdots] \end{aligned} \tag{15.49}$$

The second factor in brackets is $O(\kappa r \ln \kappa r)$ and must be handled with care in calculations. The effective bond functions for longer chains are more difficult to evaluate but we find that a chain with an odd number, $2n - 1 > 1$, of interior vertices, all third degree, has the effective bond function

$$\bar{f}_{2n-1} = [2\pi^2]^{-1} \int_0^\infty \bar{f}_1^n \bar{f}_2^n [t/r] \sin(rt)\, dt \propto \exp(-\kappa r)\kappa^{3-3n} \tag{15.50}$$

where, in the last member, terms of higher order have been neglected. A chain with an even number, $2n$, of third degree interior vertices has the effective bond function

$$\bar{f}_{2n} = [2\pi^2]^{-1} \int_0^\infty \bar{f}_1^n \bar{f}_2^{n+1} [t/r] \sin(rt)\, dt \propto \exp(-\kappa r)\kappa^{1-3n} \tag{15.51}$$

where, again, terms of higher order are neglected in the last member. If u is even, all of the vertices in Δ are third degree and therefore both a and b are even. In this case

$$J_u(\Delta) \propto \kappa^{-3(a+b)/2} = \kappa^{3-3u/2}, \qquad u \text{ even} \tag{15.52}$$

If u is odd there must be one fourth degree vertex in Δ and therefore either a or b must be odd. Then we have

$$J_u(\Delta) \propto \kappa^{2-3(a+1+b)/2} = \kappa^{(7-3u)/2}, \qquad u \text{ odd} \tag{15.53}$$

These are exactly the same orders as for the $J_u(\alpha)$, and therefore the rearrangement of bonds, $\alpha \to \Delta$, has no effect upon the singularity of the integral.

(d) *General Graphs in J_u, $u > 3$*

A complete solution to this problem has not yet been found, but a partial analysis leads to the tentative conclusion[3] that there are no terms in J_u which are more singular than the terms corresponding to cycle graphs.

(e) *The Integrals $I_B^{(u)}$*

Some of the terms of $I_B^{(u)}$ correspond to kappagraphs that are cycles of Φ'' bonds and q bonds. It is plausible that among these are some terms that are at least as singular as any other integrals in $I_B^{(u)}$, just as it is plausible that some of the terms of the same cycles are at least as singular as any other integrals in $I_C^{(u)}$.

We will call bonds with a maximum length of L, short bonds, and bonds with a minimum length of L, long bonds, Now we expand (equation (15.12)) the long Φ'' bonds of the cycles consisting of Φ'' bonds and q bonds and then, using the methods of Section 15(c), we may easily show that those terms of kappagraphs that we designate, α', are at least as singular as any other terms derived from cycles in $I_B^{(u)}$. The α' terms of $I_B^{(u)}$ are those in which there is one short Φ'' bond whose ends are connected by a chain of alternating long single and double q bonds, and in which at least one of the terminal bonds of this chain is a single bond. The integral of such a term is designated $K_u(\alpha')$. We have

$$K_u(\alpha') \propto \int_0^L \Phi''_{ik}(r)\bar{f}_{u-2}(r)r^2\,dr \tag{15.54}$$

where i and k are the vertices at the ends of the short bond and Φ''_{ik} depends on the species occupying these vertices, and where $\bar{f}$ is the effective bond function of a chain of alternate single and double q bonds (equations (15.50) and (15.51)). The approximations previously obtained for $\bar{f}$ lead to the equations

$$\begin{aligned} K_3(\alpha') &\propto \int_0^L \Phi''_{ik}(r)\exp(-\kappa r)\ln(\kappa r)r^2\,dr \\ &= \int_0^L \Phi''_{ik}\exp(-\kappa r)\ln(r)r^2\,dr + [\ln\kappa]\int_0^L \Phi''_{ik}\exp(-\kappa r)r^2\,dr \end{aligned} \tag{15.55}$$

$$K_u(\alpha') \propto \kappa^{4+(\delta-3u)/2}\int_0^L \Phi''_{ik}(r)\exp(-\kappa r)r^2\,dr, \qquad u = 4,5,6,\cdots \tag{15.56}$$

The remaining integral in each case is evidently not singular at $\kappa = 0$. On the other hand, since these integrals do depend upon the short-range interactions of the ions at i and k, it is not possible to have a cancellation of the $K_u(\alpha')$ integrals like that found for the $J_u(\alpha)$ integrals for the case, $\mu_3 = 0$.

(f) *Conclusions*

If we are correct in assuming that the cycles of Φ'' bonds and q bonds are among the most singular terms in $I_C^{(u)}$ and $I_B^{(u)}$, then the cluster theory of ionic solutions conforms to the convergence conditions, equations (15.6) and (15.7). The results for small u are summarized in Table 15.1. The contributions in the second column are both from $I_C^{(2)}$. In the third column ($u = 3$) the first row is from $I_C^{(3)}$, $J_3(\alpha)$, while the entry given for the second row is obtained from $J_3(\alpha)$ or $K_3(\alpha')$. For the rest the first row is from $J_u(\alpha)$ and the second row is from $K_u(\alpha')$, these being the most divergent integrals if $u \geq 4$ and μ_3 different from zero, or equal to zero, respectively.

The difference in orders for $\mu_3 = 0$ and $\mu_3 \neq 0$ illustrates that the effect of the long-range Coulomb forces which cause the singularities of the integrals is a function not only of κ^2/λ, the second moment of the concentration of charge types, but of the higher moments as well (equation (15.26)).

In order to compare the theory with experimental data it is in general necessary to make a detailed calculation from a model in order to include the effects of u_{ij}^* and the higher components of the direct potentials, because these may be large compared to the effects of the long-range forces in the experimental concentration range. However there are limiting laws for the concentration dependence of various thermodynamic properties at low κ which arise from the singularities of the cluster integrals and are therefore independent of the short-range forces. Of course the Debye-Hückel limiting law is the most important of these but it seems that in certain circumstances one may

TABLE 15.1. Order of $\mathfrak{S}_u$.

u	2	3	4	5	6	7
Order if $\mu_3 \neq 0$	$c^2 \ln c$	$c^{5/2}$	$c^{5/2}$	c^3	c^3	$c^{7/2}$
Order if $\mu_3 = 0$	c^2	$c^3 \ln c$	c^3	$c^{7/2}$	$c^{7/2}$	c^4

hope to observe additional limiting laws arising from the $c^2 \ln c$ term of the table.

Notes and References

1. J. E. Mayer, *J. Chem. Phys.* **18,** 1426 (1950).
2. E. Haga, *J. Phys. Soc. Japan* **8,** 714 (1953).
3. H. L. Friedman, *Molecular Phys.* **2,** 190, 436 (1959).
4. E. Meeron, *J. Chem. Phys.* **26,** 804 (1957).
5. J. G. Kirkwood and J. C. Poirier, *J. Phys. Chem.* **58,** 591 (1954).
6. Erdelyi, Magnus, Oberhettinger, and Tricomi, *Higher Transcendental Functions* (McGraw-Hill Book Co., Inc., New York, 1953).
7. Erdelyi, Magnus, Oberhettinger, and Tricomi, *Tables of Integral Transforms* (McGraw-Hill Book Co., Inc., New York, 1954).

CHAPTER 4

Application to Ionic Systems

16. Thermodynamics. Pressure as an Independent Variable

Introduction

The cluster theory of ionic solutions seems, at this stage, to be rigorously founded, internally consistent, and reasonably complete. It remains to determine whether it is useful for interpreting the observations on electrolyte solutions in terms of molecular interactions. As a rule such interpretation is achieved when good agreement is found between experimentally determined characteristics and quantities calculated from a molecular model by a statistical theory. The comparisons which have been made for the cluster theory are summarized in the following sections; they are all tentative for various reasons, as explained in each case.

In this section we prepare a basis for dealing with a difficulty that seems to be inherent in the theory based on the cluster expansion of either the grand partition function Ξ or the configuration integral Z. This is that the resulting equations for solutions all apply to a state of osmotic equilibrium, as discussed in Section 7. Furthermore the corresponding thermodynamic quantities, for example

$$\mathfrak{F}^{\text{ex}} = -kT\mathcal{V}\mathfrak{S} \tag{16.1}$$

are best defined in terms of equations in which the volume $\mathcal{V}$ is an independent variable (Section 6). On the other hand experimental work with solutions in condensed phases is mostly done near a pressure of one atmosphere and is most conveniently related to thermodynamic equations in which the pressure P is an independent variable. In addition, experimenters have usually reduced their data to partial molar excess functions while, for comparison with the cluster theory, total excess functions are much more convenient to calculate, except in the investigation of limiting laws, in which case it is easy to perform the differentiation required to go from total to partial molar quantities.

In this section we first review the formulation of thermodynamic excess functions for the concentration scales and standard states customarily employed in work with electrolyte solutions and then develop the relation of these excess functions to $\mathfrak{S}$.

An electrolyte solution is an example of a mixture with an essentially unsymmetrical relation among its components; in such a case the designation of one of the species as solvent and the others as solutes is not completely arbitrary but reflects the physical realities. The appropriate thermodynamic excess functions for such solutions are not the same as those suitable for essentially symmetrical mixtures, such as benzene–carbon tetrachloride, although the same notation is employed for both. Furthermore both kinds of total excess functions were introduced by Scatchard[1, 2] and both kinds of partial molar excess functions were introduced even earlier by Lewis and Randall.[3] The use of total excess functions for symmetrical mixtures has only recently gained wide acceptance[4, 5] and it has been proposed that there are corresponding advantages in using total rather than partial molal excess functions for unsymmetrical mixtures.[6]

Formulation of Excess Functions for Unsymmetrical Mixtures[6]

We designate one component of the mixture as the solvent w and choose as composition variables the molalities m_s of each of the solute species ($s = 1, 2, \cdots$). The chemical potential of any solute species s may be expressed as

$$\mu_s = \mu_s^0 + RT \ln m_s\gamma_s \tag{16.2}$$

We specify that the activity coefficient, γ_s, is unity when all of the m_s are zero. Then μ_s^0 is the chemical potential of solute species s in the usual hypothetical one molal reference state at the same T and P as the solution. By applying the Gibbs-Duhem equation we find that equation (16.2) implies that the chemical potential of the solvent, μ_w, is related to the total molality of solute species,

$$m = \sum m_s, \qquad s = 1, 2, \cdots \tag{16.3}$$

by the equation

$$\mu_w = \mu_w^0 - RTM_w m\phi/1000 \tag{16.4}$$

where μ_w^0 is the chemical potential of the pure solvent at the same T and P as the solution, M_w is the molecular weight of the solvent, and ϕ is the usual osmotic coefficient, given by

$$\phi = 1 + [1/m] \int_{m=0}^{m} \sum m_s \, d \ln \gamma_s \qquad (16.5)$$

A dictionary of the present nomenclature in relation to the more familiar Harned and Owen[7] nomenclature is given in Table 16.1. Note especially the distinction between the two m's.

At this point, we introduce an innovation which simplifies many of the equations. This is to use as the basic quantity of solvent not the mole, but the kilogram, and to express all specific total solution functions as the extensive function for the quantity of solution con-

TABLE 16.1. Dictionary of nomenclature.‡

This work	Description	Harned and Owen
m_s	Molality (moles per kg of solvent) of sth solute species (ion or molecule)	m_s
m	$\sum m_s$	$\nu m^{a,b}$
I	Molal ionic strength, $\frac{1}{2}\sum m_s z_s^2$	μ
x_s	m_s/m, solute fraction	ν_+/ν or ν_-/ν^a
γ_s	Practical activity coefficient of species s	γ_+ or γ_-
ϕ	Practical osmotic coefficient	ϕ
$\gamma_\pm$	$\Pi\gamma_s^{x_s}$, mean solute activity coefficient	$\gamma_\pm^a$
G_w	Partial specific (per kg) Gibbs free energy of solvent	$1000\bar{F}_1 M_1$
μ_s	Partial molal Gibbs free energy of solute species s	$\bar{F}_s$
$G_\pm$	$\sum x_s \mu_s$, mean partial molal Gibbs free energy of solutes	$\bar{F}_2/\nu^a$
H^{ex}	Excess enthalpy	$m\phi_L = -m\Delta H_D^{a,b}$
H_s^{ex}	Excess partial molal enthalpy of the solute species s	$\bar{L}_s$
$H_\pm$	$\sum H_s x_s$	$\bar{L}_2/\nu^a$
V^{ex}	Excess volume	$m(\phi_V - \bar{V}_2^0)^{a,b}$
V_s^{ex}	Excess partial molal volume of solute species s	$\bar{V}_s - \bar{V}_s^0$
$V_\pm^{ex}$	$\sum x_s V_s^{ex}$	$(\bar{V}_2 - \bar{V}_2^0)/\nu^a$

‡ From H. L. Friedman, *J. Chem. Phys.*, **32**, 1351 (1960).

[a] This correspondence exists only if the solution has exactly one electrolytic solute.

[b] This m in the Harned and Owen nomenclature is the molality of *electrolyte*. If we represent it by M, then for a solution of a single electrolyte, $M = m/\nu$, where m is the total molality of solute ions as used here.

taining a kilogram of solvent. We retain the mole as the basic quantity of each solute, and continue to employ partial molal quantities for the solutes. This asymmetry of the notation is merely a logical consequence of the use of the molal rather than mole ratio or mole fraction concentration scale. Also, we use not the partial molal, but the partial specific (per kg) free energy of the solvent, G_w :

$$G_w = [\partial \mathcal{G}/\partial \mathcal{W}]_{T,P,n_s} \qquad (16.6)$$

where $\mathcal{W}$ is the mass of solvent.

Equation (16.4) then becomes

$$G_w = G_w^0 - RTm\phi \qquad (16.4')$$

We also introduce a hypothetical reference mixture in which all of the activity coefficients are unity at any temperature, pressure, and composition. For such a mixture the osmotic coefficient is also unity (equation 16.5), and the partial free energies are given by

$$\mu_s^* = \mu_s^0 + RT \ln m_s \qquad (16.7)$$

$$G_w^* = G_w^0 - RTm \qquad (16.8)$$

It is apparent that γ_s and ϕ are simply related to partial excess functions, which express the difference in properties of a real solution and of a hypothetical reference mixture of the same composition, temperature, and pressure.

$$\mu_s^{\text{ex}} \equiv \mu_s - \mu_s^* = RT \ln \gamma_s \qquad (16.9)$$

$$G_w^{\text{ex}} \equiv G_w - G_w^* = RTm(1 - \phi). \qquad (16.10)$$

The total excess Gibbs free energy of a quantity of a real mixture containing a kilogram of solvent is

$$G^{\text{ex}} = G - G^* = G_w^{\text{ex}} + \sum m_s \mu_s^{\text{ex}} \qquad (16.11)$$

$$= RTm[1 - \phi + \sum x_s \ln \gamma_s] \qquad (16.12)$$

where

$$x_s = m_s/m \qquad (16.13)$$

is the *solute fraction* of species *s*. It is convenient to define a mean solute quantity.

$$Y_\pm = \sum x_s Y_s \qquad (16.14)$$

corresponding to any partial molar solute quantity, Y_s. This reduces to the usual mean ionic quantity[7] when the solute is obtained by the dissociation of a single strong electrolyte. Equation (16.14) may be applied to $Y_s = \ln \gamma_s$ and then we obtain

$$G^{ex} = RTm[1 - \phi + \ln \gamma_{\pm}], \tag{16.15}$$

which is valid in general and which, for an electrolyte solution with a single solute, provides a way to calculate G^{ex} from the usual[7, 8] tabulated values of ϕ and $\gamma_{\pm}$.

We now consider several derivatives of G^{ex}. The partial molal derivatives are

$$(\partial G^{ex}/\partial m_s)_{T,P,m_t \neq m_s} = \mu_s^{ex} \tag{16.16}$$

$$(\partial G^{ex}/\partial m)_{T,P,\mathbf{x}} = \mu_{\pm}^{ex} = RT \ln \gamma_{\pm} \tag{16.17}$$

and

$$[\partial(G^{ex}/m)/\partial(1/m)]_{T,P,\mathbf{x}} = G_w^{ex} \tag{16.18}$$

where $\mathbf{x} \equiv x_1, x_2, \cdots, x_s, \cdots x_\sigma$ is the set of solute fractions. The excess enthalpy is

$$H^{ex} = [\partial(G^{ex}/T)/\partial(1/T)]_{P,m,\mathbf{x}} = H_w^{ex} + \sum m_s H_s^{ex} \tag{16.19}$$

and equations for the excess entropy, excess volume, etc., are obtained in analogous fashion.

In the case of solutions of a single electrolyte, there are functions discussed in Harned and Owen[7] which correspond to some of the excess functions which we define in this way. Some relations between the two sets of enthalpy and volume functions are given in Table 16.1.

It is apparent that the excess functions, G^{ex}, H^{ex}, S^{ex}, and V^{ex} for unsymmetrical mixtures are interrelated in the same way as the fundamental thermodynamic variables $\mathcal{G}$, $\mathcal{H}$, $\mathcal{S}$, and $\mathcal{V}$. For instance, we have

$$TS^{ex} = H^{ex} - G^{ex}$$

The same interrelation exists for the excess functions which are convenient for the description of symmetrical mixtures, but not for the functions $\gamma_{\pm}$, ϕ, ϕ_L, ϕ_V, etc., which have most often been used to describe the properties of electrolyte solutions. On the other hand, the excess functions for symmetrical mixtures include the effects of solvation as well as solute-solute interaction and give a complete description of the formation of a solution from its components, while the excess functions for unsymmetrical mixtures include only that part of the solute-solvent interaction which changes with concentration. The new excess functions must, therefore, be supplemented by the thermodynamics of the formation of the solutes in their reference states in solution from pure forms, in order to give a complete description of the formation of a solution from its components.

There is one other aspect in which unsymmetrical excess functions compare unfavorably with symmetrical ones. A mixture in which the latter vanish over the entire composition range, a symmetrical ideal solution (cf. equation (10.1)), is an ideal which is closely approached by certain real systems, notably mixtures of isotopic species. On the other hand there is probably no real system which is even approximately an unsymmetrical ideal solution, that is, a mixture that conforms to equations (16.7) and (16.8) for $0 \leq m_s \leq \infty$ for every solute species. These observations serve to emphasize that unsymmetrical mixtures are harder to understand than symmetrical ones, for these difficulties cannot in any real sense be surmounted by formulating the properties of unsymmetrical mixtures in terms of symmetrical excess functions.

Excess Functions for Electrolyte Solutions

The equations developed above are applicable at once to electrolyte solutions, but for some purposes it is convenient to introduce different composition variables. For instance the molality m is often replaced by the molal ionic strength

$$I = \tfrac{1}{2} \sum m_s z_s^2 = \tfrac{1}{2} m \sum x_s z_s^2 \qquad (16.20)$$

where z_s is the ionic charge. If and only if all of the solute species are ionic (i.e., no $z_s = 0$) then equations (16.17) and (16.18) may be written as

$$(\partial G^{\text{ex}}/\partial I)_{T,P,\mathbf{x}} = [2/\sum x_s z_s^2]\mu_{\pm}^{\text{ex}} \qquad (16.21)$$

$$\left(\frac{\partial(G^{\text{ex}}/I)}{\partial(1/I)}\right)_{T,P,\mathbf{x}} = G_w^{\text{ex}} \qquad (16.22)$$

In the special case of a solution of a single binary electrolyte we have

$$\sum x_s z_s^2 = x_+ z_+^2 + x_- z_-^2 = |z_+ z_-|$$

For solutions of several electrolytes it is often convenient to reduce the thermodynamic data by subtracting the corresponding properties of component solutions of single electrolytes.[6] The remaining contributions then correspond to changes in the thermodynamic properties in a mixing process, usually arranged to be a process at constant T, P, and I. When this is done another composition variable in addition to ionic strength is needed for each electrolyte after the first.

We consider here only the simple case of a solution containing electrolyte A (ions 1 and 3) and electrolyte B (ions 2 and 3). For composition variables, we choose the total ionic strength of the solution, I, and the mixing fraction y, defined as the fraction of the ionic strength contributed by ions from electrolyte A. The excess free energy of the solution may be expressed as

$$G^{ex}(y,I) = \Delta_m G^{ex}(y,I) + yG^{ex}(1,I) + [1 - y]G^{ex}(0,I) \quad (16.23)$$

where $G_{ex}(1,I)$ applies to a solution of pure A and $G^{ex}(0,I)$ applies to a solution of pure B. Each term of the equation pertains to a solution of the same ionic strength, temperature, and pressure, although the last two variables are not exhibited. The quantity $\Delta_m G^{ex}$ is clearly the increase in excess free energy on forming the mixture from the component solutions at the same I, T, and P and is a measure of the change in molecular interactions in this process. By appropriate differentiation of (16.23) we may obtain expressions for the corresponding enthalpy and volume changes, $\Delta_m H^{ex}$ and $\Delta_m V^{ex}$, which are the directly observable increases in enthalpy and volume, respectively, in this same process. However, $\Delta_m G^{ex}$ is not directly observable but must be calculated from observed changes in partial molal free energies. We next discuss methods of performing this calculation.

In order to keep the following equations compact we now introduce the mean ionic partial excess free energy of electrolyte A, G_A^{ex}, and a reduced form of this, g_A^{ex}. These are related to the usual mean ionic activity coefficient by the equation

$$\begin{aligned} | z_1 z_3 | \, g_A^{ex} &\equiv G_A^{ex} \\ &= RT \ln \gamma_A = -\frac{z_3}{z_1 - z_3} RT \ln \gamma_1 + \frac{z_1}{z_1 - z_3} RT \ln \gamma_3 \end{aligned} \quad (16.24)$$

and there is a corresponding equation for component B. By differentiating (16.11), collecting the terms corresponding to G_A^{ex} and G_B^{ex}, and finally changing variables from m_1, m_2, m_3 to y,I we get

$$dG^{ex}(y,I) = 2g_A^{ex}(y,I)\ d(yI) + 2g_B^{ex}(y,I)\ d([1 - y]I) \quad (16.25)$$

which is the basis of several methods of calculating $\Delta_m G^{ex}$ from measured partial free energies. We summarize all of these here, classified according to the components whose partial free energies are observed in the mixtures.

Components A *and* B. For constant I we have, from (16.25),

$$(\partial G^{ex}/\partial y) = 2I[g_A^{ex} - g_B^{ex}] \quad (16.26)$$

Integration at constant I gives, with (16.23),

$$\Delta_m G^{ex}(y,I)/2I = \int_0^y [g_A^{ex}(y',I) - g_B^{ex}(y',I)]\ dy' - y \int_0^1 [g_A^{ex}(y',I) - g_B^{ex}(y',I)]\ dy' \quad (16.27)$$

Although there are few systems for which g_A^{ex} and g_B^{ex} are both independently measureable, the combination $g_A^{ex} - g_B^{ex}$ can, in principle, always be determined with the help of an ion-exchange membrane[9] or, for certain cations, by making use of the ion-exchange properties of amalgams.[10]

Component A. Now we must integrate (16.25) at constant

$$[1 - y]I \equiv t.$$

The path of integration is shown in Fig. 16.1. The result is

$$G^{ex}(y,t/[1 - y]) - G^{ex}(0,t) = 2\int_0^y g_A^{ex}(y',t/[1 - y'])\ d(y't/[1 - y']) \quad (16.28)$$

so we have

$$\Delta_m G^{ex}(y,I) = G^{ex}(0,t) - [1 - y]G^{ex}(0,I) - yG^{ex}(1,I) + 2t\int_0^y g_A^{ex}(y',t/[1 - y'])[1 - y']^{-2}\ dy' \quad (16.29)$$

This is roughly equivalent to the method of integration first proposed by McKay[11] but the present treatment is both simpler and more general.

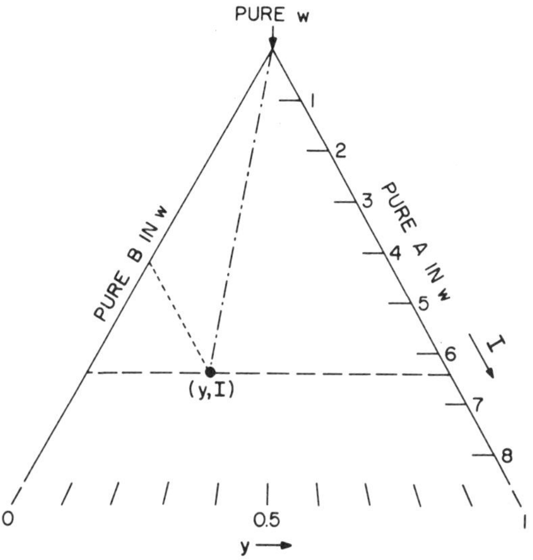

Fig. 16.1. Different paths of integration used to calculate $\Delta_m G^{ex}(y,I)$ from various partial molar free energies. The scale of I is arbitrary. - - - -, from partial molar free energies of A and B or A and w, constant I; -·-·-·, from partial molar free energy of w alone, constant y; ·····, from partial molar free energy of A alone, constant $(1 - y)I$. (This is parallel to the $y = 1$ axis.)

The solvent. We define $\Delta_m G_w^{ex}$ in the obvious way,

$$\Delta_m G_w^{ex}(y,I) \equiv G_w^{ex}(y,I) - [1 - y]G_w^{ex}(0,I) - yG_w^{ex}(1,I) \quad (16.30)$$

Given the osmotic coefficients over a suitable composition range one may calculate $\Delta_m G_w^{ex}$ from equations (16.10) and (16.30). For symmetrical mixtures $z_1 = z_2$ this calculation simplifies to

$$\Delta_m G_w^{ex}(y,I) = 2RTI[\phi(y,I) - [1 - y]\phi(0,I) - y\phi(1,I)] \quad (16.31)$$

By repeating the determination of $\Delta_m G_w^{ex}$ at various I one can get it as a function of I at given y and then integrate (16.22) at constant y to get

$$\Delta_m G^{ex}(y,I) = I \int_0^I G_w^{ex}(y,I')\, d(1/I') \quad (16.32)$$

This is equivalent to an integration first proposed by McKay and Perring.[12] The lower limit of the integral causes some difficulty (see Section 18), but this is unavoidable if only the solvent partial free energy is measured in the mixture.

The solvent and component A. Here as in the other case in which two components are directly observed the only integration required is at fixed ionic strength.

In the notation of equation (16.25) the Gibbs-Duhem equation is

$$dG_w^{ex} + 2yI\,dg_A^{ex} + 2[1 - y]I\,dg_B^{ex} = 0 \tag{16.33}$$

and in the same notation equation (16.11) is

$$G^{ex}(0,I) = G_w^{ex}(0,I) + 2Ig_B^{ex}(0,I) \tag{16.34}$$

for one of the component solutions and

$$\begin{aligned}\Delta_m G^{ex}(y,I) = \Delta_m G_w^{ex}(y,I) &+ 2Iy[g_A^{ex}(y,I) - g_A^{ex}(1,I)] \\ &+ 2I[1 - y][g_B^{ex}(y,I) - g_B^{ex}(0,I)]\end{aligned} \tag{16.35}$$

for the mixture. The third term on the right of (16.35), the only term with g_B^{ex}, may now be replaced by

$$\begin{aligned}2I[1 - y]\int_0^y dg_B^{ex} = -[1 - y]\int_0^y \frac{\partial G_w^{ex}}{\partial y}\frac{dy'}{[1 - y']} \\ - 2I[1 - y]\int_0^y \frac{\partial g_a^{ex}}{\partial y}\frac{y'}{[1 - y']}\,dy'\end{aligned} \tag{16.36}$$

giving an expression for $\Delta_m G^{ex}$ that requires only the y dependence of G_w^{ex} and g_A^{ex} at given ionic strength.

Expansions for Thermodynamic Functions for Electrolyte Mixtures

The treatment up to this point is perfectly general but now, in order to allow comparison with other methods of treating the thermodynamics of this system as well as to facilitate the comparison with the cluster theory, it is convenient to make some assumptions about the form of the y dependence of G_A^{ex}, G_B^{ex}, and $\Delta_m G^{ex}$, namely,

$$\left.\begin{aligned}G_A^{ex}(y,I) - G_A^{ex}(1,I) &= -2.303RT\sum \alpha_{An}I^n[1 - y]^n \\ G_B^{ex}(y,I) - G_B^{ex}(0,I) &= -2.303RT\sum \alpha_{Bn}I^n y^n,\end{aligned}\right\} \quad n = 1, 2, \cdots \tag{16.37}$$

$$\Delta_m G^{ex}(y,I) = I^2RTy[1 - y]\sum g_p[1 - 2y]^p, \quad p = 0, 1, 2, \cdots \tag{16.38}$$

If the terms for $n > 1$ are omitted, (16.37) reduces to the original form of Harned's rule (Harned and Owen,[7] equations (14-4-3) and (14-4-4)). The inclusion of terms for $n = 2$ is equivalent to the definition of the Harned coefficients α_{An} and α_{Bn} currently in use for the representation of the properties of mixed electrolyte solutions. The inclusion of terms of higher n increases the generality of the equations. The form of (16.38) is suggested by the corresponding expansion of G^{ex} for mixtures of two pure substances and by the experimental results[13, 14] for $\Delta_m H^{ex}$ and $\Delta_m V^{ex}$ for mixtures of solutions of 1-1 electrolytes. In these equations, the coefficients α_{An}, α_{Bn}, and g_p are independent of y, but are in general functions of I, T, and P.

The following relation of the g_p to the Harned coefficients is obtained by substituting (16.37) and (16.38) in (16.27), and collecting coefficients of $[1 - 2y]^p$.

$$g_p = -2.303 \sum_{n>p} \left\{ \frac{\alpha_{An}}{|z_1 z_3|} + [-1]^p \frac{\alpha_{Bn}}{|z_2 z_3|} \right\} \cdot \frac{I^{n-1}}{(n+1)} \sum_{m=p}^{n-1} \binom{m}{p} 2^{1-m} \tag{16.39}$$

where

$$\binom{m}{p}$$

is the binomial coefficient. The experimental work in the current literature is only accurate enough to determine the first two Harned coefficients, i.e., α_{A1} and α_{A2}. The g_p coefficients to this accuracy are

$$g_0 = -2.303 \left[\frac{\alpha_{A1}}{|z_1 z_3|} + \frac{\alpha_{B1}}{|z_2 z_3|} + I \left[\frac{\alpha_{A2}}{|z_1 z_3|} + \frac{\alpha_{B2}}{|z_2 z_3|} \right] \right] \tag{16.40}$$

$$g_1 = -\frac{2.303}{3} I \left[\frac{\alpha_{A2}}{|z_1 z_3|} - \frac{\alpha_{B2}}{|z_2 z_3|} \right] \tag{16.41}$$

We also note that in general we have, from (16.38),

$$g_0(I) = 4\Delta_m G^{ex}(\tfrac{1}{2}, I)/I^2 RT$$

So to determine $g_0(I)$ one need have $\Delta_m G^{ex}(y,I)$ only at $y = \frac{1}{2}$. A corollary of interest is that if only the partial molal free energy of component A is measured, then to determine $g_0(I)$ one must have data for g_A^{ex} from $\frac{1}{2}I$ to I, as may be easily deduced from Fig. 16.1.

It is well known[11, 15] that there are restrictions on the Harned coefficients which are obtained by the application of the cross-differentiation relations to the partial molal free energies, and therefore these coefficients cannot be independently adjusted to give the best fit with experimental data. There are no such relations among the g_p coefficients, and this is one reason why they are more suitable for the representation of the free energy relations in these systems. This is an advantage in the use of total, rather than partial molal, excess functions here, but in order to exploit it most completely it is necessary to calculate $\Delta_m G^{ex}$ more directly from the measured partial molal free energy than via the Harned coefficients. This is particularly true when the measurements relate directly to the partial free energy of the solvent.

Another advantage in the use of total functions here is most simply illustrated by considering a mixture of two 1-1 electrolytes with a common ion in a concentration range where Harned's rule is a good approximation. Then we have, to this approximation,

$$g_0 = -2.303[\alpha_{A1} + \alpha_{B1}], \qquad g_p = 0 \text{ for } p > 0 \qquad (16.42)$$

It is also true in this case that the difference in partial excess free energy of solvent in the end solutions is given by

$$G_w^{ex}(1,I) - G_w^{ex}(0,I) = 2.303RTI^2[\alpha_{A1} - \alpha_{B1}] \qquad (16.43)$$

Then the sum of Harned coefficients is a measure of interactions characterizing the mixture, but the difference in Harned coefficients merely reflects a difference in properties of the end solutions!

In general, the individual Harned coefficients include a measure of the difference in properties of the end solutions and must be combined according to (16.39) to obtain quantities that characterize the changes in molecular interaction that occur when the mixture is formed from the end solutions. Therefore, it is advantageous to formulate the thermodynamic properties of these systems in terms of total excess functions if the objective is to represent and correlate those properties of mixed electrolyte solutions which cannot be deduced by thermodynamic methods from measurements on the end solutions alone.

Some of the experimental results for mixed electrolyte solutions have been expressed in terms of total excess functions in contributions by Young,[13, 14] McKay,[11] Harned,[16, 13] and Friedman.[18]

The Relation of $\mathfrak{S}$ to G^{ex}

The following procedure is simpler, though more formal, than that originally proposed.[6] All of the mathematical operations are assumed to be at constant $\mathbf{x}$ and T. The solution under its own osmotic pressure is characterized by c,Π or m,Π where c is the particle number density and m is the molality. We have

$$mN_a = cV(m,\Pi) \tag{16.44}$$

where N_a is Avogadro's number and $V(m,\Pi)$ is the volume of solution per kilogram of solvent in the osmotic state m,Π.‡

We also have the following equations for the extensive Helmholz and Gibbs free energies.

$$\mathfrak{F}(m,\Pi) = \mathfrak{F}^{ex}(m,\Pi) + \mathfrak{F}^{id}(c) \tag{16.45}$$

$$= \mathcal{G}(m,\Pi) - \Pi\mathcal{V}(m,\Pi) \tag{16.46}$$

$$\mathcal{G}(m,\Pi) = \mathcal{G}(m,0) + \int_0^{\Pi} \mathcal{V}(m,P)\, dP \tag{16.47}$$

where in the last equation not only $\mathbf{x}$ and m but also the absolute amount of each component is unchanged in the integration. The terms on the right of (16.45) appear in equations (6.4) and (6.5). The desired excess Gibbs free energy appears in the equation (cf. equation (16.11))

$$\mathcal{G}(m,0) = \mathcal{G}^{ex}(m,0) + \mathcal{G}^{*}(m,0) \tag{16.48}$$

Now we combine the above equations with (16.1) and multiply through by $N_a/cV(m,\Pi)$ to get

$$-RT\mathfrak{S}/c = G^{ex}(m,0)/m + g^{*}(m,0) - f^{id}(c) - RT - m^{-1}\int_0^{\Pi} P\,\frac{\partial V(m,P)}{\partial P}\, dP \tag{16.49}$$

where g^{*} and f^{id} are $\mathcal{G}^{*}$ and $\mathfrak{F}^{id}$, respectively, per mole of solute. We have

$$g^{*}(m,0) = \sum_s x_s\mu_s^0 + RT\sum_s x_s \ln x_s + RT\ln m \tag{16.50}$$

$$f^{id}(c) = N_a\sum_s x_s\mu_s^{\dagger} + RT\sum_s x_s \ln x_s + RT\ln c - RT \tag{16.51}$$

‡ In equations for numerical calculation we shall always associate V with the units liters per kilogram of solvent. However at this stage, in order to avoid conversion factors, we assume the same unspecified volume units for V and $1/c$.

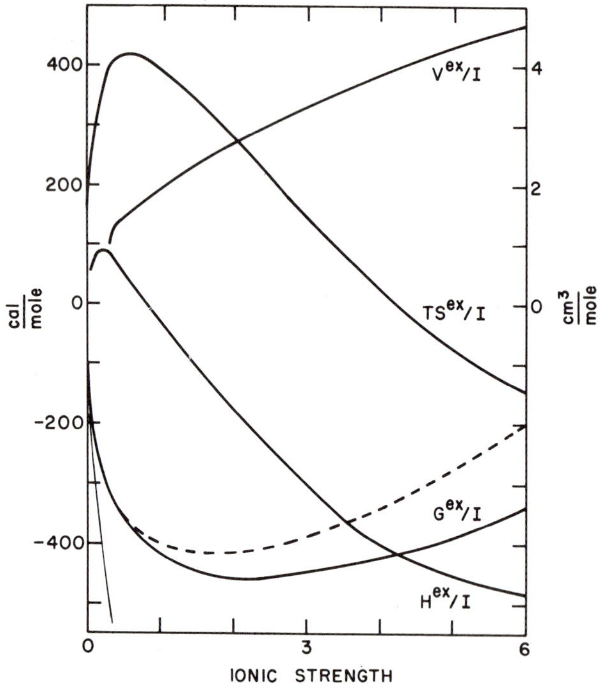

Fig. 16.2. Excess functions for aqueous NaCl solutions at 25°C. The ordinate scale for V^{ex} is on the right, the others are on the left. The broken curve is $-2RT\lambda\mathfrak{S}/\kappa^2$ calculated from equation (16.56). The light curve is the Debye-Hückel limiting law for G^{ex}/I. (From H. L. Friedman, *J. Chem. Phys.* **32,** 1351 (1960).)

The chemical potential μ_s^0 is per mole and for hypothetical one molal standard state (cf. equation (16.2)). The chemical potential $\mu_s^\dagger$ is per molecule and is for hypothetical $c_s = 1$ standard state (cf. equation (6.5)). We then find

$$\mu_s^0 - N_a\mu_s^\dagger = RT \ln (c^\dagger/m^0) = RT \ln (N_a/V(0,0)) \qquad (16.52)$$

and

$$g^*(m,0) - f^{\text{id}}(c) = RT \ln (V(m,\Pi)/V(0,0)) + RT \qquad (16.53)$$

With this result equation (16.49) becomes

$$-\mathfrak{S}/c = G^{\text{ex}}/mRT + \ln\,(V(m,\Pi)/V(0,0)) - [mRT]^{-1}\int_0^{\Pi} P[\partial V(m,P)/\partial P]\,dP \tag{16.54}$$

$$= G^{\text{ex}}/mRT + \ln\,(V(m,0)/V(0,0)) + [\Pi^2/2mRT][\partial V(m,P)/\partial P] + \cdots \tag{16.55}$$

where the last member has been obtained by using a Taylor series

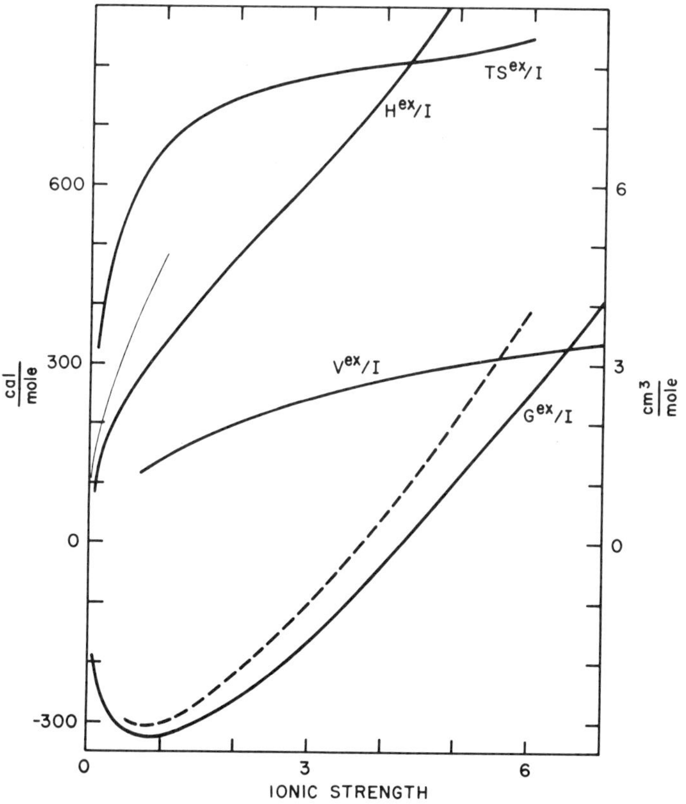

Fig. 16.3. Excess functions for aqueous LiCl solutions at 25°C. The ordinate scale for V^{ex}/I is on the right, the others are on the left. The broken curve is $-2RT\lambda\mathfrak{S}/\kappa^2$ calculated from equation (16.56). The light curve is the Deybe-Hückel limiting law for H^{ex}/I. (From H. L. Friedman, *J. Chem. Phys.* **32,** 1351 (1960).)

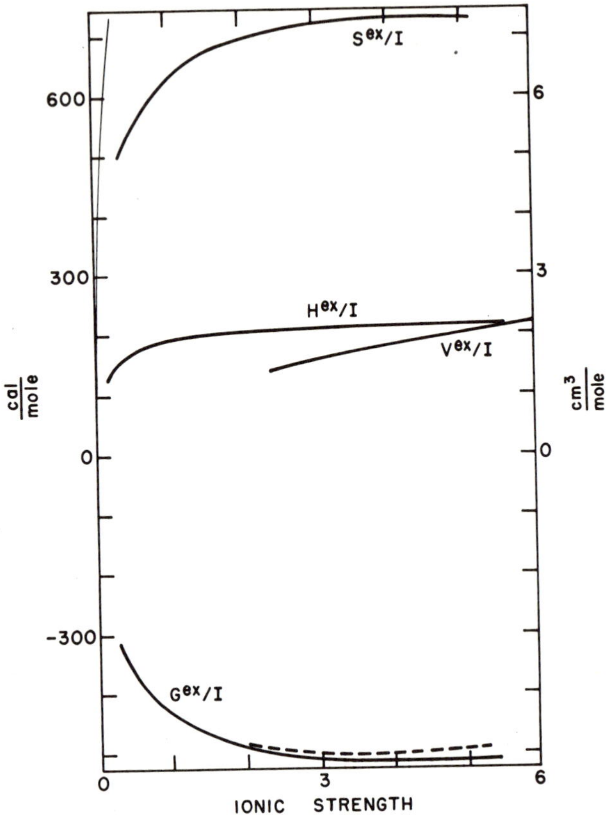

Fig. 16.4. Excess functions for aqueous $BaCl_2$ solutions at 25°C. The ordinate scale for V^{ex}/I is on the right, the others are on the left. The broken curve is $-2RT\lambda\mathfrak{S}/\kappa^2$ calculated from equation (16.56). The light curve is the Debye-Hückel limiting law for TS^{ex}/I. (From H. L. Friedman, *J. Chem. Phys.* **32,** 1351 (1960).)

expansion of $V(m,P)$ in powers of P. For many purposes the third term in the last member can be neglected (Table 16.2). Procedures for estimating Π and $\partial V(m,P)/\partial P$ by means of the Gibson-Tait equation are discussed in the original paper.[6] The preceding result is quite general but for electrolyte solutions the equivalent form

$$-2\lambda\mathfrak{S}/\kappa^2 = G^{ex}/IRT + [2/\sum_s x_s z_z^2] \ln (V(m,0)/V(0,0)) \qquad (16.56)$$

$$+ [\Pi^2/2IRT]\partial V(m,P)/\partial P + \cdots$$

TABLE 16.2. The correction terms in equation (16.55). Aqueous solutions of several electrolytes at 25°C and $I = 6$ molal.

	LiCl	NaCl	CsCl	NaI
$\ln \frac{V(m,0)}{V(0,0)}$	0.113	0.120	0.234	0.205
$\frac{\Pi^2}{2mRT} \frac{\partial V(m,P)}{\partial P}$	0.000	−0.002	−0.004	0.000

may be more convenient. A comparison of $-2\lambda\mathfrak{S}/\kappa^2$ with G^{ex}/IRT and other excess functions for typical strong electrolyte solutions is given in Figs. 16.2, 16.3 and 16.4.

In some cases, particularly for the investigation of limiting laws, it is of interest to compare the theory directly with partial molal quantities. We write (16.55) as

$$G^{ex}/mRT = -\mathfrak{S}/c - R_1(m) \tag{16.57}$$

and then differentiate according to (16.18) to get

$$G_w^{ex}/RT = m[1 - \phi] = -[V_w(m,\Pi)/N_a] \frac{\partial(\mathfrak{S}/c)}{\partial(1/c)} + m^2 \frac{\partial R_1}{\partial m} \tag{16.58}$$

while differentiation according to (16.17) gives

$$G_\pm^{ex}/RT = \ln \gamma_\pm = -\left[1 - \frac{mV_\pm(m,\Pi)}{V(m,\Pi)}\right] \frac{\partial \mathfrak{S}}{\partial c} - \frac{V_\pm(m,\Pi)}{N_a} \mathfrak{S} - \frac{\partial(mR_1)}{\partial m} \tag{16.59}$$

These equations find application mainly at low m so it is of interest to note the order in m of the various terms near $m = 0$. In (16.58) the first term on the right is $O(m^{3/2})$ while the second is $O(m^2)$. In (16.59) the first term on the right is $O(m^{1/2})$, the second is $O(m^{3/2})$, and the third is $O(m)$.

Notes and References

1. G. Scatchard, *Chem. Revs.* **8**, 321 (1931); **44**, 7 (1949).
2. G. Scatchard and S. Prentiss, *J. Am. Chem. Soc.* **56**, 1486 (1934).
3. G. N. Lewis and M. Randall, *Thermodynamics* (McGraw-Hill Book Co., Inc., New York, 1923).
4. I. Prigogine and R. Defay, *Chemical Thermodynamics* (Longmans, Green, & Co., Inc., New York, 1952).

5. E. A. Guggenheim, *Thermodynamics* (Interscience Publishers, New York, 1957), third ed.
6. H. L. Friedman, *J. Chem. Phys.* **32,** 1351 (1960).
7. H. Harned and B. Owen, *The Physical Chemistry of Electrolyte Solutions* (Reinhold Publishing Corporation, New York, 1958), third ed.
8. R. Robinson and R. Stokes, *Electrolyte Solutions* (Butterworth & Co. (Publishers) Ltd., London, 1955).
9. F. Helfferich, *Ionenaustaucher* (Verlag Chemie GMBH, Weinheim/Bergstr., 1959) B.Î.
10. H. L. Friedman and K. Schug, *J. Am. Chem. Soc.* **78,** 3881 (1956).
11. H. A. C. McKay, *Nature* **169,** 464 (1952).
12. H. A. C. McKay and J. K. Perring, *Trans. Faraday Soc.* **49,** 163 (1953).
13. T. F. Young and M. B. Smith, *J. Phys. Chem.* **58,** 716 (1954).
14. T. F. Young, Y. C. Wu, and A. A. Krawetz, *Discussions Faraday Soc.* **24,** 37 (1957).
15. E. Glueckauf, H. McKay, and A. R. Matheson, *Trans. Faraday Soc.* **47,** 428 (1951).
16. H. S. Harned, *J. Am. Chem. Soc.* **63,** 1299 (1959).
17. H. S. Harned, *J. Phys. Chem.* **64,** 112 (1960).
18. H. L. Friedman, *J. Chem. Phys.* **32,** 1134 (1960).

17. Solutions of a Single Electrolyte

(a) *Limiting Laws*

From Section 15 we find that we have in general

$$\mathfrak{S} = \kappa^3/12\pi + O(\kappa^4 \ln \kappa) \tag{17.1}$$

The statement that $\mathfrak{S}$ approaches $\kappa^3/12\pi$ to any desired accuracy as $\kappa \to 0$ is the Debye-Hückel limiting law for $\mathfrak{S}$. In the original paper of Debye and Hückel[1] the quantity that we designate as $-kT\mathfrak{S}$ is identified as the electrical work of dilution.

By applying (16.56) we see that (17.1) implies the limiting law for G^{ex}

$$\begin{aligned} G^{\text{ex}}/RT &= -I\kappa\lambda/6\pi + O(I^2 \ln I) \\ &= -\tfrac{4}{3}AI^{3/2}V^{-1/2} + O(I^2 \ln I) \end{aligned} \tag{17.2}$$

where A is defined in terms of the dimensionless product $\kappa\lambda$:

$$\kappa\lambda/8\pi \equiv A[I/V(I,\Pi)]^{1/2} \tag{17.3}$$

with $V(I,\Pi)$ the specific volume (liters of solution per kilogram of solvent) of the solution under its own osmotic pressure. The volume appearing in (17.2) may be either $V(I,\Pi)$ or $V(0,0)$, for the difference between them is only $O(I)$ and changing from one to the other makes a change of only $O(I^{5/2})$ in this equation. It is of course simplest to use $V(0,0)$, the volume of the pure solvent.

We also have

$$A = 4.2030 \times 10^6/[DT]^{3/2}$$

The notation for A varies, so to help identify it we note its appearance in the equation for the mean ionic activity coefficient of a single electrolyte according to the Debye-Hückel limiting law:

$$\ln \gamma_\pm = -\,|\,z_+z_-\,|\,A[I/V]^{1/2} \qquad (17.4)$$

For water $A = 1.1243$ at 0°C and $A = 1.1723$ at 25°C.

The temperature and pressure derivatives of (17.2) are

$$\frac{\partial(G^{ex}/T)}{\partial(1/T)} = H^{ex} = \frac{4}{3}R\,\frac{\partial(AV^{-1/2})}{\partial(1/T)}\,I^{3/2} + O(I^2 \ln I) \qquad (17.5)$$

$$\frac{\partial G^{ex}}{\partial P} = V^{ex} = -\frac{4}{3}RT\,\frac{\partial(AV^{-1/2})}{\partial P}\,I^{3/2} + O(I^2 \ln I) \qquad (17.6)$$

The Debye-Hückel limiting law functions for aqueous solutions at 25°C are

$$G^{ex} = -926 I^{3/2} \text{ cal}$$

$$H^{ex} = 472 I^{3/2} \text{ cal}$$

$$TS^{ex} = 1398 I^{3/2} \text{ cal}$$

$$V^{ex} = 2.5 I^{3/2} \text{ cm}^3$$

The positive signs of H^{ex} and TS^{ex} are worth noting.[2]

Comparisons of the various forms of the Debye-Hückel limiting law with thermodynamic data have often been reviewed[3, 4] and can be omitted here. It is of interest, however, to note in Figs. 16.2 through 16.4 the immense concentration range in which the Debye-Hückel limiting law is useless as an approximation to G^{ex}, H^{ex}, or S^{ex}. Also in Fig. 17.1 we compare the Debye-Hückel limiting law for an aqueous 3-1 electrolyte with G^{ex} calculated from the cluster theory on the basis of the primitive model. The form of the graphs is chosen for comparison with the next figure but even so it is clear that for this system one must work at $I < 10^{-3}$ molal in order for the Debye-

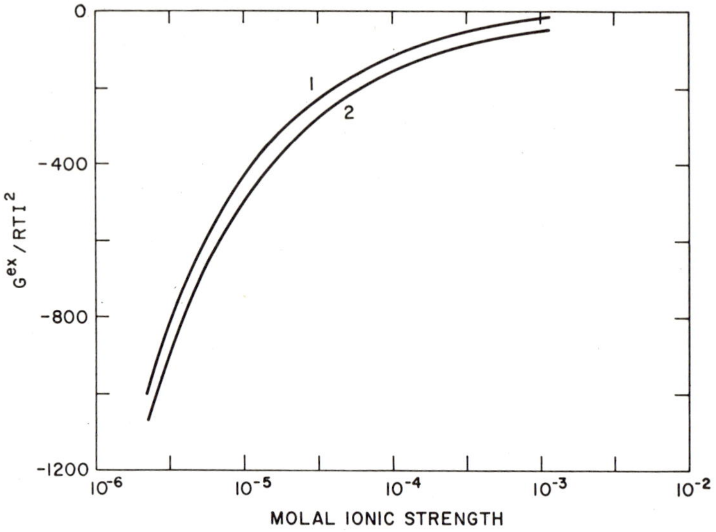

Fig. 17.1. Accuracy of the Debye-Hückel limiting law. Curve 1 is calculated from cluster theory (terms through $\mathfrak{S}_2$) for a 3-1 electrolyte in water at 25°C, using the primitive model with all ionic diameters equal to 5A (A = angstrom unit). Curve 2 is calculated for a 3-1 electrolyte in water from the Debye-Hückel limiting law.

Hückel limiting law to be the dominant term in $\mathfrak{S}$. As we shall see there are hardly any data for 3-1 electrolytes at such low ionic strengths.

Again from Section 15 it is clear that in general

$$\mathfrak{S} - \kappa^3/12\pi = \frac{\lambda^3}{12[4\pi]^2}\mu_3^2 \ln \kappa + O(\kappa^4) \tag{17.7}$$

This is a higher order limiting law in the sense that at low enough κ the first term on the right is arbitrarily large compared with the remainder. Furthermore, the first term on the right is independent of the model but depends only on the charge type of the electrolyte and the macroscopic properties of the solvent, just as in the case of the Debye-Hückel limiting law. We note that for solutions of symmetrical electrolytes ($z_+ + z_- = 0$) and their mixtures the third moment, μ_3, vanishes and the right side of (17.7) is only $O(\kappa^4)$.

It is of interest to examine the behavior of (17.7) as a limiting law for solutions of unsymmetrical electrolytes because this bears on the problem of making extrapolations to infinite dilution when one reduces partial free energy data or applies the Gibbs-Duhem equation to calculate one partial molal quantity from another.

The equations corresponding to (17.7) for G^{ex} and the partial free energies are listed here. By *X-DHLL* we abbreviate the quantity X(real) − X(Debye-Hückel limiting law). For example, the left side of (17.7) may be written $\mathfrak{S} - DHLL$. We have

$$\frac{G^{\text{ex}} - DHLL}{RT} = -\frac{1}{3V}\left[\frac{n_3}{n_2}\right]^2 A^2 I^2 \ln I + O(I^2) \qquad (17.8)$$

$$\ln \gamma_\pm - DHLL = -\frac{n_3^2}{3Vn_2} A^2 I \ln I + O(I) \qquad (17.9)$$

$$\phi - DHLL = -\frac{n_3^2}{6Vn_2} A^2 I \ln I + O(I) \qquad (17.10)$$

where ϕ is the molal osmotic coefficient and

$$n_m \equiv \mu_m / c \qquad (17.11)$$

is a reduced mth moment of the concentration of charge types. Thus n_m is unchanged when solvent is added to a solution.

It is interesting to note that the same coefficients of the higher order limiting law have been obtained from the Gronwall, LaMer, Sandved extension of the Debye-Hückel theory[5] and from Kirkwood's ionic solution theory.[6, 7]‡ On the other hand the higher order limiting law has never been employed for the reduction of data. The reason for this becomes apparent when we examine Fig. 17.2 which is based on the calculation of $\mathfrak{S}_2$ for a system favorable for observing the higher order limiting law. We observe that only for $I < 10^{-4}$ molal is this law the dominant term in G^{ex}-*DHLL*. Moreover experiments in the range $I < 10^{-5}$ molal would be required to establish the limiting slope by experiment.

The most suitable experiments for comparison with the higher order limiting law extend only down to 10^{-3} M ionic strength. The emf data of Shedlovsky and MacInnes[8] for $LaCl_3$(aq) are shown in Fig. 17.3. The point at lowest concentration tends off in the right direction, but, by comparison with Fig. 17.2, the approach to the higher order limiting law is not expected at so high a concentration. Data for the partial free energy of a solute are hard to test in this

‡ This observation was also made by J. G. Kirkwood (private communication). See also the comments pertaining to footnote 8 in reference 8a.

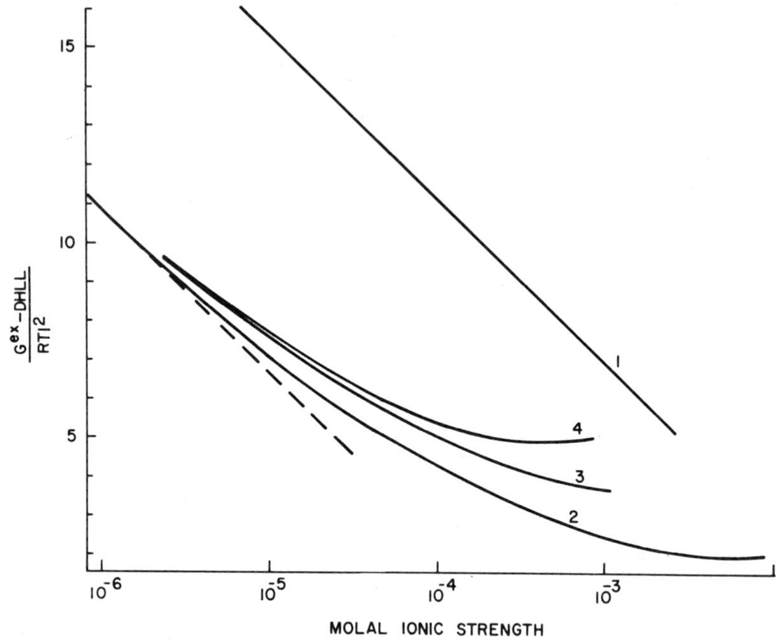

Fig. 17.2. The higher order limiting law. The slope of curve 1 is the higher order limiting law for a 3-1 or 1-3 electrolyte in water at 25°C. Curves 2, 3, and 4 are calculated in various ways from the restricted primitive model for such a solution with $a = 5$ A. Curve 2 is the result of evaluating $\mathfrak{S}_2$ exactly, while for curve 3 only terms of $\mathfrak{S}_2$ whose order is not greater than $c^{5/2}$ are retained. For curve 4 all contributions to $\mathfrak{S}$ of order not greater than $c^{5/2}$ are included. Thus curve 4 includes contributions from $\mathfrak{S}_3$ and $\mathfrak{S}_4$ as well as from $\mathfrak{S}_2$. The basis for all of these calculations is developed later in this section.

way because of the necessity of extrapolating to infinite dilution to get activity coefficients. However, if the data were more numerous and more precise this uncertainty might be avoided by differentiating the data graphically[9] and comparing the derivative with the theory. As it is, the ordinate in Fig. 17.3 should really be labeled

$$\frac{\log_{10} \gamma_{\pm} - DHLL - \epsilon}{I}$$

where ϵ is the extrapolation error. An extrapolation error that is negligible for most purposes may make a large contribution to this quotient at small I.

The freezing-point data of Scatchard, Vonnegut, and Beaumont,[10] Fig. 17.4, do not suffer from this difficulty of extrapolation but the

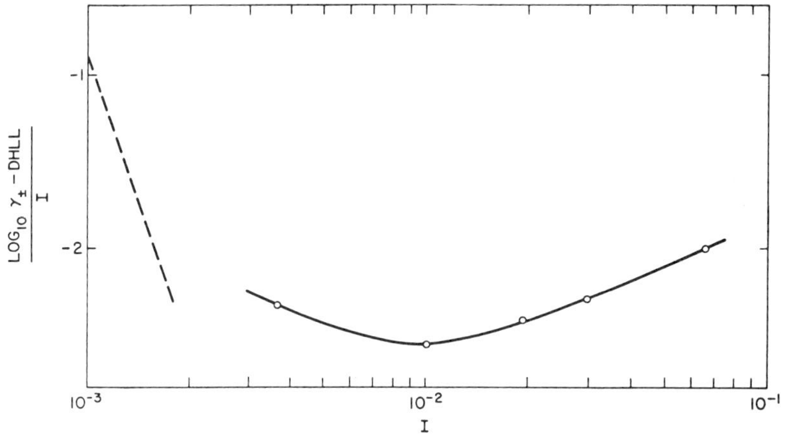

Fig. 17.3. Activity coefficients of aqueous $LaCl_3$ solutions at 25°C. Shedlovsky and MacInnes derived these data from their emf measurements on cells with transference combined with independent determinations of the transport numbers. The broken line is the slope of the higher order limiting law.

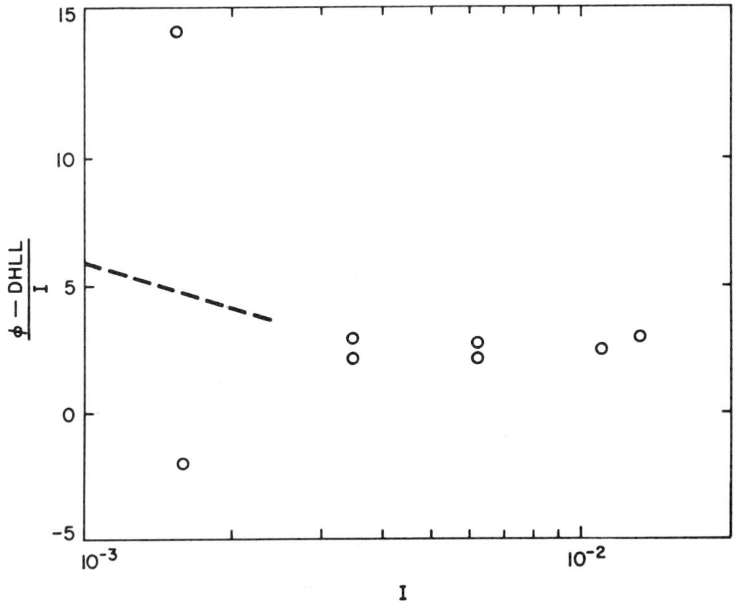

Fig. 17.4. Osmotic coefficients of aqueous $LaCl_3$ solutions at 0°C. These data were derived by Scatchard, Vonnegut, and Beaumont from their freezing-point measurements. The broken line is the slope of the higher order limiting law.

precision of the data at lowest ionic strength is not all that one might hope for.

(b) *The Numerical Evaluation of* $\mathfrak{S}_2$

This term of $\mathfrak{S}$ is defined in equation (15.4). We have the problem of the numerical evaluation of

$$B_{ij}(\kappa) = \frac{4\pi}{[i,j]!} \int_0^\infty \Phi_{ij}'''(r) r^2 \, dr \tag{17.12}$$

In the case of the most general $u_{ij}^*(r)$ each integral is a new problem, but we shall describe here a procedure that is applicable if u_{ij}^* is sectionally uniform:

$$\begin{aligned} u^*(r) &= \infty, & 0 \le r < a^{(1)} \\ &= h^{(1)}, & a^{(1)} \le r < a^{(2)} \\ &= h^{(2)}, & a^{(2)} \le r < a^{(3)} \\ &\;\vdots \\ &= 0, & a^{(\omega)} \le r < \infty \end{aligned} \tag{17.13}$$

where the $h^{(n)}$ and $a^{(n)}$ are parameters calculated from a model. The simplest example of a sectionally uniform u^* is that given by the primitive model, in which case $\omega = 1$. The next simplest example is a square well form of u^*, in which case $\omega = 2$ and there are three parameters, $a^{(1)}$, $a^{(2)}$, and $h^{(1)}$. A potential of this form has often been used in treating the properties of imperfect gases but has only recently been applied to ionic solutions in a paper by Kelbg[11] who does not, however, employ the cluster theory

To proceed we first assume that u_{ij}^* corresponds to the primitive model for which (17.12) may be written

$$B_{ij}(\kappa) = \frac{4\pi}{[i,j]!}\left[\int_{a_{ij}}^\infty [e^{q_{ij}} - Q_{ij}] r^2 \, dr - \int_0^{a_{ij}} Q_{ij} r^2 \, dr\right] \tag{17.14}$$

where

$$q_{ij} = -z_i z_j \lambda e^{-\kappa r}/4\pi r \tag{17.15}$$

$$Q_{ij} \equiv 1 + q_{ij} + \tfrac{1}{2} q_{ij}^2 \tag{17.16}$$

Now we change to dimensionless variables. We define

$$L_{ij} = -z_i z_j \lambda / 4\pi a_{ij} \tag{17.17}$$

$$K_{ij} = \kappa a_{ij} \tag{17.18}$$

$$x_{ij} = r/a_{ij} \tag{17.19}$$

In terms of these variables (17.15) becomes

$$q = Le^{-Kx}/x \tag{17.20}$$

and (17.14) becomes

$$B_{ij}(\kappa) = \frac{-z_i z_j \lambda}{[i,j]!\kappa^2} J(L_{ij}, K_{ij}) \qquad \text{(primitive model)} \tag{17.21}$$

where

$$J(L,K) \equiv K^2 L^{-1} \left[\int_1^\infty [e^q - Q]x^2 \, dx - \int_0^1 Qx^2 \, dx \right] \tag{17.22}$$

It is easily seen that for sectionally uniform u^* we may also express $B_{ij}(\kappa)$ in terms of $J(L,K)$. For example, for the square well potential we deduce

$$B_{ij}(\kappa) = \frac{-z_i z_j \lambda}{[i,j]!\kappa^2} [[1 - t]J(L_{ij}^{(2)}, K_{ij}^{(2)}) - tJ(L_{ij}^{(1)}, K_{ij}^{(1)})] \tag{17.23}$$

where

$$t \equiv e^{-h/kT}$$
$$L_{ij}^{(1)} \equiv -z_i z_j / 4\pi a_{ij}^{(1)}$$
$$K_{ij}^{(1)} \equiv -\kappa a_{ij}^{(1)}$$

In general we may define a quantity $\bar{J}$ by the equation

$$B_{ij}(\kappa) = \frac{-z_i z_j \lambda}{[i,j]!\kappa^2} \bar{J}_{ij} \tag{17.24}$$

In the case of the primitive model $\bar{J}$ is given by (17.22). In the case of the square well model it is given by the bracketed quantity in (17.23). Then we have in general, for a solution of a single binary electrolyte (cf. equation (15.4)),

$$\begin{aligned} \mathfrak{S}_2 &= c_+^2 B_{++} + c_+ c_- B_{+-} + c_-^2 B_{--} \\ &= \frac{c \mid z_+ z_- \mid}{2[z_+ - z_-]^2} [2\bar{J}_{+-} - \bar{J}_{--} - \bar{J}_{++}] \end{aligned} \tag{17.25}$$

which is convenient for numerical calculation.

A table of $J(L,K)$ is given in the Appendix. In order to calculate this function the first integral in (17.22) may be evaluated numerically and the second analytically or the two integrals may be combined after expanding e^q in a power series in q, to get

$$J(L,K) = K^2L^{-1} \sum_{p\geq 0} \frac{L^p}{p!} e_{p-2}(pK) \tag{17.26}$$

where

$$e_n(x) \equiv -\int_0^1 e^{-xt}t^{-n}\,dt \qquad \text{if } n \leq 0 \tag{17.27}$$

$$e_n(x) \equiv E_n(x) = \int_1^\infty e^{-xt}t^{-n}\,dt \quad \text{if } n \geq 1 \tag{17.28}$$

The generalized exponential integrals $E_n(x)$ (whose appearance here was first noticed by Meeron[12]) have been extensively investigated[13] and tabulated[14] while the integrals in (17.27) are elementary. Nevertheless the series in (17.26) is convenient for calculation of $J(L,K)$ only if $|L|$ is small or K is large; otherwise the convergence of the series is very slow.

The series in (17.26) corresponds closely to that used by Poirier[15] to calculate activity coefficients from the Mayer theory. He tabulates the functions

$$g_n(x) \equiv x^3 e_{n-3}(nx)/4[n-1]! \tag{17.29}$$

and

$$b_n(x) \equiv x^2\left[\frac{e_{n-2}(nx)}{n!} - \frac{xe_{n-3}(nx)}{4[n-1]!}\right] \tag{17.29$'$}$$

for which we have the equations

$$\frac{\partial(J/K^2)}{\partial K} = \frac{-4}{LK^3}\sum_{p>0} L^p g_p(K) \tag{17.30}$$

$$\frac{\partial \mathfrak{S}_2}{\partial c} = \frac{|z_+z_-|}{[z_+ - z_-]^2}[2J^a_{+-} - J^a_{++} - J^a_{--}]$$

$$J^a \equiv \frac{1}{2}J + \frac{1}{4}K\frac{\partial J}{\partial K} \tag{17.31‡}$$

$$= L^{-1}\sum_{p\geq 0} L^p b_p(K)$$

‡ For the $p = 0$ terms note that $[-1]! = \infty$. Also note that Poirier's functions b_n and g_n have no relation to the functions for which these symbols are used elsewhere in this book.

and

$$\frac{\partial(\mathfrak{S}_2/c)}{\partial(1/c)} = -\frac{c}{2}\frac{|z_+z_-|}{[z_+ - z_-]^2}[2J^b_{+-} - J^b_{++} - J^b_{--}]$$

$$J^b \equiv \frac{1}{2}K\frac{\partial J}{\partial K} = L^{-1}\sum_{p\geq 0} L^p[b_p(K) - g_p(K)] \qquad (17.32)‡$$

Equations (17.31) and (17.32) may be combined, respectively, with (16.59) and (16.58) to get the contributions of $\mathfrak{S}_2$ to the partial free energy of the solute and solvent. Poirier[15] also gives equations and tables for calculating several other derivatives of $\mathfrak{S}_2$. Scatchard has given some examples of the slowness of convergence of these series.[7]

Since the transcendental functions J, J^a, and J^b that are required to calculate $\mathfrak{S}_2$ and its derivatives do not have convenient series expansions it seems advisable to construct tables of each of them to facilitate calculations. The table of $J(L,K)$ given in the Appendix is only a first step in this direction because for many purposes a denser table of this function or tables of the derivative functions will be required. To get $J(L,K)$ at small K it may be more practical to proceed in the following way. We use the identity (obtained by partial integration of the defining integral for $E_n(x)$)

$$E_n(x) = \frac{[-x]^{n-1}}{[n-1]!}E_1(x) + e^{-x}\sum_{m=0}^{n-2}\frac{[n-2-m]![-x]^m}{[n-1]!} \qquad (17.33)$$

to get

$$J(L,K) = K^2L^{-1}[A(L,K) + B(L,K) + C(L,K)] \qquad (17.34)$$

$$A(L,K) \equiv \sum_{p=0}^{2} L^p e_{p-2}(pK)/p! \qquad (17.35)$$

$$B(L,K) \equiv \sum_{p\geq 4}\frac{[Le^{-K}]^p}{p![p-3]!}\sum_{m=0}^{p-4}[p-4-m]![-pK]^m \qquad (17.36)$$

$$C(L,K) \equiv L^3\sum_{p\geq 3}\frac{[-LpK]^{p-3}}{p![p-3]!}E_1(pK) \qquad (17.37)$$

Now A is just three terms, the singularity is in C which is a rapidly convergent series at small K, while B, unlike J and C, has derivatives of all orders at all $K \geq 0$ and may be expanded as a power series around $K = 0$:

$$B(L,K) = B(L,0) + KB^{(1)}(L,0) + \tfrac{1}{2}K^2B^{(2)}(L,0) + \cdots \qquad (17.38)$$

where

$$B^{(n)}(L,0) \equiv \lim_{K=0} \frac{\partial^n B(L,K)}{\partial K^n} \tag{17.39}$$

We find

$$B(L,0) = \sum_{p \geq 4} L^p/p!(p-3) \tag{17.40}$$

$$B^{(1)}(L,0) = -L[L^3/3! + B(L,0)] \tag{17.41}$$

The higher coefficients are also readily obtained but they involve infinite sums that are not arithmetically reducible to $B(L,0)$. For instance we have

$$B^{(2)}(L,0) = L[\tfrac{2}{3}L^3 - \tfrac{5}{16}L^4 + LB(L,0) - \sum_{p \geq 5} L^p/p![p-4]] \tag{17.42}$$

The sum $B(L,0)$ has been studied by Kirkwood and Scatchard.[7, 10] It seems that their results, which are summarized here, could readily be extended to evaluate the new sums appearing in the higher $B^{(n)}$.

$$\begin{aligned} B(L,0) &= L^3 \int_0^L t^{-4}(e^t - 1 - t - t^2/2 - t^3/6)\, dt \\ &= L^3[Ei(L) - \ln |L| - \gamma]/6 - e^L[2 + L + L^2]/6 \\ &\qquad + [2 + 3L + 3L^2 + 11L^3/6]/6 \end{aligned}$$

where γ is Euler's constant, 0.5772 $\cdots$ and $Ei(x)$ is the often tabulated function

$$Ei(x) \equiv \int_{-\infty}^{x} t^{-1}e^t\, dt$$

If $L < 0$ we have

$$B(L,0) = \frac{1}{3} + \frac{L}{2} + \frac{L^2}{2} - \frac{L^3}{6}\left[\gamma + \ln|L| - \frac{11}{6}\right] - E_4(|L|)$$

while if L is a large positive number we have the asymptotic expansion

$$B(L,0) = \frac{e^L}{L}\left[1 + \frac{4}{L} + \frac{20}{L^2} + \cdots\right]$$

TABLE 17.1. The function $B(L,0)$.‡

	$\log_{10}(B(L,0)/L^4)$	
$\lvert L \rvert$	$L > 0$	$L < 0$
0	$\bar{8}.61979$	$\bar{8}.61979$
1	$\bar{8}.66607$	$\bar{8}.57882$
2	$\bar{8}.71888$	$\bar{8}.54228$
3	$\bar{8}.87979$	$\bar{8}.50932$
4	$\bar{8}.85058$	$\bar{8}.47946$
5	$\bar{8}.93352$	$\bar{8}.45192$
6	$\bar{9}.03104$	$\bar{8}.42671$
7	$\bar{9}.14569$	$\bar{8}.40329$
8	$\bar{9}.27974$	$\bar{8}.38148$
9	$\bar{9}.43489$	$\bar{8}.36110$
10	$\bar{9}.61190$	$\bar{8}.34193$
11	$\bar{9}.81051$	$\bar{8}.32385$
12	0.02951	$\bar{8}.30683$
13	0.26703	$\bar{8}.29052$
14	0.52909	$\bar{8}.27510$
15	0.78995	$\bar{8}.26038$

‡ From G. Scatchard B. Vonnegut and D. W. Beaumont *J. Chem. Phys.* **33,** 1292 (1960).

A convenient table of $B(L,0)$, given in Table 17.1, is also taken from Scatchard, Vonnegut and Beaumont.[10]

Restricted primitive model. Most of the calculations of $\mathfrak{S}_2$ and derived thermodynamic quantities that have so far been made are based on the primitive model combined with the following additional restrictive assumption:

$$a_{++} = a_{--} = a_{+-} \equiv a \tag{17.43}$$

In this case all $K_{ij} = K \equiv \kappa a$ and the bracketed quantity in (17.25) becomes

$$\begin{aligned} 2J_{+-} - J_{++} - J_{--} \\ &= K^2 \sum_{p\geq 0} [2L_{+-}^{p-1} - L_{++}^{p-1} - L_{--}^{p-1}]\, e_{p-2}(pK)/p! \\ &= \frac{K^2[z_+ - z_-]^2}{\lvert z_+z_- \rvert^2 L} \sum_{p\geq 0} \frac{[-L]^p n_p^2 e_{p-2}(pK)}{p!} \end{aligned} \tag{17.44}$$

where

$$L \equiv \lambda/4\pi a > 0$$

$$n_p \equiv \mu_p/c = \sum_s x_s z_s^p$$

$$= \frac{| z_+ z_- |}{[z_+ - z_-]} [z_+^{p-1} - z_-^{p-1}]$$

for a simple binary electrolyte. Then for the restricted primitive model we have

$$\mathfrak{S}_2 = 2\pi c^2 a^3 \sum_{p \geq 0} [-L]^p n_p^2 e_{p-2}(pK)/p! \tag{17.45}$$

$$\partial \mathfrak{S}_2/\partial c = [L \mid z_+ z_- \mid]^{-1} \sum_{p \geq 0} [-L]^p b_p(K) n_p^2 \tag{17.46}$$

$$\partial(\mathfrak{S}_2/c)/\partial(1/c) = - c[2 \mid z_+ z_- \mid L]^{-1} \sum_{p \geq 0} [-L]^p [b_p(K) - g_p(K)] n_p^2 \tag{17.47}$$

The expression used by Poirier[15] to calculate mean ionic activity coefficients for comparison with experiment is equivalent to what may be obtained by combining (17.46) with (16.59). His procedure was to calculate ln $\gamma_\pm$ as a function of concentration for each valence type of interest and for several values of the parameter a. Then for each salt a value was selected to give agreement of calculated and experimental values at some particular concentration. This concentration is indicated by an arrow in Fig. 17.5 where Poirier's results are summarized.

The agreement in Fig. 17.5 is quite good for ionic strengths less than 0.1 molal but allowance must be made for the fact that a graph of this form tends to suppress differences between computed and experimental values of $\mathfrak{S}_2$ at low concentration. However, the "experimental" values used here all involve an extrapolation to infinite dilution. For a more careful comparison this extrapolation must be made with the help of the foregoing theory.

It is to be noted[16] that the approach of the curve for $ZnSO_4$ to the Debye-Hückel limiting law from the low side, a phenomenon that has often been adduced as evidence for the association of Zn^{++} with SO_4^{--} in aqueous solution, is consistent with the restricted primitive model without auxiliary assumptions. Guggenheim[17] has recently shown that the partial free energies of electrolytes of this charge type at low concentration can also be calculated from the restricted primi-

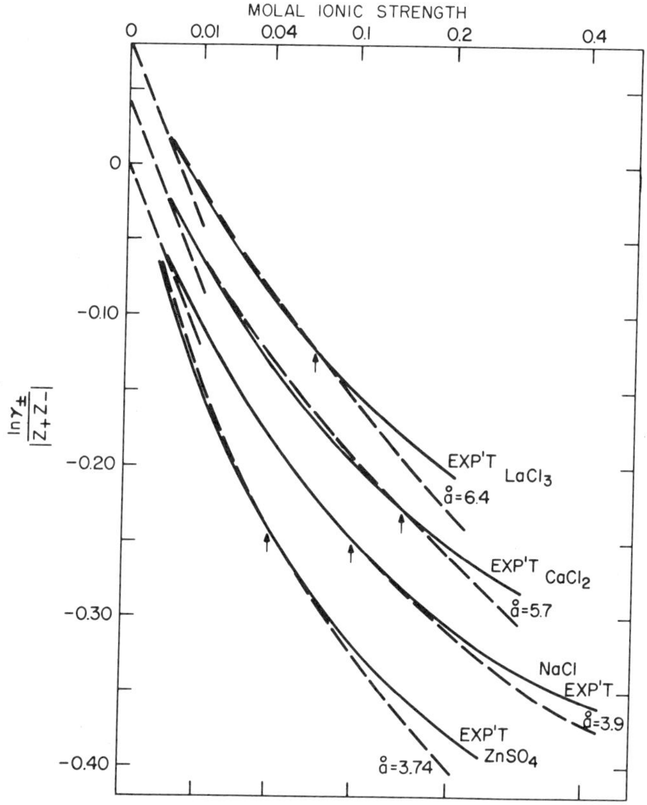

Fig. 17.5. Comparison of experimental results with Poirier's calculations from the primitive model. The ordinate scale has been shifted for $CaCl_2$ and $LaCl_3$.

tive model on the basis of the extended Debye-Hückel theory without auxiliary assumptions involving ion association.

While it is clear on intuitive grounds that a_{+-} must have a large effect in determining the calculated properties compared to a_{++} and and a_{--} it is also of interest to examine this generalization quantitatively and so to get some idea of the restrictiveness of (17.43). This is easy to do, at least for calculations that are limited to $\mathfrak{S}_2$.[12] We begin with (17.14) for a pair of ions of the same charge, say z_+. This may then be written as

$$B_{++}(\kappa) = 2\pi \int_0^\infty [e^{q_{++}} - Q_{++}]r^2\, dr - 2\pi \int_0^{a_{++}} e^{q_{++}} r^2\, dr \quad (17.48)$$

From this we find at once

$$\frac{\partial B_{++}}{\partial a_{++}} = -2\pi a_{++}^2 \exp\left(q_{++}(a_{++})\right) \tag{17.49}$$

This is put into a form for practical calculations by combining it with (16.56) and (17.3) to get

$$\begin{aligned}\delta_{++} &\equiv \partial(G^{\text{ex}}/I^2RT)/\partial \ln a_{++} \\ &= 4\left[\frac{A}{z_+[z_+ - z_-]}\right]^2 \frac{\exp(-z_+^2 Le^{-K})}{VL^3}\end{aligned} \tag{17.50}$$

where $K = \kappa a_{++}$ and $L = \lambda/4\pi a_{++}$. This equation is readily applied to any particular system to see whether δ_{++} (or δ_{--}) is negligible compared to the experimental or calculated G^{ex}/I^2RT. For example this is readily seen to be the case for $LaCl_3$(aq) for the range of compositions covered in Fig. 17.2. (In this case $A \sim 1$, $\lambda/4\pi \sim 7$ A and $a_{++} \sim 5$ A.) On the other hand for mixed electrolyte solutions it does not follow that if δ_{++} is negligible then a_{++} has a negligible effect on $\Delta_m G^{\text{ex}}/RTI^2$. Indeed in Section 18 it becomes clear that $\Delta_m G^{\text{ex}}$ depends strongly on a_{++} in some cases.

(c) *Numerical Calculation of Higher Terms*

The only calculations in the literature of terms higher than $\mathfrak{S}_2$ for solutions of a single electrolyte are those of Haga.[18] His approach was to proceed on the basis of calculations like those in Section 15 and to select all terms whose contribution to $\mathfrak{S}$ is of order at most $c^{5/2}$. It is of interest here to proceed along the same lines because this is the easiest way to get some idea of the importance of the higher terms.

We find from Section 15 that to include all terms of order not more than $c^{5/2}$ we need only consider part of the expansion of $\mathfrak{S}$ as follows:

$$\mathfrak{S} = \kappa^3/12\pi + \mathfrak{S}_2 + \mathfrak{S}_3 + \mathfrak{S}_4 \tag{17.51}$$

The part of the contribution of $\mathfrak{S}_2$ that is of order at most $c^{5/2}$ is readily deduced from (17.25) combined with (17.34). It is apparent that we can neglect all of the contributions to A, B, and C of (17.34) that are of order higher than $c^{1/2}$, or what is the same, of order in K higher than K^1.

From Section 15 we learn that the lowest order terms of $\mathfrak{S}_3$ and $\mathfrak{S}_4$ are of order $c^{5/2}$ and that these come from the terms of kappagraphs represented in Fig. 17.6. Haga calculated the contributions of 3α and 4ω but neglected 4α.

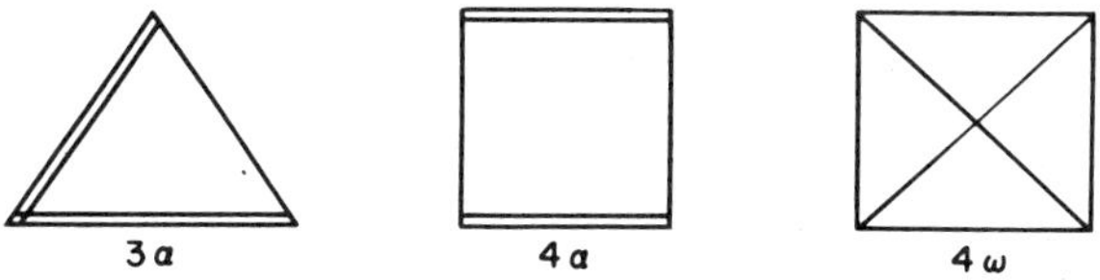

Fig. 17.6. All of the bonds are q bonds.

The term of $\mathfrak{S}_3$ derived from 3α is $(T(3\alpha) = 1)$

$$\mathfrak{S}_3(\alpha) = I_\alpha^{(3)}\lambda^5\mu_4\mu_3^2/3![2!]^2$$

$$\begin{aligned} I_\alpha^{(3)} &= \int q^2(r_{ij})q^2(r_{ik})q(r_{jk})\ d\{i,j\} \\ &= \frac{2}{\pi^2}\int_0^\infty \tilde{f}_1\tilde{f}_2^2t^2\ dt \qquad (17.52) \\ &= \frac{1}{32\pi^4}\int_0^\infty \frac{[\tan^{-1}(t/2\kappa)]^2}{\kappa^2+t^2}\ dt \\ &= \frac{1}{128\pi\kappa}\int_0^1 \frac{\phi^2\sec^2(\pi\phi/2)}{1+4\tan^2(\pi\phi/2)}\ d\phi \\ &= 2.59_0 \times 10^{-4}\kappa^{-1} \qquad (17.53) \end{aligned}$$

where, following Haga,[18] we have first transformed the integral by convolution and then done the final integral numerically. So we have

$$\mathfrak{S}_3(\alpha) = -1.07_9 \times 10^{-5}\lambda^2\kappa^5[z_+ + z_-]^2[z_+^3 - z_-^3]/[z_+ - z_-]. \qquad (17.54)$$

The term of $\mathfrak{S}_4$ derived from 4α is $(T(4\alpha) = 6)$,

$$\mathfrak{S}_4(\alpha) = 6I_\alpha^{(4)}\lambda^6\mu_3^4/4![2!]^2$$

$$\begin{aligned} I_\alpha^{(4)} &= \int q(r_{ij})q^2(r_{jh})q(r_{hl})q^2(r_{il})\ d\{i,j,k\} \qquad (17.55) \\ &= \frac{1}{128\pi\kappa^3}\int_0^1 \frac{\phi^2\ \sec^2(\pi\phi/2)\ d\phi}{[1+4\tan^2(\pi\phi/2)]^2} \\ &= 3.25_3 \times 10^{-5}\kappa^{-3} \qquad (17.56) \end{aligned}$$

$$\mathfrak{S}_4(\alpha) = 2.033 \times 10^{-6}\lambda^2\kappa^5[z_+ + z_-]^4 \qquad (17.57)$$

Finally we have $(T(4\omega) = 1)$

$$\mathfrak{S}_4(\omega) = I_\omega^{(4)}\lambda^6\mu_3^4/4! \qquad (17.58)$$

$$\begin{aligned} I_\omega^{(4)} &= \int q(r_{ij})q(r_{jk})q(r_{kl})q(r_{il})q(r_{ik})q(r_{jl})\ d\{i,j,k\} \qquad (17.59) \\ &= 1.3 \times 10^{-5}\kappa^{-3} \end{aligned}$$

according to Haga's estimation of this difficult integral. It follows that

$$\mathfrak{S}_4(\omega) = 5.41 \times 10^{-7}\lambda^2\kappa^5[z_+ + z_-]^4 \qquad (17.60)$$

The combined effect of these higher terms is shown in Fig. 17.2. It is clear that the effect of the higher terms is not negligible even at the lower end of the experimental concentration range ($I = 10^{-3}\,M$); but neither are the terms considered here large compared to $\mathfrak{S}_2$. The contribution of $\mathfrak{S}_3(\alpha)$ is much larger than that of $\mathfrak{S}_4(\alpha)$ and $\mathfrak{S}_4(\omega)$.

From consideration of these results it seems that inclusion of some of the higher terms, at least $\mathfrak{S}_3$, is required to get physically significant parameters from comparing calculations from a model with experimental data for unsymmetrical electrolytes. More refined calculations are required to investigate this point for symmetrical electrolytes because in this case $\mathfrak{S}_3(\alpha)$, $\mathfrak{S}_4(\alpha)$, and $\mathfrak{S}_4(\omega)$ all vanish.

Notes and References

1. P. Debye and E. Hückel, *Physik. Z.* **24,** 185 (1923).
2. H. L. Friedman, *J. Chem. Phys.* **32,** 1351 (1960).
3. H. S. Harned and B. B. Owen, *The Physical Chemistry of Electrolyte Solutions* (Reinhold Publishing Corporation, New York, 1958), third ed.
4. R. A. Robinson and R. H. Stokes, *Electrolyte Solutions* (Butterworth & Co. (Publishers) Ltd., London, 1955).
5. T. H. Gronwall, V. K. LaMer, and K. Sandved, *Physik Z.*, **29,** 558 (1928).
6. J. G. Kirkwood, *Chem. Revs.* **19,** 275 (1936).
7. G. Scatchard, *U. S. National Bureau of Standards Circular 524,* p. 185 (1953).
8a. T. Shedlovsky and D. A. MacInnes, *J. Am. Chem. Soc.* **61,** 200 (1939).
8b. T. Shedlovsky, *J. Am. Chem. Soc.* **72,** 3680 (1950).
9. T. F. Young and W. L. Groenier, *J. Am. Chem. Soc.* **58,** 187 (1936).
10. G. Scatchard, B. Vonnegut, and D. W. Beaumont, *J. Chem. Phys.* **33,** 1292 (1960).
11. G. Kelbg, *Z. physik. Chem. (Leipzig*) **214,** 141 (1960).
12. E. Meeron, *J. Chem. Phys.* **26,** 804 (1957).
13a. J. LeCaine, *A Table of Integrals Involving the Functions $E_n(x)$* (National Research Council of Canada, Ottawa, 1948) N.R.C. No. 1553.
13b. A. Erdelyi, W. Magnus, F. Oberhettinger, and F. Tricomi, *Higher Transcendental Functions* (McGraw-Hill Book Co., Inc., New York, 1953), Vol. 2, p. 134.
14a. G. Placzek, *The Functions $E_n(x)$* (National Research Council of Canada, Ottawa, 1946) N.R.C. No. 1547.
14b. *Tables of Functions and Zeros of Functions,* U. S. National Bureau of Standards, Applied Math. Series, No. 37 (1954).

15. J. Poirier, *J. Chem. Phys.* **21**, 965, (1953); **21**, 972 (1953).
16. H. S. Frank and M. S. Tsao, *Ann. Rev. Phys. Chem.* **5**, 43 (1954).
17. E. A. Guggenheim, *Trans. Faraday Soc.* **56**, 1152 (1960).
18. E. Haga, *J. Phys. Soc. Japan* **8**, 714 (1953).

18. Mixed Electrolyte Solutions

(a) *Introduction*

As we have emphasized, the cluster expansion for ionic systems

$$\mathfrak{S} = \kappa^3/12\pi + \sum{}'' \mathbf{c}^{\mathrm{n}} B_{\mathrm{n}}(\kappa) \tag{13.44}$$

is different from the virial expansion, not only because of the leading term but also because $B_{\mathrm{n}}(\kappa)$ depends on the composition of the system through κ. Now we observe that in the isochoric, isothermal mixing process in which we mix two electrolyte solutions with the same κ in the same solvent, each $B_{\mathrm{n}}(\kappa)$ behaves only as a constant coefficient. On this basis it is possible to learn something about the change in $\mathfrak{S}$ in this process, $\Delta_m\mathfrak{S}$, without evaluating the $B_{\mathrm{n}}(\kappa)$. In this section we exploit this observation and then go on to evaluate $\Delta_m\mathfrak{S}$ on the basis of the primitive model for comparison with experiment. The treatment here closely follows that previously given.[1]

The mixing process that is most simply related to the theory is described with reference to Fig. 18.1 in which two different solutions with the same solvent are in osmotic equilibrium with a reservoir of solvent which is maintained in the state, T, P_0. The (extensive) volumes of the two solutions are $y\mathfrak{V}$ and $[1 - y]\mathfrak{V}$.

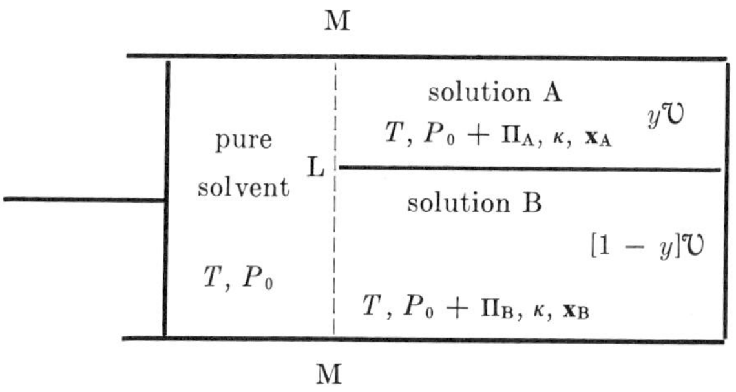

Fig. 18.1. Experiment corresponding to the definition of $\Delta_m\mathfrak{S}$. The partition, MLM, is permeable only to solvent species. The piston is adjusted during the experiment to maintain P_0 at a predetermined value. (From H. L. Friedman, *J. Chem. Phys.* **32**, 1134 (1960).)

In the mixing process in which the partition LL is removed and osmotic equilibrium is reestablished with the original solvent state, T, P_0, there is a change in $\mathfrak{S}$ that is given by

$$\Delta_m\mathfrak{S}(y,\kappa) = \mathfrak{S}(y,\kappa) - y\mathfrak{S}(1,\kappa) - [1 - y]\mathfrak{S}(0,\kappa) \tag{18.1}$$

where $\mathfrak{S}(1,\kappa)$ applies to solution A and $\mathfrak{S}(0,\kappa)$ to solution B. We assume that κ is unchanged in the process. It is clear that if we substitute the cluster expansion for each $\mathfrak{S}$ on the right of (18.1) we get

$$\Delta_m\mathfrak{S}(y,\kappa) = \sum_{u\geq 2} B_{\mathfrak{u}}(\kappa)\Delta_m\mathbf{c}^{\mathfrak{u}}(y,\kappa) \tag{18.2}$$

where

$$\Delta_m\mathbf{c}^{\mathfrak{u}} \equiv [\mathbf{c}(y,\kappa)]^{\mathfrak{u}} - y[\mathbf{c}(1,\kappa)]^{\mathfrak{u}} - [1 - y][\mathbf{c}(0,\kappa)]^{\mathfrak{u}} \tag{18.3}$$

is a purely stoichiometric quantity.

Just as $\mathfrak{S}$ is closely related to G^{ex}, so must $\Delta_m\mathfrak{S}$ be closely related to $\Delta_m G^{\text{ex}}$, the increase in free energy in an isobaric, isothermal process in which we mix two solutions of the same molal ionic strength. (Cf. equation (16.23).) In particular if we express the experimental results in terms of the coefficients of the expansion (equation (16.38))

$$\Delta_m G^{\text{ex}}(y,I) = I^2RTy[1 - y][g_0 + g_1Y + g_2Y^2 + \cdots] \tag{18.4}$$

where

$$Y \equiv 1 - 2y$$

(where g_p is independent of Y at fixed I), then it is natural to express the theoretical calculations as the coefficients of the expansion

$$\Delta_m\mathfrak{S}(y,\kappa) = -\tfrac{1}{2}[\kappa^2/\lambda]^2y[1 - y][\mathfrak{s}_0 + \mathfrak{s}_1Y + \mathfrak{s}_2Y^2 + \cdots] \tag{18.6}$$

(where $\mathfrak{s}_p$ is independent of Y at fixed κ) and it is clear that $\mathfrak{s}_0$ is closely related to g_0.

By comparison with (18.2) one finds that in the limit of low concentration $\mathfrak{s}_0$ is given by

$$\sum_{u=2} B_{\mathfrak{u}}(\kappa)\Delta_m\mathbf{c}^{\mathfrak{u}}(y,\kappa) \tag{18.7}$$

since the contribution of this term does not vanish and since (Section 15) the contributions of the higher terms become negligible at low κ. However, if a system conforms closely to Harned's rule, *even at high ionic strength*, then so is (18.7) the dominant term in $\mathfrak{s}_0$ even at high ionic strength. This important, although tentative, conclusion is developed in parts (b) and (c) of this section.

(b) *Thermodynamic Considerations*

In order to find the exact relations among the first few $\mathfrak{s}_p$ and g_p we begin with the thermodynamic relation of $\mathfrak{S}$ to G^{ex}, the excess free energy of the same solution in the state, T, P_0 :

$$2\lambda\mathfrak{S}/\kappa^2 = -G^{ex}/IRT + B \tag{18.8}$$

where we find from (16.56)

$$B = -\frac{2}{\sum x_s z_s^2} \ln \frac{V(m,0)}{V(0,0)} - \frac{\Pi^2}{2IRT} \frac{\partial V(m,P)}{\partial P} + \cdots \tag{18.8a}$$

As discussed in Section 16, B is of the nature of a small correction term whose main contribution comes from the difference in concentration scales of $\mathfrak{S}$ and G^{ex}.

Although the three solution states corresponding to the three terms on the right side of equation (18.1) have the same value of κ, they do not in general have the same value of I, the molal ionic strength. This is because $\mathfrak{S}$ is a property of a solution of definite‡ $\mathbf{x}$, κ, and T, but the amount of solvent in the solution of given volume is determined by the osmotic equilibrium with the solvent at P_0 and is therefore not an independent variable in the formulation of $\mathfrak{S}$. The situation is illustrated in Fig. 18.2. The coordinates of a point in this diagram are y, I and therefore we will express the excess free energy as a function of two variables: $G^{ex}(y,I)$. The other variables, namely T, P_0, $\mathbf{x}_A$, and $\mathbf{x}_B$, are constant in this problem. We designate the value of I at a particular y of the κ-constant curve as I_y and obtain

$$\begin{aligned} 2\lambda\Delta_m\mathfrak{S}(y,\kappa)/\kappa^2 = & -\Delta_m G^{ex}(y,I_y)/I_y RT \\ & + \Delta_m B(y,I_y) + E(y,I_y) \end{aligned} \tag{18.9}$$

where $\Delta_m B$ like $\Delta_m G^{ex}$ applies to the isobaric mixing process at constant I:

$$\Delta_m B(y,I_y) \equiv B(y,I_y) - yB(1,I_y) - [1 - y]B(0,I_y) \tag{18.10}$$

The remaining term in equation (18.9) is found by difference to be

‡ $\mathbf{x}$ is the set of solute fractions. See equation 16.13.

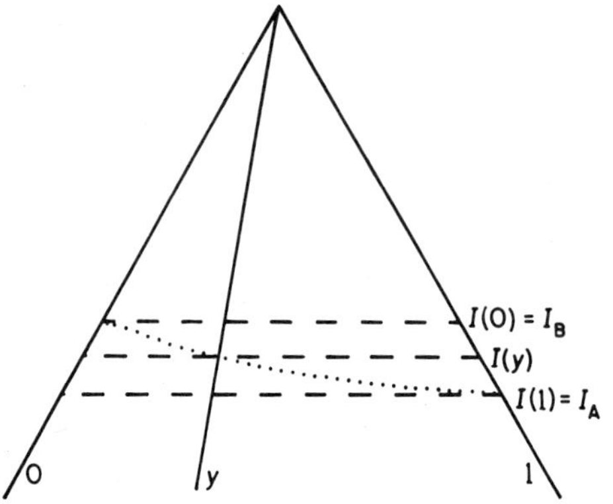

Fig. 18.2. Schematic relation of I to y at constant κ. ———, line of constant y; -------, line of constant I; ······, line of constant κ. For a given solution corresponding to a point, y, I, of this graph, the function κ pertains to the solution at the pressure $P_0 + \Pi(y,I)$, where Π is the osmotic pressure. (From H. L. Friedman, *J. Chem. Phys.* **32,** 1134 (1960).)

$$\begin{aligned} E(y,I_y) \equiv {} & [y/RT]\{[G^{ex}(1,I_1)/I_1] - [G^{ex}(1,I_y)/I_y]\} \\ & + [[1 - y]/RT]\{[G^{ex}(0,I_0)/I_0] - [G^{ex}(0,I_y)/I_y]\} \\ & -y\{B(1,I_1) - B(1,I_y)\} - [1 - y]\{B(0,I_0) - B(0,I_y)\} \end{aligned} \tag{18.11}$$

Each of the differences in braces in this equation refers to a change in a function of I, either from I_y to I_1 on the $y = 1$ line of Fig. 18.2, or from I_y to I_0 on the $y = 0$ line. Each such function may be expanded in a Taylor's series in I about I_y. For instance,

$$G^{ex}(1,I_1)/I_1 = [G^{ex}(1,I_y)/I_y] + [I_1 - I_y]\frac{\partial(G^{ex}/I)}{\partial I} + \frac{1}{2}[I_1 - I_y]^2\frac{\partial^2(G^{ex}/I)}{\partial I^2} + \cdots$$

where each derivative is evaluated at I_y. We may also represent the variation of I_y on the constant κ curve of Fig. 18.2 as

$$I_y = I_0 + y[I_1 - I_0] + \Delta_m I(y,\kappa) \tag{18.12}$$

where this Δ_m applies to the process at constant κ. With these substitutions, 18.9 becomes

$$\begin{aligned} E(y,I_y) = {} & y[1 - y][(I_0 - I_1)/I_y^2 RT] \\ & \cdot[G_w^{ex}(1,I_y) - G_w^{ex}(0,I_y)] + \cdots \end{aligned} \tag{18.13}$$

where $G_w^{\text{ex}}(y,I)$ is the excess partial free energy of the solvent in the state y, I, and where the higher terms which are omitted are negligible compared to the first, at least for mixtures of solutions of $1-1$ electrolytes. We expand $\Delta_m B$ and E in the same way as $\Delta_m G^{\text{ex}}$:

$$\Delta_m B(y,I) = y[1 - y]I[b_0 + b_1 Y + \cdots] \tag{18.14}$$

$$E(y,I) = y[1 - y]I[e_0 + e_1 Y + \cdots] \tag{18.15}$$

The coefficients of (18.14) and (18.15) can also be determined by experiment although at this time there is no system for which all of the necessary data are available.

Now we wish to express the $\mathfrak{g}_p$ coefficients in terms of the g_p coefficients by substituting these expansions for the Δ_m quantities in equation (18.9) and collecting coefficients of Y^p. However, first it must be recalled that the $\mathfrak{g}_p$ coefficients are independent of Y for variations at constant κ at the osmotic pressure (i.e., along the dotted curve of Fig. 18.2), but the coefficients g_p, b_p, and e_p are independent of Y for variations at constant I at the pressure P_0 (i.e., along a horizontal line of Fig. 18.2. We take this into account by introducing more Taylor's series expansions. Thus we have for variation along the curve of constant κ:

$$\begin{aligned} I(Y) &= I(Y = 0) + (\partial I/\partial Y)_\kappa Y + \tfrac{1}{2}(\partial^2 I/\partial Y^2)_\kappa Y^2 + \cdots \\ g_0(Y) &= g_0(Y = 0) + (\partial g_0/\partial Y)_\kappa Y + (\partial^2 g_0/\partial Y^2)_\kappa \tfrac{1}{2} Y^2 + \cdots \\ (\partial g_0/\partial Y)_\kappa &= (\partial g_0/\partial I)(\partial I/\partial Y)_\kappa , \\ (\partial^2 g_0/\partial Y^2)_\kappa &= (\partial g_0/\partial I)(\partial^2 I/\partial Y^2)_\kappa + (\partial^2 g_0/\partial I^2)(\partial I/\partial Y)_\kappa^2 \end{aligned} \tag{18.16}$$

and similarly for the other experimental coefficients. In this way we obtain the desired relations:

$$[\kappa^2/\lambda I]\mathfrak{g}_0 = g_0 - b_0 - e_0 \tag{18.17}$$

$$\begin{aligned} [\kappa^2/\lambda I]\mathfrak{g}_1 &= g_1 - b_1 - e_1 \\ &\quad + (\partial \ln I/\partial Y)_\kappa(\partial/\partial I)\{I[g_0 - b_0 - e_0]\} \end{aligned} \tag{18.18}$$

$$\begin{aligned} [\kappa^2/\lambda I]\mathfrak{g}_2 &= g_2 - b_2 - e_2 + \tfrac{1}{2}[\partial \ln I/\partial Y]_\kappa^2(\partial/\partial I) \\ &\quad \{I^2(\partial/\partial I)[g_0 - b_0 - e_0]\} \\ &\quad + [1/2I][\partial \ln I/\partial Y]_\kappa(\partial/\partial I)\{I^2[g_1 - b_1 - e_1]\} \\ &\quad + [1/2I][\partial^2 I/\partial Y^2]_\kappa(\partial/\partial I)\{I[g_0 - b_0 - e_0]\} \end{aligned} \tag{18.19}$$

In these equations, $\kappa^2\mathfrak{s}_p$ is independent of Y and the remaining functions pertain to the ionic strength characteristic of $Y = 0$ at the given κ. The dependence of I on Y at constant κ contributes not only the $\partial I/\partial Y$ terms in these equations but also the e_p terms, as may be seen from (18.13).

Now we consider the special case in which there is a single electrolyte with two species of ions in solution A and another binary electrolyte in solution B, with either the anion species or the cation species the same in both solutions. In the majority of such systems, which have been investigated experimentally, it is found that Harned's rule is nearly obeyed in the mixing process. In our nomenclature, this means that in such systems $|g_0|$ is much larger than $|g_1|$, $|g_2|$, etc. (equation (16.42)). We also find in the cases for which we have the required data that the right side of equation (18.17) is nearly equal to g_0, the right side of equation (18.18) is nearly equal to g_1, etc. When a system conforms to all these conditions then it must also be true that $|\mathfrak{s}_0|$ is large compared to $|\mathfrak{s}_1|$, $|\mathfrak{s}_2|$, etc. We use this result in the following section to simplify the theoretical calculation of $\mathfrak{s}_0$, and the usefulness of these calculations therefore depends upon the generality of this behavior. Although it seems likely that it is as general as the approximate validity of Harned's rule, further work is required to establish this.

(c) *The Calculation of* $\mathfrak{s}_p$

We wish to collect the coefficients of $y(1 - y)Y^p$ in the equation

$$\Delta_m\mathfrak{S}(y,\kappa) = \sum_{u\geq 2} B_{\mathrm{u}}(\kappa)\Delta_m\mathbf{c}^{\mathrm{u}}(y,\kappa) \tag{18.20}$$

The functional dependence on y must come from $\Delta_m\mathbf{c}^{\mathrm{u}}$ which we then need to expand in terms of the form $y(1 - y)Y^p$. To proceed we represent the concentration set in solution A as

$$\mathbf{c}(1,\kappa) = c_1, 0, c_3[1 + h] \tag{18.21}$$

and in solution B as

$$\mathbf{c}(0,\kappa) = 0, c_2, c_3[1 - h] \tag{18.22}$$

Thus, species 3 is the *common* ion and the *hetero* ion is species 1 in solution A and species 2 in solution B. The term, h, is to allow for the fact that unless species 1 and 2 have the same charge, the concentration of species 3 will not be the same in the two end solutions. The arithmetic mean of the concentration of species 3 in the end solutions is c_3 and from the condition of constant κ we find

$$h = [z_1 - z_2]/[2z_3 - z_1 - z_2] \tag{18.23}$$

Species 3 may be either a cation or an anion, but we note that both z_1z_3 and z_2z_3 are always negative.

It follows that in a mixture the concentration set is

$$\mathbf{c}(y,\kappa) = yc_1\,,\ [1 - y]c_2\,,\ [1 - hY]c_3 \tag{18.24}$$

We also introduce the special notations

$$[x]_0^p \equiv 1 + x + \cdots + x^p \quad \text{if } p \geq 0, \tag{18.25}$$
$$\equiv 0 \quad \text{if } p < 0$$

$$\langle p\rangle \equiv \tfrac{1}{2}[1 - [-1]^p] = 0 \quad \text{if } p \text{ is even} \tag{18.26}$$
$$= 1 \quad \text{if } p \text{ is odd}$$

Then we have four special cases to consider for $\Delta_m\mathbf{c}^{\mathbf{u}}$ depending on whether neither, either, or both $\mathbf{c}(0,\kappa)^{\mathbf{u}}$ and $\mathbf{c}(1,\kappa)^{\mathbf{u}}$ vanish.

$$u_1 > 0, \quad u_2 > 0, \quad u_3 \geq 0$$
$$\Delta_m\mathbf{c}^{\mathbf{u}} = y(1 - y)c_1^{u_1}c_2^{u_2}c_3^{u_3}2^{2-u_1-u_2}[1 - Y]^{u_1-1} \cdot[1 + Y]^{u_2-1}[1 - hY]^{u_3} \tag{18.27}$$

$$u_1 = 0, \quad u_2 > 0, \quad u_3 \geq 0$$
$$\Delta_m\mathbf{c}^{\mathbf{u}} = -y[1 - y]c_2^{u_2}c_3^{u_3}\sum_{p=0}^{u_3}\binom{u_3}{p}[-h]^p \cdot\left\{2\left[\frac{1+Y}{2}\right]^{u_2-1}[Y]_0^{p-1} + \left[\frac{1+Y}{2}\right]_0^{u_2-2}\right\} \tag{18.28}$$

$$u_1 > 0, \quad u_2 = 0, \quad u_3 \geq 0$$
$$\Delta_m\mathbf{c}^{\mathbf{u}} = -y[1 - y]c_1^{u_1}c_3^{u_3}\sum_{p=0}^{u_3}\binom{u_3}{p}h^p \cdot\left\{2\left[\frac{1-Y}{2}\right]^{u_1-1}[-Y]_0^{p-1} + \left[\frac{1-Y}{2}\right]_0^{u_1-2}\right\} \tag{18.29}$$

$$u_1 = 0, \quad u_2 = 0, \quad u_3 \geq 2$$
$$\Delta_m\mathbf{c}^{\mathbf{u}} = -4y[1 - y]c_3^{u_3}\sum_{p=2}^{u_3}\binom{u_3}{p}h^pY^{\langle p\rangle}[Y^2]_0^{p/2-1-\langle p\rangle/2} \tag{18.30}$$

In these equations

$$\binom{n}{m}$$

is the binominal coefficient and $h^0 = 1$ even if $h = 0$.

It is apparent upon examining these results that every set **u**, makes a contribution to $\mathfrak{s}_0$ but that most sets contribute also to some of the higher $\mathfrak{s}_p$. It is convenient to classify the terms of $\mathfrak{s}_0$ on the basis of this observation. Let $\mathfrak{s}_{0u[p]}$ be the sum of the contributions to $\mathfrak{s}_0$ of all sets that contain u particles and that do not contribute to any of the terms $\mathfrak{s}_{p+1}$, $\mathfrak{s}_{p+2}$, etc. Then we deduce from equations (18.27) to (18.30) that

$$\mathfrak{s}_0 = \sum_{u>1} \sum_{p=0}^{u-2} \mathfrak{s}_{0u[p]} \tag{18.31}$$

Inspection of these equations also shows that if a given set, **u**, contributes to $\mathfrak{s}_p$, $p > 0$, as well as to $\mathfrak{s}_0$ then its contribution to $\mathfrak{s}_p$ is at least of the same order of magnitude as its contribution to $\mathfrak{s}_0$. Therefore, if we are to have the usual relation for a Harned's rule system

$$|\mathfrak{s}_0| > |\mathfrak{s}_p|, \qquad p > 0$$

as discussed in the last paragraph of Section 2, then there are certain restrictions on the contributions of the various sets to the $\mathfrak{s}_p$ coefficients. We may have one or both of the following restrictions: (1) There is a nearly complete cancellation of the contributions of various sets, **u**, to each $\mathfrak{s}_p$, except $\mathfrak{s}_0$. (2) Those sets that contribute to any $\mathfrak{s}_p$ in addition to $\mathfrak{s}_0$ make negligible contributions to $\mathfrak{s}_0$ and every other $\mathfrak{s}_p$.

Now we assume that restriction (2) correctly describes the relations in Harned's rule systems. Of course, the deduction of either restriction from the theory would be equivalent to a deduction of the approximate validity of Harned's rule.

The assumption that we have made is equivalent to the approximation

$$\mathfrak{s}_0 \simeq \sum \mathfrak{s}_{0u[0]} \equiv \sum \mathfrak{s}_{0u} \qquad u = 2, 3, 4, \cdots \tag{18.32}$$

where, in the third member, we have indicated that the subscript, [0], will be omitted in the sequel.

The sets that contribute to the first few $\mathcal{S}_{0u}$ are readily found from equations (18.27) to (18.30) with the results shown in Table 18.1. Some details of the contributions to $\mathcal{S}_{02}$ are shown in Table 18.2. We also use the stoichiometric relations,

$$c_1 = [\kappa^2/\lambda]/z_1[z_1 - z_3] \tag{18.33}$$

$$c_2 = [\kappa^2/\lambda]/z_2[z_2 - z_3] \tag{18.34}$$

$$2hc_3 = [\kappa^2/\lambda][z_1 - z_2]/z_3[z_2 - z_3][z_1 - z_3] \tag{18.35}$$

and so obtain the equation

$$-\frac{\mathcal{S}_{02}}{2} = \frac{B(1,1,0)}{z_1z_2[z_1 - z_3][z_2 - z_3]} - \frac{B(2,0,0)}{z_1^2[z_1 - z_3]^2} - \frac{B(0,2,0)}{z_2^2[z_2 - z_3]^2} + \frac{z_1 - z_2}{z_3[z_1 - z_3][z_2 - z_3]}\left[\frac{B(0,1,1)}{z_2[z_2 - z_3]} - \frac{B(1,0,1)}{z_1[z_1 - z_3]} - \frac{[z_1 - z_2]B(0,0,2)}{z_3[z_1 - z_3][z_2 - z_3]}\right] \tag{18.36}$$

Note that the integral $B_{\mathbf{u}}(\kappa)$ is written $B(\mathbf{u})$ when the set $\mathbf{u}$ is written out, as in the preceding equation where $\mathbf{u} = u_1, u_2, u_3$.

TABLE 18.1 Sets that contribute to $\mathcal{S}_{0u}$.‡

	$\mathcal{S}_{02}$		$\mathcal{S}_{03}$	$\mathcal{S}_{04}$
$h \neq 0$	1, 1, 0 2, 0, 0 0, 2, 0	0, 1, 1 1, 0, 1 0, 0, 2	none	none
$h = 0$	1, 1, 0 2, 0, 0 0, 2, 0		1, 1, 1 2, 0, 1 0, 2, 1	1, 1, 2 2, 0, 2 0, 2, 2

‡ From H. L. Friedman, *J. Chem. Phys.*, **32**, 1134 (1960).

TABLE 18.2. Contributions to $\mathcal{S}_{02}$.‡

$\mathbf{u}$	1, 1, 0	2, 0, 0	0, 2, 0	0, 1, 1	1, 0, 1	0, 0, 2
$\Delta_m\mathbf{c}^{\mathbf{u}}/y[1 - y]$	c_1c_2	$-c_1^2$	$-c_2^2$	$2hc_2c_3$	$-2hc_1c_3$	$-4h^2c_3^2$
$\mathbf{u}!$	1	2	2	1	1	2

‡ From H. L. Friedman, *J. Chem. Phys.*, **32**, 1134 (1960).

For the case, $h \neq 0$ (unsymmetrical mixtures), $\mathcal{S}_{02}$ is the only non-vanishing $\mathcal{S}_{0u}$. For the case $h = 0$ (symmetrical mixtures) the last three terms in equation (18.36) vanish, but the higher $\mathcal{S}_{0u}$ do not vanish. In fact, we have

$$-\mathcal{S}_{0,n+2} = 2\left[\frac{\kappa^2}{\lambda}\right]^n \frac{1}{n!} \frac{B(1,1,n) - B(2,0,n) - B(0,2,n)}{z_1^2 z_3^n [z_3 - z_1]^{n+2}} \qquad \text{if } h = 0,\ n \geq 0 \tag{18.37}$$

$$= 0 \qquad \text{if } h \neq 0 \text{ and } n > 0$$

In the remainder of this section we discuss several applications of these equations.

(d) *Symmetrical Mixtures, General Relations*

Symmetrical mixtures are those for which $z_1 = z_2$.

We will consider here the first two terms of $\mathcal{S}_0$, namely $\mathcal{S}_{02}$ and $\mathcal{S}_{03}$. It is convenient for the purposes of this and the following section to introduce a different set notation. The set, ij, written without separating commas, denotes one ion of species i and one of species j. Then, the set consisting of two ions of species i is designated as ii. The same principle will also be employed for larger sets.

The definition of $B_u(\kappa)$ for $u = 2$ is

$$B_{ij}(\kappa) = \frac{4\pi}{[i,j]!} \int_0^\infty \Phi_{ij}''' r^2\, dr \tag{18.38}$$

where (equation (13.42))

$$\Phi_{ij}''' = [1 + k_{ij}] \exp(q_{ij}) - 1 - q_{ij} - \tfrac{1}{2} q_{ij}^2$$

Now we substitute equation (18.38) in (18.37) with $n = 0$, and note that in this case ($h = 0$) we have $q_{11} = q_{12} = q_{22}$ and therefore,

$$-\mathcal{S}_{02} = 4\pi z_1^{-2} [z_1 - z_3]^{-2} \int_0^\infty \delta k e^{q_{11}} r^2\, dr \tag{18.39}$$

where we have introduced the operator δ that in this case denotes

$$\delta k = 2k_{12}(r) - k_{11}(r) - k_{22}(r) \tag{18.40}$$

It is of interest to examine the behavior of (18.39) at $\kappa = 0$. In the first place we find that the integral approaches a finite value and, indeed, attains the same form as the mixing parameter characteristic of the regular solution theory. Furthermore we have

$$-d\mathcal{S}_{02}/d\kappa = 4\pi z_1^{-2}[z_1 - z_3]^{-2} \int_0^\infty \delta k[z_1^2\lambda/4\pi]\cdot\exp(q_{11} - \kappa r)r^2\,dr \quad (18.41)$$

because k_{ij} is independent of κ. In the limit of $\kappa = 0$, neither $\mathcal{S}_{02}$ nor $d\mathcal{S}_{02}/d\kappa$ vanishes in general, but we have the limiting relation

$$d\mathcal{S}_{02}/d\kappa = [z_1^2\lambda/4\pi]\mathcal{S}_{02}\,, \qquad \kappa = 0 \quad (18.42)$$

At higher concentrations, the concentration dependence of $\mathcal{S}_{02}$ depends on the nature of the k_{ij} functions in a more complicated way. The behavior for the primitive model is indicated in Fig. 18.6.

Now it is easy to deduce from (18.37) and Section 15 that

$$\lim_{\kappa\to 0} \mathcal{S}_0 = \lim_{\kappa\to 0} \mathcal{S}_{02} \quad (18.43)$$

It also follows that

$$\lim_{\kappa\to 0} d\mathcal{S}_0/d\kappa = \lim_{\kappa\to 0} d\mathcal{S}_{02}/d\kappa \quad (18.44)$$

Therefore, the following limiting law for symmetrical electrolyte mixtures at vanishing I may be deduced from equation (18.42),

$$\begin{aligned} d\ln g_0/d(I)^{1/2} &= [z_1^2\lambda/4\pi][\kappa^2/I]^{1/2} \\ &= 2z_1^2A/V_w^{1/2} \end{aligned} \quad (18.45)$$

where V_w is the specific volume (liters/kg) of pure solvent and A is defined in (17.3). This is a general result of Mayer's ionic solution theory (with the assumptions of Section 15) and does not depend on the choice of a model for the calculation of the direct potentials nor on the validity of Harned's rule at finite concentrations. By appropriate differentiation, corresponding limiting laws may be obtained for $\Delta_m H^{ex}$, $\Delta_m V^{ex}$, etc. Unfortunately, the experimental techniques which are needed to investigate the validity of equation (18.45) or its derivatives seem to be far out of the range of present-day attainments. This is shown by the approach of the calculated $\mathcal{S}_{02}$ to the limiting law only at ionic strengths less than 0.01 molal in Fig. 18.6. Nevertheless it is interesting to observe another example of a limiting law that owes its existence to the long range on the interionic forces. The common characteristic of the Debye-Hückel and higher order limiting laws is that each is a relation that is independent of the details of the short-range ionic interactions and that cannot be deduced from the laws of thermodynamics alone.

The higher $\mathcal{S}_{0u}$ are much more difficult to calculate. Reference may be made to the original article[1] for an investigation of $\mathcal{S}_{03}$. In the same place some procedures for the numerical evaluation of $\mathcal{S}_{02}$ and $\mathcal{S}_{03}$ for the primitive model are also discussed.

(e) *Symmetrical Mixtures. Comparison with Experiment*

The calculations from the cluster theory that are summarized here are based on the primitive model. Of all of the mixed electrolyte solutions that have been investigated experimentally[2] the most suitable for comparison with the primitive model are the mixed alkali chlorides.[3] In each case the activity of the solvent has been measured (by the isopiestic method), and therefore to calculate g_0 from the data requires an integration to infinite dilution (Fig. 16.1). In fact if we expand $\Delta_m G_w^{ex}$ as

$$\Delta_m G_w^{ex}(y,I) = RTI^2 y[1 - y][w_0 + w_1 Y + \cdots] \qquad (18.46)$$

then the w_p coefficients may be determined by measurement of the activity of the solvent as a function of y at constant I, but we have

$$Ig_0(I) = -\int_0^I w_0(I')\,dI' \qquad (18.47)$$

The Harned α coefficients commonly used to represent the results of the isopiestic measurements (not the same as the Harned coefficients in Section 16) are related to w_0 by

$$w_0 = 2.303(\alpha_{12} + \alpha_{21}) \qquad (18.48)$$

Water activity measurements by the cryoscopic method have been found to be accurate enough to yield reliable values of w_0 down to $I = 0.1$ molal[4] but usually the isopiestic method is useful only to determine w_0 down to about $I = 1$ molal. An extrapolation is required to perform the integration (18.47) in either case, but of course a reliable extrapolation to zero concentration from 1 molal is difficult to make. In Fig. 18.3 we show the results of several alternative ways of making the extrapolation for LiCl, CsCl, H_2O, 25°C, for which the data extend only down to $I = 2$ molal. Although the differences among the "experimental" g_0 curves are appreciable, they are not so severe as to completely interfere with our objective, which is to show

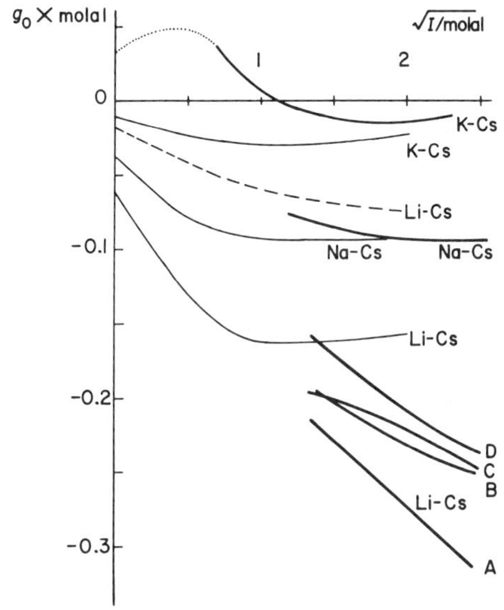

Fig. 18.3. Comparison of g_0 values calculated from the primitive model and from experiment for several mixtures of aqueous alkali-metal chloride solutions at 25°C. ———, calculated; ━━━, experimental; ······, a possible extrapolation to conform with the limiting law; -------, calculated on the assumption that the ionic radii are 1.5× the crystal radii. If 1× the crystal radii is used, the curve obtained is similar in shape to the 1.5× curve, but with about $\frac{1}{3}$ the ordinate at each I. The four "experimental" curves for the LiCl–CsCl system illustrate the effect of calculating g_0 from the w_0 data by using different approximations[1] in applying equation (18.47):

A: $g_0 = -w_0$ (used for the other systems here and in Fig. 18.4).

B: $g_0 = -w_0 + \frac{1}{2}\langle I\, dw_0/dI\rangle_{\text{AV}}$.

C: $g_0 = (5/I)[-w_0 + (I/2)(dw_0/dI) - (I^2/6)(d^2w_0/dI^2)]_{I=5} - (1/I)\displaystyle\int_5^I w_0\, dI$

D: $g_0 = (2/I)[-\frac{3}{4}w_0]_{I=2} - (1/I)\displaystyle\int_2^I w_0\, dI.$

(From H. L. Friedman, *J. Chem. Phys.* **32,** 1134 (1960).)

To understand the notation on the scales note that I has the units molal. Then I/molal is dimensionless and can be represented by a pure number. For example if $I = 5$ molal then $I/\text{molal} = 5$. In the same way g_0 has the units molal^{-1} and $g_0 \times$molal is dimensionless. For other examples of this natural system of representing the units of physical quantities see E. A. Guggenheim, *Thermodynamics*, Third ed. (Interscience, New York, 1957).

what sort of agreement with experiment is obtained on the basis of the primitive model with the assumption $\mathcal{S}_0 = \mathcal{S}_{02}$. The reason is that we can obtain equivalent agreement of the calculated g_0 with any of the "experimental" g_0 curves by rather minor adjustments of the a_{ij} parameters of the primitive model.

We could avoid the extrapolation problem by calculating w_0 from the theory. This would be particularly appropriate if our purpose were the precise comparison of theoretical calculations with data obtained at low ionic strength. However, the relation of the model to the calculated thermodynamic quantity becomes less transparent when one calculates w_0 instead of g_0 because the equations are less simple.

For the rest we shall employ the approximation

$$g_0(\text{expt}) = -w_0 \tag{18.49}$$

For the preliminary calculations we shall also employ the approximation

$$g_0(\text{calc}) = \mathcal{S}_{02}\kappa^2/\lambda I \tag{18.50}$$

For numerical calculations based on the primitive model or other models giving a sectionally uniform u_{ij}^* it may be convenient to employ the $J(L,K)$ functions introduced in Section 17 and tabulated in the appendix. In terms of this function equation (18.37) becomes, for $n = 0$,

$$\kappa^2\mathcal{S}_{02}/\lambda = [2J_{12} - J_{11} - J_{22}]/[z_1 - z_3]^2 \tag{18.51}$$

where we have abbreviated $J(L_{ij}, K_{ij}) = J_{ij}$. The results reported here, however, were obtained by a somewhat different numerical method.[1]

It remains to choose values of the a_{ij} parameters of the primitive model. The parameters that are obtained by fitting the primitive model to the data for solutions of a single electrolyte (e.g., Fig. 17.5) are essentially a_{+-} and do not seem appropriate as a basis for estimating a_{ij} when i and j have the same sign. Thus if we consider the primitive model as an idealization of a real electrolyte solution, the parameter a_{+-} must be more sensitive to ion solvation than a_{++} or a_{--}: When two ions of opposite sign approach each other there is a region of high electric field between them which tends to stabilize coordinated solvent molecules, but when two ions of the same sign approach each other the opposite is true. Thus the inverse trend of a_{+-} with cation crystal radius for the aqueous alkali halides is commonly attributed to the effect of hydration[5] and is not expected for a_{ij} if both i and j are cations.

Therefore we have assumed that if i and j are cations then

$$a_{ij} = C[r_i + r_j] \tag{18.52}$$

where r_i and r_j are the Pauling crystal radii and C is a constant. We find that $C = 2$ gives rough agreement between g_0(calc) and g_0 (expt). This rough agreement is not obtained with very different values of C as shown for one example in Fig. 18.3. More comparisons on the same approximate basis for $C = 2$ are shown in Fig. 18.4, and Fig. 18.5 serves to emphasize that the calculated g_0 values change from system to system in the same way as the experimental values. The calculated curves in Figs. 18.3 and 18.4 show that in each case the primitive model leads to negative values of $\mathcal{S}_{02}$ and, therefore, when this is the principle term in g_0, to a net attractive effect on mixing. This result for $\mathcal{S}_{02}$ must always be obtained for the primitive model if

$$a_{12} = \tfrac{1}{2}[a_{11} + a_{22}]$$

The positive values of g_0 shown by some of the systems represented

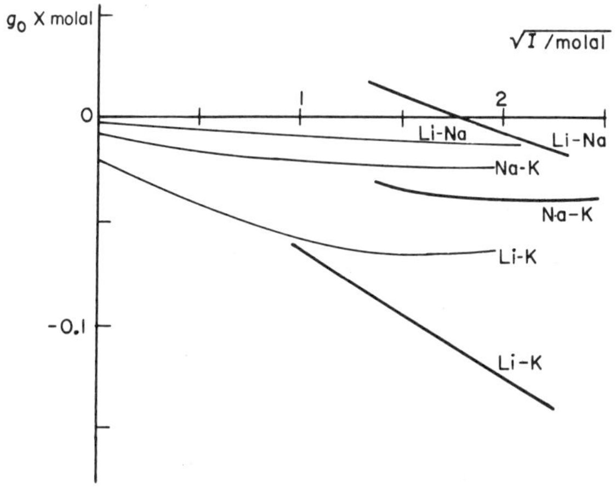

Fig. 18.4. Comparison of g_0 values calculated from the primitive model and from experiment for several mixtures of aqueous alkali–metal chloride solutions at 25°C. ———, calculated; ———, experimental. (From H. L. Friedman, *J. Chem. Phys.* **32,** 1134 (1960).)

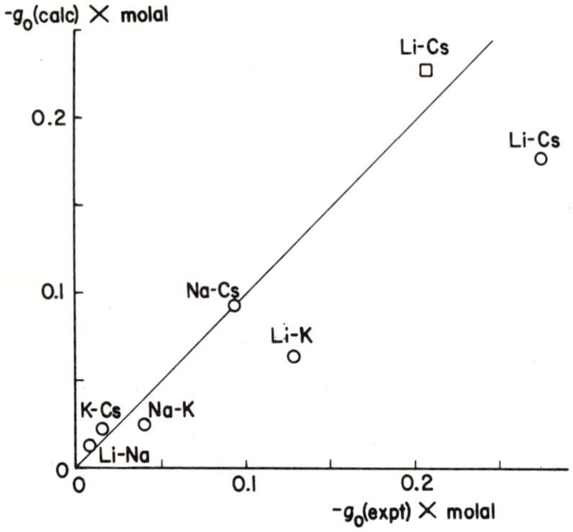

Fig. 18.5. Survey of the comparison of theory with experiment at $I = 4$ molal using the model and approximations described in the text. The circles are based on Figs. 18.3 and 18.4, while the square is based on the more refined comparison of Fig. 18.6.

in these figures as well as by the systems HCl–LiCl, HCl–NaCl, and HCl–KCl therefore demonstrate that if $\mathcal{S}_{02}$ is the principal term, then such systems cannot be adequately represented by the primitive model, even for a qualitative purpose. Of course it is not surprising that the primitive model is less successful in representing the differences in interaction of sets of ions including hydrogen ion than sets of alkali metal ions alone. However, it is also observed that when alkali metal ions are most nearly the same size the differences in interaction are again not accounted for with significant accuracy by the primitive model. In this connection, it is interesting that the system HCl–CsCl behaves in a way that is qualitatively in accord with the primitive model: g_0 is negative.

In one respect, the general theory, equation (18.45), which does not depend on a model at all, also tends to be inconsistent with some of the data. This equation requires that g_0 (or w_0) have the same sign as dg_0/dI (or dw_0/dI) in the limit of $I = 0$. As an example, the KCl–CsCl data are inconsistent with the theory unless the sign of the slope changes at low I as suggested by the dotted line in Fig. 18.3.

The theoretical calculations for the system LiCl, CsCl, H_2O, 25°C.

have been made in more detail[1] with the results shown in Fig. 18.6. Method B (Fig. 18.3) has been used to obtain the g_0 values shown here. Some approximations are also needed to estimate b_0 and e_0 but their effect is relatively minor. In the calculation of $\mathfrak{S}_{03}$ one requires a_{LiCl} and a_{CsCl}. For these it is appropriate to use the a_{ij} parameters obtained from single electrolyte solutions by the modified Debye-Hückel theory.[2] The best theoretical curve (solid line) agrees with the highly corrected experimental data (squares) within 20% and shows a qualitatively similar concentration dependence. The agreement could doubtless be improved by adjustment of the various a_{ij} parameters but it seems pointless to do so because of the uncertainty in the experimental g_0 values and the shortcomings of the primitive model. The important point is that making various corrections to the calculation leading to Fig. 18.3 tends to improve the agreement with experiment. It is especially gratifying that the effect of $\mathfrak{S}_{03}$ appears only as a small correction to $\mathfrak{S}_{02}$, even at high concentration, exactly in accordance with our assumption (2) in part (c). However, in the absence of a more extensive investigation we must allow the possibility that this result is fortuitous.

The results of the calculations at low I are included in Fig. 18.6

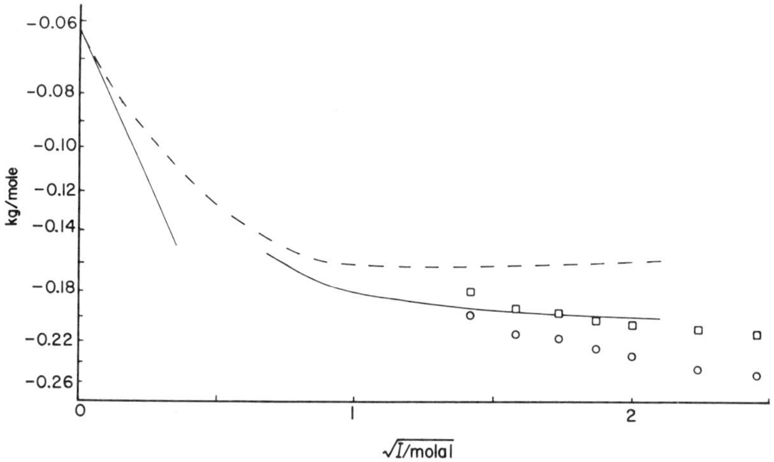

Fig. 18.6. The system LiCl, CsCl, H_2O, at 25°C. ○, g_0, experimental; □, $g_0 - b_0 - e_0$, experimental;------, $[\kappa^2/\lambda I]\mathfrak{S}_{02}$; ———, $[\kappa^2/\lambda I][\mathfrak{S}_{02} + \mathfrak{S}_{03}]$. The scale of ordinates is logarithmic. (From H. L. Friedman, *J. Chem. Phys.* **32,** 1134 (1960).)

to illustrate two of the novel features of the theory. These do not depend on the primitive model and they are noted again here only to allow us the benefits of a definite example. The first of these features is the non-zero intercept of g_0 at $I = 0$. This intercept depends on the short-range pairwise interactions, u^*_{ij}, of the hetero ions. The other feature is the approach of g_0 to its limiting value at $I = 0$, as governed by the higher order limiting law, equation (18.45). Of course, the Debye-Hückel limiting law does not appear here because g_0 is a parameter of the process in which mixing occurs at constant I. Note also that these calculations indicate that only for $I < 0.01$ molal may g_0 be expected to conform reasonably closely to the higher order limiting law.

One feature of the present theory is that g_0 is determined mainly by $\mathcal{S}_{02}$ and is, therefore, nearly independent of the common ions, species 3. Unfortunately, the systems which are similar to those discussed above, except in the use of Br^- or I^- instead of Cl^-, have not been investigated except for the enthalpy determinations summarized below. However, free energy data are available[7] for the systems, KOH, KCl, H_2O and NaOH, NaCl, H_2O at 25°C and $I = 1$ molal. The g_p parameters that we calculate from these data (using equation (16.40)) are given in Table 18.3. This discrepancy between the two g_0 values is clearly outside of the experimental error but this may be an unfavorable case for this comparison because $\mathcal{S}_{03}$ may be more important compared to $\mathcal{S}_{02}$ in this system than in alkali halide mixtures because of the unsymmetrical charge distribution in the ion OH^-. Note that for these systems the hetero ions are OH^- and Cl^- and the common ion is Na^+ or K^+. The contributions to g_1 also seem to be unusually large in these systems, although not so large as might be expected on the basis of a casual inspection of the original Harned coefficients.[7] (For the system NaOH, NaCl at $I = 0.5$ molal, $g_0 = -0.088$, and $g_1 = 0.031$—so the comparison at $I = 1$ molal may be misleading in suggesting that g_1 is much less important if Na^+ rather than K^+ is the common ion.) An examination of the integrals that

TABLE 18.3 g_p parameters calculated from equation (16.40).

	NaOH, NaCl	KOH, KCl
g_0/molal^{-1}	−0.081	−0.105
g_1/molal^{-1}	0.003	0.017

TABLE 18.4. g_0 values obtained by freezing-point measurements.

g_0(0°C)	Electrolyte
-0.1075 molal^{-1}	$LiNO_3$, KNO_3
-0.0649	LiCl, KCl
0.0218	LiCl, $LiNO_3$
0.0000	KCl, KNO_3

contribute to $\mathfrak{s}_1$ leads us to suggest that the same factors that make $\mathfrak{s}_{03}$ so large compared to $\mathfrak{s}_{02}$ also make $\mathfrak{s}_1$ large.

Another example of this sort of disagreement is found in the results that Scatchard and Prentiss[8] obtained by freezing-point measurements on mixed electrolyte solutions. The g_0 function in terms of the coefficients of their equations is, for a mixture of electrolytes 1 and 2,

$$g_0 = 12I[D_{112} + D_{122} - D_{111} - D_{222}] + 12\sqrt{2}I^{3/2}[E_{112} + E_{122} - E_{111} - E_{222}]$$

The g_0 values at $I = 1$ molal and 0°C are given in Table 18.4. Again the effect of the common ion indicates that $\mathfrak{s}_{03}$ is not negligible compared to $\mathfrak{s}_{02}$. Moreover, the above equation for g_0 is not consistent with the limiting law (equation (18.45)), although unfortunately the data are not suitable to determine whether g_0 does indeed tend to vanish as $I \to 0$, as these authors assumed.

It is much more difficult to make calculations for comparison with the enthalpy and volume data[1] although the appropriate equations are readily written down. Thus we define h_p coefficients, analogous to the g_p coefficients, as follows:

$$\Delta_m H^{ex}(y) = I^2 y[1 - y]RT \sum h_p Y^p, \qquad p = 0,1,2,\cdots \qquad (18.53)$$

where $\Delta_m H^{ex} = \Delta_m H$, the increase in enthalpy on mixing at constant pressure and ionic strength. Comparison with equation (18.4) shows that

$$h_p = -T[\partial g_p/\partial T] \qquad (18.54)$$

Substitution of equation (18.17) yields

$$h_0 = T[\kappa^2/\lambda I]\{\partial \mathfrak{s}_0/\partial T - \mathfrak{s}_0[\partial \ln V(I, \Pi)/\partial T]\} - (T/I)[(\partial b_0/\partial T) + (\partial e_0/\partial T)] \qquad (18.55)$$

TABLE 18.5. Values of h_0 in mixed-electrolyte solutions.

Salt pair	1000 h_0
LiCl, LiBr	5.40 molal^{-1}
NaCl, NaBr	5.25
KCl, KBr	5.32 ± 0.07

The difficulty is that in numerical calculations from the primitive model the various terms have different signs and the result appears as a small difference among large quantities which is very sensitive to an assumed da_{ij}/dT. However, we note one respect in which the enthalpy data tend to support the present theory. If we assume that the term in $\partial \mathcal{S}_0/\partial T$ is the principal one in equation (18.55), and that the principal term in this derivative is $\partial \mathcal{S}_{02}/\partial T$, then we must conclude that h_0 is nearly independent of the common ion. Some measurements of Young, Wu, and Krawetz[9] show that h_0 is indeed nearly independent of the common ion when this is changed from Li^+ to Na^+ to K^+ for the pair of hetero ions Cl^-, Br^-. They find the results given in Table 18.5 for 1 M aqueous solutions at 25°C.

Equations for the enthalpy and volume effects having the same form as (18.53) but with only the $p = 0$ term seem to have been given for the first time by Young and Smith.[10] Recently Rush and Scatchard have presented similar equations and applied them to the volume effects in aqueous HCl, $BaCl_2$ mixtures at 25°C.

(f) *Unsymmetrical Mixtures*

We start with equation (18.36) which cannot be simplified for unsymmetrical mixtures. However, we will consider only the particular case $z_2 = 3$, $z_1 = -z_3 = 1$. This corresponds to the systems HCl, $AlCl_3$, H_2O and HCl, $CeCl_3$, H_2O which have been investigated experimentally.[3] The systems HCl, $BaCl_2$, H_2O and HCl, $SrCl_2$, H_2O have also been investigated, but they do not exhibit any features that are qualitatively different from those of the 1,1 − 1,3, H_2O mixtures. It is unfortunate that there are no data for systems with an alkali metal chloride in place of HCl, because at the present stage we must use the primitive model for comparison of theory and experiment and it seems likely that this is less satisfactory for solutions of H^+ than for any other simple ion. However, it is of interest to demonstrate that the theory correctly yields the feature which qualitatively dis-

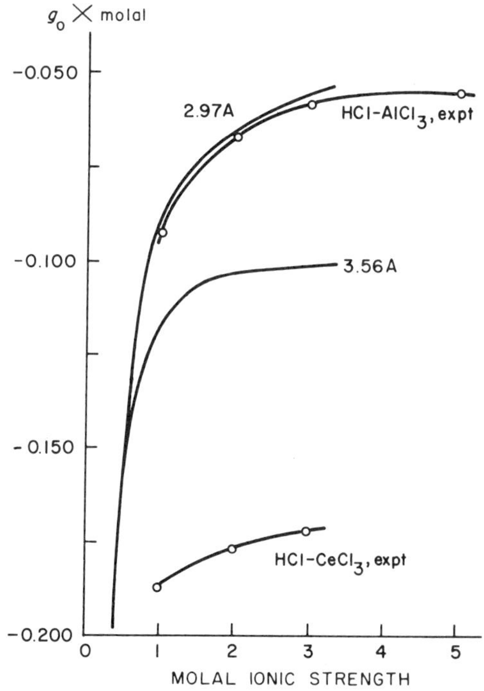

Fig. 18.7. Comparison of g_0 values for two unsymmetrical mixtures in water at 25°C with values calculated from the primitive model with all a_{ij} = 2.97 A or 3.56 A.

tinguishes the unsymmetrical mixtures from the symmetrical mixtures. This is the sign of the derivative, dg_0/dI. (Compare Fig. 18.7 with Figs. 18.3 and 18.4.)

For calculations with the primitive model it is convenient to introduce the $J(L,K)$ function into (18.36) to get the equation

$$\frac{\kappa^2}{\lambda}\,\mathfrak{g}_{02} = \frac{2J_{12}}{[z_1 - z_3][z_2 - z_3]} - \frac{J_{11}}{[z_1 - z_3]^2} - \frac{J_{22}}{[z_2 - z_3]^2} + \frac{z_1 - z_2}{[z_1 - z_3][z_2 - z_3]}\left[\frac{2J_{23}}{z_2 - z_3} - \frac{2J_{13}}{z_1 - z_3} - \frac{[z_1 - z_2]J_{33}}{[z_1 - z_3][z_2 - z_3]}\right] \qquad (18.56)$$

which is valid for either symmetrical or unsymmetrical mixtures. In this equation we have abbreviated $J(L_{ij},K_{ij}) = J_{ij}$. Of course the same equation applies for more general models if $\bar{J}$ is used instead of J. Now we specialize to the particular case mentioned above and also employ the approximation, equation (18.51), to get

$$Ig_0\,(\text{calc}) = \frac{J_{12}}{4} - \frac{J_{11}}{4} - \frac{J_{22}}{16} - \frac{J_{23}}{8} + \frac{J_{13}}{4} - \frac{J_{33}}{16} \qquad (18.57)$$

This expression does not vanish for the case in which all of the a_{ij} are equal so we may expect that for unsymmetrical mixtures the first order contribution to $\Delta_m G^{\text{ex}}$ comes from higher terms in the long-range electrostatic interactions (failure of the ionic strength principle) rather than from differences in the specific interaction of ions. Partly in order to illustrate this point, and partly because there is not much basis for assigning individual values to the a_{ij} for these systems, we only present numerical results for the case in which the a_{ij} are all the same (Fig. 18.7).

The magnitudes of the contributions of various sets to g_0 are illustrated for the case in which all $a_{ij} = 2.97$ A, $I = 1.060$ molal, by writing out the numerical values of the terms in equation (18.57) in the same order: $1.060\ g_0 = 0.354 - 0.101 - 0.302 + 0.209 - 0.227 - 0.025$. The prominence of the contributions of sets of like-charged ions is especially noteworthy. This may at first seem surprising but it must be recalled that in this approximation g_0 is determined by differences in long-range interactions rather than by differences in short-range specific interactions.

Evidently the theory does yield the positive values of dg_0/dI which distinguish unsymmetrical from symmetrical mixtures. The theory of unsymmetrical mixtures also requires that as $I \to 0$, g_0 goes to $-\infty$ as $\ln I$. There do not seem to be any data in dilute solution with which to compare this result.

(g) *The Primitive Model*

From the point of view of the Mayer theory the characteristics of the primitive model are that it leads to a step function for u_{ij}^* and to vanishing higher components of the direct potentials. But in addition, in order to know that thermodynamic properties calculated on the basis of the primitive model are properties of a physical model, however idealized, we must regard the a_{ij} as made up of additive contributions from the radii of the ions. That is, only hard sphere ions can lead exactly to a step function for u_{ij}^* and hard sphere ions have a_{ij} additive in the ion radii. Conversely, if a set of ionic radii can be found to fit all of the thermodynamic data for the simplest solutes, the alkali halides, then we may conclude that the equilibrium properties of these systems are indeed understandable in terms of the primitive model.

The a_{++} values used in this section do form a set that is additive in the ionic radii. The a_{+-} values from the extended Debye-Hückel theory, which are not very different from those required by the Mayer theory plus primitive model,[10] form a set that is nearly additive in the ionic radii. The combined set of these a_{++} and a_{+-} values is clearly very far from additive in the ionic radii, and it seems unlikely that one can find a combined set that is additive and is in even rough agreement with the data for both single and mixed electrolytes. The main difficulty is that the large g_0 values for the systems LiCl, CsCl; LiCl, KCl; and NaCl, CsCl require considerably larger differences in radii among the cations than are consistent with the a_{+-} values. (Note that the a_{++} values used here correspond to an increase in radius from Li^+ to Cs^+, contrary to the trend of the a_{+-} values, but that the calculation of g_0 is only affected to a small degree if we assume instead that the radii increase from Cs^+ to Li^+. The reason for this is that δk for equation (18.39) is unchanged if we interchange the designations of the ions.)

Although there are these difficulties with the primitive model it is apparent that the free energy data for moderately dilute electrolyte solutions would be reasonably consistent with a model that led to potentials u_{ij}^* with the following characteristics:

(1) For the purpose of calculating the cluster integrals one may, with reasonable accuracy, replace u_{ij}^* by a smoothed potential; $u_{ij}^* = \infty$ if $r < a_{ij}$, $u_{ij}^* = 0$ if $r > a_{ij}$.

(2) Each of the sets of parameters, a_{++}, a_{+-}, and a_{--} is composed of additive contributions from a set of "radii" but there are three different sets of such "radii", one each for $++$, $+-$, and $--$. It will be interesting to see to what extent the solvent structure must be invoked in models to obtain this behavior.

Very recently Scatchard has made another criticism of the primitive model, namely that it leads to a repulsive effect that is independent of charge and that it would lead to deviations from ideality even for "an isotope of water" in aqueous solution. The difficulty referred to here is discussed in the first part of Section 10 and is not relevent to the calculations for ionic solutions. It does not really imply a shortcoming of the primitive model but is related to the awkwardness of treating symmetrical mixtures such as mixtures of isotopes by the McMillan-Mayer theory.

Notes and References

1. H. L. Friedman, *J. Chem. Phys.* **32,** 1134 (1960).
2. H. S. Harned and B. B. Owen, *The Physical Chemistry of Electrolyte Solutions* (Reinhold Publishing Corporation, New York, 1958), Third ed.
3. R. A. Robinson and R. H. Stokes, *Electrolyte Solutions* (Butterworth & Co. (Publishers) Ltd., London, 1955). The measurements on the mixed alkali chloride solutions were mostly made by Robinson. See the original paper[1] for references to his work.
4. G. Scatchard and S. Prentiss, *J. Am. Chem. Soc.* **56,** 1486 (1934).
5. H. S. Frank and M. S. Tsao, *Ann. Rev. Phys. Chem.* **4,** 43 (1953).
6. H. S. Harned and B. B. Owen, *The Physical Chemistry of Electrolyte Solutions,* (Reinhold Publishing Corporation, New York, 1958), Third ed., Table 12-5-2, second column.
7. H. S. Harned and M. A. Cook, *J. Am. Chem. Soc.* **59,** 1890 (1937).
8. G. Scatchard and S. S. Prentiss, *J. Am. Chem. Soc.* **56,** 2320 (1934).
9. T. F. Young, Y. C. Wu, and A. A. Krawetz, *Discussions Faraday Soc.* **24,** 37 (1957).
10. T. F. Young and M. B. Smith, *J. Phys. Chem.* **58,** 716 (1954).
11. R. M. Rush and G. Scatchard, *J. Phys. Chem.* **65,** 2240 (1961).
12. G. Scatchard, *J. Am. Chem. Soc.* **83,** 2636 (1961).

Appendix

TABLE OF $J(L,K)$

[*This table was prepared by R. P. Kelisky and C. E. Shanesy IBM Research Computing Center* (see equation 17.26 for definitions).]

L	$10^4\,J$	$10^4\,J$	$10^4\,J$	$10^3\,J$	$10^3\,J$	$10^3\,J$	$10^3\,J$
	K = 0.02	0.04	0.06	0.08	0.10	0.15	0.20
−30.0	566.8	1494.	2565.	371.8	492.8	810.9	1143.
−28.0	513.1	1360.	2343.	340.4	452.0	745.9	1053.
−26.0	460.7	1229.	2124.	309.5	411.6	681.5	964.5
−24.0	409.8	1101.	1910.	279.0	371.8	617.8	876.4
−22.0	360.6	975.3	1699.	249.0	332.7	554.9	789.2
−20.0	313.1	853.5	1494.	219.6	294.2	492.8	702.9
−18.0	267.6	735.5	1294.	191.0	256.5	431.8	617.8
−16.0	224.1	621.8	1101.	163.0	219.6	371.8	534.1
−14.0	182.9	513.1	913.9	136.0	183.9	313.3	452.0
−12.0	144.4	409.8	735.5	110.1	149.4	256.5	371.8
−10.0	108.8	313.1	566.8	85.35	116.4	201.6	294.2
−9.5	100.4	290.1	526.4	79.40	108.5	188.3	275.2
−9.0	92.26	267.6	486.7	73.55	100.6	175.2	256.5
−8.5	84.32	245.5	447.8	67.81	92.92	162.2	237.9
−8.0	76.61	224.1	409.8	62.18	85.35	149.4	219.6
−7.5	69.16	203.2	372.8	56.68	77.93	136.9	201.6
−7.0	61.96	182.9	336.6	51.31	70.67	124.5	183.9
−6.5	55.03	163.3	301.5	46.07	63.58	112.4	166.5
−6.0	48.39	144.4	267.6	40.99	56.68	100.6	149.4
−5.5	42.05	126.2	234.7	36.06	49.98	89.12	132.7
−5.0	36.03	108.8	203.2	31.31	43.51	77.93	116.5
−4.5	30.34	92.27	173.1	26.76	37.28	67.11	100.7
−4.0	25.01	76.63	144.4	22.41	31.32	56.70	85.40
−3.5	20.07	61.98	117.5	18.30	25.67	46.76	70.76
−3.0	15.54	48.44	92.37	14.46	20.35	37.36	56.86
−2.8	13.85	43.36	82.91	13.01	18.34	33.78	51.54
−2.6	12.24	38.48	73.80	11.60	16.39	30.31	46.37
−2.4	10.71	33.82	65.08	10.26	14.52	26.96	41.39
−2.2	9.262	29.40	56.76	8.973	12.73	23.75	36.60
−2.0	7.902	25.23	48.90	7.755	11.03	20.69	32.04

(*continued*)

TABLE OF J (L,K)—*Continued*

L	$10^4 J$	$10^4 J$	$10^4 J$	$10^3 J$	$10^3 J$	$10^3 J$	$10^3 J$
	K = 0.02	0.04	0.06	0.08	0.10	0.15	0.20
−1.8	6.636	21.32	41.53	6.611	9.436	17.82	27.74
−1.6	5.472	17.72	34.72	5.552	7.958	15.16	23.77
−1.4	4.421	14.47	28.55	4.595	6.621	12.76	20.22
−1.2	3.500	11.62	23.17	3.762	5.464	10.71	17.22
−1.0	2.736	9.282	18.81	3.096	4.550	9.146	15.02
1.0	−3.595	−15.92	−37.51	−6.841	−10.85	−24.77	−43.94
1.2	−2.981	−14.25	−34.75	−6.467	−10.40	−24.21	−43.39
1.4	−2.260	−12.33	−31.63	−6.056	−9.917	−23.71	−43.11
1.6	−1.396	−10.01	−27.86	−5.558	−9.338	−23.13	−42.82
1.8	−.3638	−7.203	−23.23	−4.942	−8.611	−22.37	−42.37
2.0	.8591	−3.827	−17.60	−4.180	−7.698	−21.34	−41.63
2.2	2.293	.1875	−10.82	−3.250	−6.566	−19.99	−40.53
2.4	3.958	4.912	−2.746	−2.130	−5.185	−18.27	−38.97
2.6	5.876	10.42	6.764	−.7980	−3.524	−16.13	−36.90
2.8	8.071	16.79	17.86	.7700	−1.551	−13.51	−34.24
3.0	10.57	24.11	30.71	2.599	.7663	−10.36	−30.92
3.5	18.32	47.20	71.72	8.498	8.322	.2182	−19.24
4.0	28.68	78.64	128.3	16.72	18.96	15.55	−1.644
4.5	42.38	120.8	205.0	27.96	33.60	37.00	23.53
5.0	60.41	177.1	308.0	43.13	53.45	66.37	58.43
5.5	84.16	252.1	445.9	63.51	80.16	106.1	105.9
6.0	115.6	352.1	630.5	90.83	116.0	159.4	169.6
6.5	157.5	486.2	878.6	127.5	164.1	230.7	254.7
7.0	213.7	667.2	1213.	177.0	228.9	326.0	367.9
7.5	290.0	913.4	1668.	244.1	316.4	453.8	518.3
8.0	394.7	1251.	2291.	335.6	435.2	625.5	718.3
8.5	539.7	1719.	3150.	461.2	597.5	857.0	984.8
9.0	742.8	2371.	4344.	634.8	820.6	1171.	1341.
9.5	1030.	3291.	6015.	876.4	1129.	1598.	1819.
10.0	1440.	4596.	8373.	1215.	1559.	2182.	2463.

L	$10^3 J$	$10^3 J$	$10^3 J$	$10^3 J$	$10^3 J$	$10^3 J$	$10^3 J$
	K = 0.25	0.30	0.35	0.40	0.45	0.50	0.55
−30.0	1483.	1830.	2181.	2534.	2891.	3249.	3609.
−28.0	1369.	1691.	2016.	2345.	2677.	3010.	3345.
−26.0	1256.	1552.	1853.	2157.	2463.	2772.	3082.
−24.0	1143.	1415.	1691.	1970.	2251.	2534.	2819.
−22.0	1031.	1278.	1529.	1783.	2040.	2298.	2558.
−20.0	920.4	1143.	1369.	1598.	1830.	2063.	2298.
−18.0	810.9	1009.	1210.	1415.	1621.	1830.	2040.
−16.0	702.9	876.4	1053.	1233.	1415.	1598.	1783.
−14.0	596.8	745.9	898.4	1053.	1210.	1369.	1529.
−12.0	492.8	617.8	745.9	876.4	1009.	1143.	1278.

TABLE OF *J* (*L*,*K*)—*Continued*

L	$10^3 J$	$10^3 J$	$10^3 J$	$10^3 J$	$10^3 J$	$10^3 J$	$10^3 J$
	K = 0.25	0.30	0.35	0.40	0.45	0.50	0.55
−10.0	391.7	492.8	596.8	702.9	810.9	920.4	1031.
−9.5	366.9	462.2	560.1	660.2	762.1	865.5	970.0
−9.0	342.4	431.8	523.8	617.8	713.7	810.9	909.4
−8.5	318.1	401.6	487.7	575.8	665.6	756.7	849.1
−8.0	294.2	371.8	452.0	534.1	617.9	703.0	789.2
−7.5	270.5	342.4	416.7	492.8	570.6	649.6	729.8
−7.0	247.2	313.3	381.8	452.0	523.8	596.8	671.0
−6.5	224.2	284.7	347.3	411.7	477.5	544.6	612.7
−6.0	201.6	256.5	313.3	371.9	431.8	492.9	555.1
−5.5	179.5	228.8	280.0	332.7	386.8	442.0	498.2
−5.0	157.9	201.7	247.2	294.3	342.6	392.0	442.3
−4.5	136.9	175.2	215.3	256.7	299.3	342.9	387.4
−4.0	116.5	149.6	184.2	220.0	257.0	294.9	333.7
−3.5	96.92	124.8	154.1	184.5	216.0	248.4	281.6
−3.0	78.24	101.1	125.3	150.5	176.7	203.8	231.6
−2.8	71.07	92.04	114.2	137.4	161.6	186.6	212.4
−2.6	64.11	83.20	103.5	124.7	146.9	169.9	193.6
−2.4	57.37	74.65	93.03	112.4	132.6	153.7	175.5
−2.2	50.90	66.42	83.01	100.5	119.0	138.2	158.2
−2.0	44.74	58.59	73.47	89.29	106.0	123.5	141.8
−1.8	38.94	51.24	64.54	78.77	93.88	109.9	126.7
−1.6	33.59	44.48	56.37	69.19	82.93	97.57	113.1
−1.4	28.84	38.52	49.21	60.89	73.53	87.15	101.8
−1.2	24.89	33.65	43.49	54.40	66.38	79.45	93.63
−1.0	22.12	30.43	39.94	50.68	62.68	75.95	90.54
1.0	−67.95	−96.41	−129.0	−165.3	−205.0	−248.0	−293.8
1.2	−67.50	−96.10	−128.8	−165.2	−204.9	−247.7	−293.3
1.4	−67.62	−96.76	−130.1	−167.1	−207.5	−251.0	−297.1
1.6	−67.89	−97.80	−132.0	−170.1	−211.6	−256.2	−303.4
1.8	−68.08	−98.90	−134.3	−173.6	−216.5	−262.5	−311.3
2.0	−68.02	−99.84	−136.5	−177.3	−221.8	−269.5	−319.9
2.2	−67.58	−100.4	−138.4	−180.8	−227.1	−276.6	−329.0
2.4	−66.67	−100.6	−140.0	−184.1	−232.2	−283.8	−338.3
2.6	−65.21	−100.2	−141.0	−186.9	−237.0	−290.7	−347.5
2.8	−63.09	−99.15	−141.5	−189.1	−241.3	−297.3	−356.5
3.0	−60.25	−97.35	−141.2	−190.7	−245.0	−303.4	−365.1
3.5	−49.38	−88.95	−136.6	−191.1	−251.1	−315.8	−384.3
4.0	−31.92	−73.73	−125.3	−185.0	−251.3	−323.0	−398.8
4.5	−6.082	−49.86	−105.5	−170.9	−244.0	−323.3	−407.3
5.0	30.43	−15.05	−75.00	−146.6	−227.3	−315.1	−408.3

(*continued*)

TABLE OF J (L,K)—*Continued*

L	$10^3 J$	$10^3 J$	$10^3 J$	$10^3 J$	$10^3 J$	$10^3 J$	$10^3 J$
	K = 0.25	0.30	0.35	0.40	0.45	0.50	0.55
5.5	80.57	33.58	−31.09	−109.7	−199.1	−296.5	−399.9
6.0	148.2	99.73	29.62	−57.20	−156.6	−265.2	−380.3
6.5	238.4	188.2	111.5	14.79	−96.55	−218.3	−347.1
7.0	357.9	305.3	220.1	111.1	−14.86	−152.3	−297.3
7.5	515.6	459.2	362.7	237.9	93.67	−63.12	−227.5
8.0	723.5	660.7	548.7	403.0	235.5	54.73	−133.4
8.5	997.7	924.0	790.0	616.6	419.0	207.8	−9.590
9.0	1360.	1268.	1103.	891.7	654.5	404.5	150.4
9.5	1839.	1718.	1508.	1245.	955.2	655.2	354.8
10.0	2475.	1307.	2032.	1698.	1338.	972.9	613.5

L	$10^2 J$	$10^2 J$	$10J$	$10J$	$10J$	$10J$	$10J$
	K = 0.60	0.65	0.70	0.75	0.80	1.0	1.2
−30.0	397.1	433.3	46.97	50.62	54.27	68.94	83.67
−28.0	368.2	401.9	43.58	46.97	50.37	64.04	77.77
−26.0	339.3	370.6	40.19	43.33	46.69	59.15	71.88
−24.0	310.6	339.3	36.82	39.71	42.61	54.27	66.00
−22.0	281.9	308.2	33.45	36.09	38.74	49.40	60.13
−20.0	253.4	277.2	30.10	32.49	34.89	44.55	54.27
−18.0	225.1	246.3	26.77	28.91	31.06	39.71	48.43
−16.0	197.0	215.7	23.45	25.34	27.24	34.89	42.61
−14.0	169.1	185.3	20.16	21.81	23.45	30.10	36.82
−12.0	141.5	155.2	16.91	18.30	19.70	25.35	31.06
−10.0	114.3	125.6	13.69	14.84	15.98	20.64	25.35
−9.5	107.6	118.2	12.90	13.98	15.06	19.47	23.93
−9.0	100.9	110.9	12.10	13.12	14.15	18.30	22.52
−8.5	94.24	103.7	11.32	12.27	13.24	17.14	21.12
−8.0	87.65	96.46	10.53	11.43	12.33	15.99	19.71
−7.5	81.10	89.30	9.757	10.59	11.43	14.84	18.32
−7.0	74.60	82.19	8.986	9.759	10.54	13.70	16.94
−6.5	68.17	75.15	8.221	8.933	9.651	12.57	15.56
−6.0	61.81	68.19	7.464	8.115	8.771	11.45	14.20
−5.5	55.53	61.31	6.716	7.307	7.904	10.34	12.85
−5.0	49.34	54.53	5.978	6.510	7.047	9.246	11.52
−4.5	43.26	47.86	5.253	5.726	6.204	8.169	10.21
−4.0	37.32	41.35	4.544	4.959	5.380	7.116	8.934
−3.5	31.56	35.02	3.855	4.214	4.579	6.094	7.698
−3.0	26.02	28.94	3.194	3.499	3.811	5.119	6.525
−2.8	23.89	26.61	2.939	3.224	3.516	4.746	6.080
−2.6	21.81	24.33	2.692	2.958	3.230	4.387	5.653
−2.4	19.81	22.14	2.455	2.702	2.956	4.043	5.249
−2.2	17.90	20.05	2.228	2.458	2.695	3.721	4.874
−2.0	16.10	18.09	2.015	2.230	2.453	3.425	4.536

TABLE OF J (L,K)—*Continued*

L	10^2J	10^2J	$10J$	$10J$	$10J$	$10J$	$10J$
	K = 0.60	0.65	0.70	0.75	0.80	1.0	1.2
−1.8	14.43	16.28	1.821	2.023	2.233	3.164	4.247
−1.6	12.95	14.69	1.651	1.843	2.045	2.950	4.026
−1.4	11.73	13.39	1.516	1.703	1.900	2.804	3.901
−1.2	10.89	12.54	1.431	1.620	1.821	2.759	3.920
−1.0	10.65	12.38	1.426	1.628	1.846	2.873	4.168
1.0	−34.25	−39.36	−4.472	−5.030	−5.609	−8.114	−10.89
1.2	−34.15	−39.19	−4.446	−4.992	−5.557	−7.980	−10.62
1.4	−34.57	−39.65	−4.494	−5.041	−5.604	−8.003	−10.59
1.6	−35.31	−40.49	−4.586	−5.141	−5.712	−8.123	−10.70
1.8	−36.24	−41.56	−4.707	−5.275	−5.857	−8.306	−10.90
2.0	−37.28	−42.78	−4.846	−5.431	−6.029	−8.533	−11.16
2.2	−38.39	−44.09	−4.998	−5.601	−6.219	−8.792	−11.48
2.4	−39.54	−45.46	−5.156	−5.782	−6.421	−9.073	−11.82
2.6	−40.69	−46.85	−5.319	−5.968	−6.631	−9.371	−12.20
2.8	−41.84	−48.25	−5.484	−6.159	−6.846	−9.681	−12.59
3.0	−42.96	−49.63	−5.649	−6.351	−7.064	10.000	−13.00
3.5	−45.57	−52.97	−6.055	−6.828	−7.614	−10.82	−14.07
4.0	−47.80	−55.98	−6.436	−7.289	−8.154	−11.67	−15.19
4.5	−49.50	−58.55	−6.781	−7.721	−8.671	−12.51	−16.33
5.0	−50.54	−60.55	−7.076	−8.111	−9.155	−13.35	−17.48
5.5	−50.77	−61.84	−7.331	−8.450	−9.596	−14.17	−18.64
6.0	−50.00	−62.27	−7.472	−8.727	−9.984	−14.97	−19.79
6.5	−48.06	−61.69	−7.546	−8.929	−10.31	−15.73	−20.93
7.0	−44.69	−59.88	−7.517	−9.044	−10.56	−16.47	−22.06
7.5	−39.61	−56.63	−7.364	−9.055	−10.73	−17.16	−23.17
8.0	−32.47	−51.64	−7.067	−8.944	−10.79	−17.80	−24.26
8.5	−22.85	−44.59	−6.598	−8.692	−10.73	−18.39	−25.33
9.0	−10.24	−35.06	−5.925	−8.271	−10.54	−18.92	−26.37
9.5	6.003	−22.58	−5.011	−7.654	−10.19	−19.37	−27.38
10.0	26.64	−6.540	−3.810	−6.806	−9.650	−19.75	−28.35

L	$10J$	$10J$	$10J$	J	J	J	J
	K = 1.4	1.6	1.8	2.0	2.2	2.4	2.6
−30.0	98.46	113.3	128.1	14.30	15.79	17.28	18.77
−28.0	91.56	105.4	119.2	13.31	14.70	16.09	17.48
−26.0	84.66	97.47	110.3	12.32	13.61	14.90	16.19
−24.0	77.77	89.58	101.4	11.33	12.52	13.71	14.90
−22.0	70.90	81.71	92.54	10.34	11.43	12.52	13.61
−20.0	64.04	73.84	83.68	9.354	10.34	11.33	12.33
−18.0	57.20	66.00	74.83	8.369	9.257	10.15	11.04
−16.0	50.37	58.18	55.01	7.387	8.176	8.967	9.762
−14.0	43.58	50.38	57.22	6.408	7.098	7.791	8.487
−12.0	36.82	42.63	48.47	5.434	6.025	6.620	7.219
−10.0	30.12	34.93	39.78	4.467	4.961	5.459	5.962
−9.5	28.45	33.02	37.62	4.227	4.696	5.171	5.650
−9.0	26.79	31.11	35.47	3.988	4.433	4.884	5.339
−8.5	25.14	29.21	33.33	3.749	4.171	4.598	5.030
−8.0	23.49	27.32	31.20	3.512	3.910	4.314	4.723

(continued)

TABLE OF J (L,K)—*Continued*

L	$10J$	$10J$	$10J$	J	J	J	J
	K = 1.4	1.6	1.8	2.0	2.2	2.4	2.6
−7.5	21.85	25.44	29.08	3.277	3.651	4.032	4.419
−7.0	20.23	23.57	26.97	3.042	3.394	3.752	4.117
−6.5	18.61	21.71	24.88	2.810	3.139	3.475	3.818
−6.0	17.01	19.88	22.81	2.581	2.888	3.202	3.523
−5.5	15.42	18.06	20.77	2.354	2.640	2.933	3.234
−5.0	13.86	16.27	18.76	2.132	2.397	2.670	2.951
−4.5	12.33	14.52	16.79	1.915	2.160	2.415	2.678
−4.0	10.83	12.82	14.89	1.706	1.933	2.170	2.417
−3.5	9.393	11.18	13.08	1.508	1.719	1.941	2.174
−3.0	8.035	9.655	11.39	1.325	1.523	1.734	1.958
−2.8	7.524	9.085	10.77	1.258	1.453	1.661	1.883
−2.6	7.037	8.546	10.19	1.197	1.388	1.595	1.815
−2.4	6.580	8.046	9.654	1.141	1.331	1.537	1.758
−2.2	6.163	7.597	9.185	1.093	1.284	1.491	1.714
−2.0	5.795	7.213	8.797	1.055	1.248	1.459	1.687
−1.8	5.494	6.915	8.519	1.031	1.229	1.446	1.683
−1.6	5.284	6.737	8.391	1.025	1.232	1.461	1.710
−1.4	5.205	6.728	8.478	1.046	1.268	1.513	1.782
−1.2	5.321	6.973	8.886	1.106	1.351	1.623	1.921
−1.0	5.748	7.626	9.809	1.230	1.511	1.824	2.167
1.0	−13.89	−17.11	−20.54	−2.418	−2.802	−3.207	−3.632
1.2	−13.46	−16.46	−19.62	−2.294	−2.642	−3.006	−3.386
1.4	−13.33	−16.20	−19.21	−2.233	−2.558	−2.896	−3.247
1.6	−13.40	−16.21	−19.11	−2.212	−2.523	−2.843	−3.175
1.8	−13.59	−16.38	−19.24	−2.217	−2.519	−2.829	−3.148
2.0	−13.88	−16.66	−19.51	−2.241	−2.538	−2.841	−3.152
2.2	−14.23	−17.04	−19.89	−2.279	−2.573	−2.873	−3.179
2.4	−14.63	−17.48	−20.35	−2.326	−2.621	−2.919	−3.223
2.6	−15.07	−17.96	−20.88	−2.381	−2.677	−2.977	−3.280
2.8	−15.53	−18.49	−21.45	−2.443	−2.741	−3.043	−3.347
3.0	−16.02	−19.04	−22.06	−2.509	−2.812	−3.116	−3.422
3.5	−17.31	−20.53	−23.72	−2.690	−3.006	−3.322	−3.638
4.0	−18.68	−22.12	−25.51	−2.886	−3.219	−3.550	−3.881
4.5	−20.08	−23.76	−27.38	−3.094	−3.445	−3.794	−4.141
5.0	−21.52	−25.46	−29.30	−3.308	−3.680	−4.048	−4.414
5.5	−22.97	−27.17	−31.27	−3.528	−3.922	−4.311	−4.696
6.0	−24.43	−28.91	−33.27	−3.751	−4.168	−4.578	−4.985
6.5	−25.90	−30.67	−35.28	−3.978	−4.418	−4.851	−5.279
7.0	−27.36	−32.43	−37.32	−4.206	−4.670	−5.126	−5.577
7.5	−28.82	−34.20	−39.36	−4.437	−4.925	−5.405	−5.878
8.0	−30.28	−35.97	−41.42	−4.669	−5.182	−5.686	−6.182
8.5	−31.73	−37.74	−43.48	−4.901	−5.440	−5.968	−6.488
9.0	−33.17	−39.51	−45.55	−5.135	−5.700	−6.252	−6.796
9.5	−34.59	−41.28	−47.61	−5.370	−5.960	−6.537	−7.105
10.0	−36.00	−43.05	−49.69	−5.605	−6.221	−6.824	−7.416

Index of Notation

In general, symbols which appear on only one or two consecutive pages are not indexed. In cases in which a given symbol represents different functions in different parts of the book, each use is listed separately with its range of pages.

Roman

a, a_{ij}, $a_{ij}^{(n)}$, distance parameter, 214, 215, 219–248
a_s, activity of solute species s in a solution, 82–112
$\mathbf{a}$, activity set, 22–112
A, coefficient depending on solvent, 208–211
$A_{\mathbf{n}}$, assignment operator, 175–190
b_0, b_1, coefficients in expansion of B, 229–243
$b_n(x)$, Poirier function, 216–220
$b_{\mathbf{k}}$, reducible cluster integral, 47–90
$B = B(y,I_y)$, a term in the free energy, 227–229
$B(L,K)$, term in expansion of $J(L,K)$, 217–219
B_n or $B_{\mathbf{n}}$, irreducible cluster integral, 12, 52
$B_{ij}(\kappa)$, modified irreducible cluster integral for cluster i, j, 160
$B_{\mathbf{u}}(\kappa)$, modified irreducible cluster integral for cluster $\mathbf{u}$, 148
$B(u_1,u_2,u_3)$, same as $B_{\mathbf{u}}(\kappa)$ but with $\mathbf{u}$ written out, u_1,u_2,u_3, 233–248
c_s, concentration (particle number density) of species s, 19
c, total particle number density, $N/\mathcal{V}$, 10, 19, 67
c_s^*, applies to "in" region, 105–113
C_n, convolution integral with integrand corresponding to a cycle of n bonds, 124–125, 150–151
$\mathbf{c}(y,\kappa)$, concentration set in the mixture y,κ, 226–231
D, dielectric constant of pure solvent, 106, 118
DHLL, a Debye-Hückel limiting law function, 211
$e_n(x)$, 216–220
e_0, e_1, coefficient in expansion of $E(y,I_y)$, 229–243
$E(y,I_y)$, term of free energy of mixing, 227–228
E_i, energy of quantum state, 35, 72–78

$\mathcal{E}^{ex}$, excess extensive energy, 69
$f(L,R)$, bond with cut-off, 182
$\bar{f}$, effective bond function for a chain of f bonds, 185
$\tilde{f}_1$, transform of f, 182
$\tilde{f}_2$, transform of f^2, 182
$\tilde{f}_3(t)$, transform of f^3, 185
$\mathfrak{F}$, extensive Helmholz free energy, 67
$\mathfrak{F}\dagger$, extensive Helmholz free energy in hypothetical $c_s = 1$ reference state, 71
$\mathfrak{F}^{id}$, Helmholz free energy of ideal gas, 68
$\mathfrak{F}^{ex}$, excess extensive Helmholz free energy, 12, 67, 68
$g(r)$ or $g_{ij}(r)$, distance dependence of Coulomb potential of interaction of two charges, 16, 132, 136
$g_{\mathbf{A}}^{ex}$, reduced mean ionic partial molal excess free energy of electrolyte A, 197–202
$g_n(x)$, Poirier function, 216–220
g_p, mixing coefficient, 200–202, 225–248
$g_{\mathbf{n}}$, $g_{\mathbf{r}}$, $g_{\mathbf{a}}$, generic spatial correlation function, 79–172
$G(\{\mathbf{n}\},\mathbf{z})$, modified correlation function, 80–87
G^{ex}, total excess Gibbs free energy of a quantity of a real mixture containing a kilogram of solvent, 194
G_w^{ex}, excess partial specific (per kg) free energy of solvent, 194
$\mathcal{G}$, extensive Gibbs free energy, 70
$\mathbf{h}$, $h^{(w)}$, 178–190
h, a function of ionic charges in the mixing theory, 231–234
i,j, set $\mathbf{m} = 1_i, 1_j$, 117, 149, 234
$[i,j]!$, This is unity if i and j are different species, otherwise it is two, 149
i $[-1]^{1/2}$, 126
I_B^u, 176–190
I_C^u, 176–190
I, molal ionic strength, 193, 196–248
I_y, 227–248
j, designation of a pair of vertices of a protograph, 153–190
$J_u(\tau)$, a cluster integral, 177–190
$J(L,K) \equiv J_{ij}(L_{ij}, K_{ij})$, an ionic solution integral, 215–248
J_{ij}, abbreviation, 245–248
k, Boltzmann constant, 5, 67
k, number of dimensions, 126–130

k_{ij}, cluster function of u^*_{ij}, 147–189
$K_u(a')$, cluster integral, 187–189
K_{ij}, dimensionless function of ionic strength and ionic sizes, 215–248
L, a macroscopic but small length, 123–125, 176–189
L_{ij}, dimensionless function of ionic charges and sizes, 214–248
m_s, molality of species s in solution, 192–248
m, total molality of solute species, 192–248
n, number of bonds in chain, 126–146
$n_m = \mu_m/c$, reduced moment of concentration of charge types, 211–248
n_s, total number of vertices of species s in all chains, 154–158
n_s^α, number of vertices of species s in chain α, 154–158
$\mathbf{n}^{(\alpha)}$, composition set of chain α, 154–158
$\langle n_s^{(\alpha)}\rangle$, array of $n_s^{(\alpha)}$, 154–158
$\{\mathbf{n}\}$, set of spatial coordinates of molecules of the composition set n, 34–66, 79–248
$\{\mathbf{n}\}$, complete coordinate set, 20–33, 76–78
$\{\mathbf{n}\}_i$, set of internal coordinates, 20
N, total number of molecules in system, 5, 67
N_s, number of molecules of component s in system, 68
$\mathbf{N}$, composition set of a system or of region of a grand ensemble, 72
$O(f(x))$, order of $f(x)$, 174
p_n, integral on a chain of n bonds, 16, 126
$\mathbf{p}]\mathbf{n}$, a partition set defining a partition of $\mathbf{n}$, 30
$\mathbf{p}]\{\mathbf{n}\}$, partition set, 30
P, pressure, 13
$P\mathcal{V}$, a thermodynamic potential, 69
P_0, pressure on pure solvent in osmotic equilibrium with solution, 81
P, a probability density, 74, 77
$P(\mathbf{r}:\mathbf{m})$, cluster integral for the expansion of $g_\mathbf{r}$, 96
$P_\kappa(\mathbf{k}:\mathbf{m})$, cluster integral for the expansion of $g_\mathbf{k}$ in an ionic system, 169
$q(r)$, sum over g-bond chains, 16, 108, 136, 158
$q_{ab}(r_{ab})$, a q bond from a to b, including the charge factors, 159, 214
Q_{ij}, a sum of parallel q bonds, 214, 248
$Q(\mathbf{n}:\mathbf{m})$, cluster integral appearing in density expansion of potential of average force, 97
$Q_\kappa(\mathbf{k}:\mathbf{m})$, cluster integral in the expansion of potential of average force for ionic solutions, 170
r, r_{ij}, separation of two ions or molecules, 5, 13, 117

$\mathbf{r}$, a composition set, 93–99
$\mathbf{r}$, a vector, 126–146
R, radius of containing vessel, assumed spherical, 13–65
R, gas constant, 192–248
$_s$, subscript denoting a particular species, 19
$\mathcal{s}_\mathbf{n}(\{\mathbf{n}\})$, a sum of ALSC graphs, 46, 47
$\mathcal{S}$, extensive entropy, 69, 73
$\mathrm{Si}(x)$, sine integral, 182
$S_\mathbf{k}(\{\mathbf{k}\})$, a sum of ALDC graphs, 51, 116, 152
$S_\mathbf{u}(\kappa)$, a sum of modified ALDC graphs of the ionic solution theory, 159–163
$\mathbf{t}$, a vector conjugate to $\mathbf{r}$, 126–146
T, temperature, 5
$\mathbf{t}]\mathbf{n}$, tree set, 22, 51
$T(\tau')$, combinatorial factor, 178, 223
$T(\mathbf{t}]\mathbf{n})$, tree coefficient, 53, 59
$u(r_{ij}) = U(r_{ij})$, direct potential for molecules i and j at separation r_{ij}, 5, 36, 117
u^*_{ij}, non-Coulomb part of the direct potential, 117, 119, 147, 214
$u_\mathbf{m}(\{\mathbf{m}\})$, a component of the direct potential, 36, 42, 116
$\bar{u}_{ab}$, modified two-body direct potential, 169
$U(\{N\})$, direct potential of N molecules at $\{N\}$, 5
$U_\mathbf{N}$, $U(\{\mathbf{N}\})$, $U_\mathbf{N}(\{\mathbf{N}\})$, the direct potential for set $\mathbf{N}$ at $\{\mathbf{N}\}$, 34–37
$V_w(m,P)$, partial specific volume of solvent in solution of molality m under pressure P, 207
$V(m,P)$, volume of solution per kg of solvent in the state of molality m and pressure P, 203
$\mathcal{V}$, volume of system, 5, 67
$\mathrm{W}_\mathbf{k}(\{\mathbf{k}\},\mathbf{c})$, potential of average force of set $\mathbf{k}$ at $\{\mathbf{k}\}$ in medium of composition $\mathbf{c}$, 36, 96, 169
w_p, a mixing parameter, 236–248
x, an independent thermodynamic variable, 70, 99
x, membrane potential, 106–107
$\mathbf{x}$ Set of x_i, mole fractions, 69–103
$\mathbf{x}$, set of x_s, the effective concentrations in the Donnan equilibrium, 110–111
$\mathbf{x}$, set of x_s, solute fractions, 195–248
y, an ionic strength parameter in the Donnan problem, 106–107
y, mixing fraction, 197–248

$Y = 1 - 2y$, 226
z_s, fugacity of component s, 22, 68–104, 110–111
z_s, ionic charge, 105–109, 112–248
$\mathbf{z}$, the set of z_s, fugacities, 75–90, 110–111
$\mathbf{z}$, the set of z_s, charges, 150–248
$Z(N,\mathfrak{V},T)$, configuration integral for system of N particles in volume $\mathfrak{V}$ at temperature T, 5, 42, 74, 78

German

$\mathfrak{s}_p$, coefficient in expansion of $\Delta_m\mathfrak{S}$, 266
$\mathfrak{s}_{0u}$, a sum of certain cluster integrals, 232
$\mathfrak{S}$, the reduced excess Helmholz free energy, 12, 68
$\mathfrak{S}_u$, a group of terms in the cluster expansion of the free energy of an ionic system, 174
$\mathfrak{S}_u(\tau, c)$, a part of $\mathfrak{S}_u$, 177
$\mathfrak{S}^*$, function, later to be identified as $\mathfrak{S}$, 90
$\mathfrak{S}_c$, leading term of $\mathfrak{S}$ for ionic systems, 16, 149

Greek

α, designation of a particular g-bond chain, 154–158
α, expanded kappagraphs that are cycles with a minimum number of bonds, 180–189
α_{An}, Harned coefficient, 200–248
β, partially defined irreducible cluster integral, 9–12
$\beta_{\mathbf{k}}$, irreducible cluster integral defined in terms of reducible cluster integrals, 60–65
β, expanded kappagraphs that are cycles with no vertices of order 3, 181–189
γ, expanded kappagraphs that are bicycles with a minimum number of bonds, 185–189
γ_s, activity coefficient of species s, 106–113, 192–207
$\gamma_{\mathbf{m}}$, cluster function, 42, 43, 116
γ_m, cluster function for interaction of m molecules, 43
γ_{ij}, cluster function for particles i and j, 5
$\delta(r)$, Dirac delta function, 108, 129–146
$\delta = 0$ or 1, 183–190
$\Delta_m\mathbf{c}^{\mathbf{u}}(y,\kappa)$, increase in $\mathbf{c}^{\mathbf{u}}$ in mixing at fixed κ, T, and $\mathfrak{V}$, 226
$\Delta_m\mathfrak{S}$, increase in $\mathfrak{S}$ in mixing at fixed κ, T, and $\mathfrak{V}$, 225

$\Delta_m B$, increase in B in mixing at fixed I, T, and P, 227
$\Delta_m G^{ex}$, increase in G^{ex} in mixing at fixed I, T, and P, 197
$\Delta_m H^{ex}$, increase in H^{ex} in mixing at fixed I, T, and P, 197, 243
ϵ, magnitude of electronic charge, 13
κ_0, κ of pure solvent, 115, 164
κ, Debye reciprocal length, 14, 106, 136, 150
λ, the Coulomb length, 13–106, 147
μ_s, chemical potential of component s, 68, 192–194
μ_s^{ex}, excess chemical potential of component s, 68, 194
μ_s^{id}, chemical potential of s in hypothetical ideal gas state, 68
$\mu_s^{\dagger}$, chemical potential of pure component s in hypothetical $c_s = 1$ gas state, 68, 204
μ_s^0, chemical potential of species s in hypothetical one molal reference state, 192, 201
$\boldsymbol{\mu}$, chemical potential set, 21–22, 72–78
μ_n, nth moment of concentration of charge types, 106–109, 171–189, 211
$\boldsymbol{\mu}$, set of moments of concentration, 179–180
$\boldsymbol{\nu}_j$, number of chains, 160
$\boldsymbol{\nu}$, set of numbers of chains, 153
ξ, charging parameters, 99
Ξ, grand partition function, 75
Π, osmotic pressure, 81
ρ, concentration of non-diffusing ions in "in" region, 105–107
ρ, local density of electric charge, 133–134
σ, the number of molecular species in a given system, 19
σ_3, simplest surface term, 63
$\sigma_{\mathbf{n}}$, surface term for set $\mathbf{n}$, 46, 63, 116, 164
τ, topology of a graph or protograph, 8, 153
τ_i, element of a protograph, 153
ϕ, osmotic coefficient, 193–213
$\Psi_i(\{\mathbf{N}\})$, eigenfunction, 35, 76–78
Ψ, Coulomb electrical potential, 112, 133–143

Operators

$\setminus$ difference operator for sets, 19
∇ gradient operator, 35, 133
∇^2 Laplacian operator, 134

$\cup$	union operator, 18
$\subseteq$, $\subset$	subset operator, 18
	Fourier transform operator, 126
	(For example, $\tilde{g}$ is the Fourier transform of g. However note one exception, $\tilde{\mu}$, 112.)

Miscellaneous

$\binom{\mathbf{N}}{\mathbf{m}}$	generalized binomial coefficient, 24
$\mathbf{n}!$	generalized factorial, 21
$\mathbf{c}^{\mathbf{n}}$	generalized exponential, 21
$\in$	element relation, 19
$\{\ \}$	coordinates; see $\{\mathbf{n}\}$
$\mathbf{0}$	set of zeros, 19

Subject Index

Activity, 82
Analytical continuation, 137
Association of ions, 220
At least doubly connected, ALDC, 48
At least singly connected, ALSC, 46

Bicycle graphs, 185
Binary electrolyte, 196
Binomial theorem, 24
Bond, 6, 43, 127

Cavendish experiment, 133
Chain, 45, 50, 127
Classification of sets, 22
Class of graphs, 155
Cluster
  irreducible, 48
  reducible, 46
Cluster expansion
  of configuration integral, 8, 41
  of correlation function, 96, 169
  of excess free energy, 91, 165
  of potential of average force, 96, 169
Cluster functions
  expansion of, 147
  of ionic solution theory, 163
Cluster integrals
  convergence of modified, 165
  in correlation function expansion, 96, 169
  divergence of, 13, 115, 123, 136
  irreducible, 47, 51, 60, 89
  modified irreducible, 148, 159, 173 ff.
  in potential of average force expansion, 96, 170
  reducible, 47, 51, 60, 89
  singularity of modified, 173 ff.
Common ion, 230
Component potentials, 36
  cluster expansion of, 104
  examples of higher, 38

Composition set, 19
Concentration set, 19
Connectivity, 45, 153
Convergence of cluster expansion series, 86, 91, 137, 173 ff.
Convolution approximation, 166
Convolution integral, 127
Convolution theorem, 127, 182 ff.
Coordinate set, 20
Correlation function, 78
  relation to thermodynamic excess functions, 98 ff.
  spatial, 79, 85, 93; for ionic solutions, 167
  uses, 98, 172
Coulomb potential, 115, 118
  definition of, 136, 143
  Fourier transform of, 132
Covering of a set, 28
Cumulant, 31, 47, 48, 89
Cycle, 124, 127, 149

Debye Hückel limiting law, 14, 17, 151, 174, 208, 209
Debye Hückel theory, 139 ff.
Debye potential 17, 137, 141, 142, 159
  correction to, 170
Dirac delta function, 129
  method to determine $q(r)$, 143
Directly connected, 45
Direct potential, 36
  for ionic solutions, 3, 116 ff., 117
  sectionally uniform u_{ij}^*, 214
  solvent contributions to u_{ij}^*, 119
  square-well form for u_{ij}^*, 215
Disjoint set, 20
Distinguishable graphs, 7, 46
Donnan equilibrium, 105

Edge, 46
Elements of a protograph, 152
Elements of a set, 18

Empty set, 19
Excess Gibbs free energy
  calculation from data for electrolyte mixtures, 199
  relation to $\mathfrak{S}$, 203
Excess thermodynamic functions, 67 ff., 191 ff.
  cluster expansions for, 91, 98, 163
  Debye-Hückel limiting laws for, 209
  for electrolyte solutions, 196 ff.
  expansions, for mixed electrolyte solutions, 200
  partial, 191, 194 ff.
  relation of G^{ex} to $\mathfrak{S}$, 203
  total, 191, 202

Fourier transforms, 16, 126 ff.
  convolution theorem, 127
  inversion formula, 128
  table, 131
Fuchs's derivation of the cluster theory of multicomponent systems, 59
Fugacity, 68
Functional derivative, 94

g-bond chains, 136, 148
  sum over, 136, 137 ff.
Generating function, 53
  for partitions, 32
  for tree coefficients, 56
Generic probability density, 76
Grand canonical ensemble, 72
Grand partition function, 71
  generalization of, 75
Graphs, 6, 43
  expansion of, 147, 177
Guggenheim's deduction of the ideal solution equation from the cluster theory, 101 ff.

Haga's calculation of higher terms, 222
Harned coefficients, 201, 236
Harned's rule, 201, 226, 230, 235
Hetero ion, 230
Higher-order limiting law, 210, 235, 243
Hill's theory of the Donnan equilibrium, 105 ff.
Husimi tree, 48, 90
Hypothetical states, *see* Standard states

Indirectly connected, 45
Inversion of Fourier transforms, 128

Kappagraph, 163, 177
k bonds, 147, 163
Kirkwood - Scatchard function $B(L,O)$, 218
Kirkwood superposition approximation, 103

Labeled vertices, 7, 43 ff. 151 ff.
Lebowitz and Percus method to calculate the cluster expansion of the correlation function, 94
Levine and Wrigley's calculated values of u_{ij}^*, 122
Limiting laws, 208, 235, 237, 246

McMillan-Mayer theory, 72, 84
Mayer's ionic solution theory, 147 ff.
Mayer's summation procedure (rearrangement), 14, 166, 145, 148, 173
Mean solute quantity, 194
Mixed electrolyte solutions, 225
  symmetric, 234
  unsymmetric, 85, 244
Mixing fraction, 197
Mixtures, symmetric and unsymmetric, 85, 100, 192, 195
Molal concentration scale, 193, 194
Molal ionic strength, 196
Moment of a distribution, 31
Moment of the concentration of charges, 179
Multinomial theorem, 25

Nodes, 148

Notation, 17
brackets, 22
dimensionless forms of functions for scales on figures and entries in tables, 237
for extensive thermodynamic variables, 67
for sets, 18
sums, products, and integrals, 22, 31
thermodynamics of ionic solutions, 193

Ordered products, 158, 178
Osmotic equilibrium, 81, 105 ff., 118, 191, 203, 225

Partition, 28, 47, 51, 89
ordered, 57
Partition set, 22, 29
Phase transition, 86, 98
Poirier's calculations using the cluster theory, 216, 220
Poisson-Boltzmann equation, 140, 143, 144, 145
Poisson's equation, 134, 139, 140, 143
Potential of average force, 34, 103
cluster expansion of, 96, 104, 169
Primitive model, 1, 214, 239, 246
Protograph, 151
Prototype graph, 153

Radial functions, 130

Salpeter-Mattis method to calculate $q(r)$, 143
Scatchard-Prentiss equation for electrolytic mixtures, 243
Scatchard, Vonnegut, and Beaumont freezing point data for $LaCl_3$(aq), 213
Sequences of series, 25
compared to Taylor's series, 86
Sets, 18
classification of, 22
notation for various, 19
operations with, 20
as summation indices, 22, 23, 31
Shedlovsky and MacInnes emf data for $LaCl_3$(aq), 213
Skeleton, 6, 43
Solutions, *see* Mixtures
Solvent, Solute, 36, 81, 83, 85, 192
Specific probability density, 76
Standard states, 70
hypothetical $c = 1$ gas state (†), 192, 204
hypothetical $m = 1$ solution state (^{0}), 192, 204
ideal gas state (^{id}), 68
solution reference state (*), 194
Surface terms, 46, 53, 60 ff., 91, 116, 164 ff.

Thermodynamic functions, 67, 191
Thiele seminvariant, 31
Topology, 6, 45
Tree, 22, 49
Tree coefficient, 53
generating function for, 56
recursion formula for, 53
Tree Set $\mathbf{t}]\{\mathbf{n}\}$, 51

van't Hoff model, 82
Vertex, 6
Vertex cluster, 50
Virial, generalized, 99
Virial coefficient, 13
Virial expansion, 13, 165

Yukawa potential, 115